ELECTROANALYTICAL METHODS
in Chemical and Environmental Analysis

ELECTROANALYTICAL METHODS
in Chemical
and Environmental Analysis

Edited by
Robert Kalvoda

UNESCO Laboratory of Environmental Electrochemistry
The J. Heyrovský Institute of Physical Chemistry and Electrochemistry
Czechoslovak Academy of Sciences, Prague, Czechoslovakia

PLENUM PRESS • NEW YORK AND LONDON
SNTL • PUBLISHERS OF TECHNICAL LITERATURE, PRAGUE

Distributed throughout the world with the exception of the Socialist countries by

Plenum Press
A Division of Plenum Publishing Corporation
233 Spring Street, New York, N.Y. 10013
ISBN 0-306-41799-5
Library of Congress Catalog Card Number 85-042729

© 1987 Robert Kalvoda (Editor)
Translated by Madeleine Štulíková
English edition first published in 1987 simultaneously by
Plenum Publishing Corporation and
SNTL – Publishers of Technical Literature, Prague

Printed in Czechoslovakia

"The natural resources of the earth including the air, water, land, flora, and fauna and especially representative samples of natural ecosystems must be safeguarded for the benefit of present and future generations through careful planning or management, as appropriate."

(From the DECLARATION on the HUMAN ENVIRON-MENT – The United Nations Conference on the Human Environment, Stockholm, June 1972.)

Editor:
Robert Kalvoda

Contributors:

Jaroslav Čáp   Chemoprojekt, Department of Automation, Prague

Josef Fexa   Institute of Chemical Technology, Department of Physics and Electrotechnics, Prague

Pavel Hofmann   Water Research Institute, Prague

Robert Kalvoda   UNESCO Laboratory of Environmental Electrochemistry, The J. Heyrovský Institute of Physical Chemistry and Electrochemistry, Czechoslovak Academy of Sciences, Prague

Miloslav Kopanica   UNESCO Laboratory of Environmental Electrochemistry, The J. Heyrovský Institute of Physical Chemistry and Electrochemistry, Czechoslovak Academy of Sciences, Prague

František Opekar   UNESCO Laboratory of Environmental Electrochemistry, The J. Heyrovský Institute of Physical Chemistry and Electrochemistry, Czechoslovak Academy of Sciences, Prague

Jan Pokorný   Botanical Institute, Czechoslovak Academy of Sciences, Třeboň

Lubomír Šerák   UNESCO Laboratory of Environmental Electrochemistry, The J. Heyrovský Institute of Physical Chemistry and Electrochemistry, Czechoslovak Academy of Sciences, Prague

Karel Štulík   Charles University, Faculty of Science, Department of Analytical Chemistry, Prague

Jiří Tenygl   UNESCO Laboratory of Environmental Electrochemistry, The J. Heyrovský Institute of Physical Chemistry and Electrochemistry, Czechoslovak Academy of Sciences, Prague

Antonín Trojánek   UNESCO Laboratory of Environmental Electrochemistry, The J. Heyrovský Institute of Physical Chemistry and Electrochemistry, Czechoslovak Academy of Sciences, Prague

Josef Veselý   Geological Survey of Czechoslovakia, Prague

Jaroslav Zýka   Charles University, Faculty of Science, Department of Analytical Chemistry, Prague

PREFACE

Electroanalytical methods play an important role in chemical analysis in view of their sensitivity, simplicity, broad application, adaptability to field applications, and relatively low investment outlay. They find applications in all branches of science and technology where chemical analyses are required. These qualities are of particular value in environmental protection, which involves mass application of instrumental techniques and large numbers of monitoring stations and networks around the world. The methods are applicable to an enormous number of chemical substances encountered not only in environmental monitoring but in many other fields.

This volume presents the current state of the art of electrochemical analytical methods, such as polarography, voltammetry, potentiometry with ion-selective electrodes, etc. It summarizes the potential offered by these techniques for monitoring various substances affecting health and the environment and for chemical analysis in general.

R. Kalvoda

CONTENTS

12

CHAPTER 1

INTRODUCTION

Robert Kalvoda

One of the most pressing tasks for mankind during the scientific and technical revolution, in addition to efforts to preserve peace, is the protection of the environment. This topic has been discussed so thoroughly in general that it is unnecessary to do so here.

A great variety of branches of science participate actively in the solution of environmental problems, including medicine, biology, physics, chemistry, and also electrochemistry. The latter has played an important role here in several respects. In addition to the analytical aspects treated in this book, electrochemistry is used in the removal of toxic substances or other products and wastes polluting the environment as a result of industrial activity (metallurgy, chemical and food industries, energy production, etc.), from the point of view of agriculture or of communal hygiene. Electrochemistry can also help in decreasing or eliminating the occurrence of some substances, e.g., by wasteless technologies, especially in the chemical industry and related fields. Moreover, electrochemistry can, and most probably will, affect energy production and all kinds of transport, which are among the major polluters of the environment. These questions and others are discussed in greater detail in other books [1 −4].

This book summarizes the potentialities offered by electrochemistry from an analytical point of view, in monitoring various substances polluting or affecting the environment, as a result of human activities brought about by technical progress. Electroanalysis is characterized by unusually broad applicability, from analyses for traces of heavy metals in waters or atmospheric aerosols to determinations of organic compounds with which we are in continuous contact in our surroundings. Especially polarography enables the detection of many high-priority pollutants that are listed, e.g., in Ref. 5. It is certainly interesting that more than two-thirds of the substances given in the Fourth Annual Report on Carcinogens [6] are polarographically determinable. One of the advantages of electroanalysis is the relatively simple, cheap instrumentation that can often be applied to field work and operated by technicians. This is extremely useful in the present trend to build various analytical laboratories for monitoring pollutants in the framework of national, regional, or global programs of environmental protection.

The following chapters discuss the use of selected electrochemical methods in solving the problems of environmental protection, with emphasis on tested and successful applications. At the end of the book are also discussed some possibilities of using electroanalysis in tackling ecological problems.

REFERENCES

1. Bockris J. O'M. (Ed.): Electrochemistry of Cleaner Environment. Plenum Press, New York 1971.
2. Bockris J. O'M. (Ed.): Environmental Chemistry. Plenum Press. New York 1977.
3. Balej J. (Ed.): Electrochemistry and the Environment (in Czech). Academia, Prague 1982.
4. Kalvoda R. and Parsons R. (Eds): Electrochemistry in Research and Development. Plenum Press, New York 1986
5. Keith L. H. and Telliard W. A.: *Environment. Sci. Technol.* **13,** 416 (1979).
6. Fourth Annual Report on Carcinogens, US Department of Health and Human Services, USA, 1985.

CHAPTER 2

ANALYTICAL CHEMISTRY AND ENVIRONMENTAL PROTECTION

Jaroslav Zýka

One of the most important problems to be solved at present by mankind, in addition to securing food and providing necessary sources of energy, is the protection of the environment and creation of conditions for modification of the environment in an ecologically optimal way for the further development of the human race.

Human intellect has led to the attainment of such a high level of science and technology that, in accordance with the law of equilibrium, which is valid not only in chemistry and other natural sciences but also in human life in general, the equilibrium in the relationship between man and nature has been disturbed and technical progress is beginning to turn against those who created it, i.e., people, mankind.

Most people feel, and to a certain extent objective experience has confirmed, that chemistry has contributed to this disturbance of the equilibrium; on the other hand, chemistry has yielded many favorable effects. It is paradoxical that progress in chemistry, inseparable from the development of human society and the present high standard of living, has often led to pollution of the atmosphere, water, and soil and thus also indirectly of plant and animal food with unfavorable consequences for people, plants, and the whole environment of living organisms. However, man is capable of restoring this equilibrium.

Hence, chemistry plays one of the most important roles in environmental protection and ecology, analytical chemistry being the most important field, because it points out the problems to be tackled, even though it cannot solve them itself. The results of recent medical research indicate that some inorganic and organic compounds have harmful effects in amounts substantially lower than those assumed a few decades ago, and thus it is necessary to develop analytical procedures capable of detecting and determining these substances with better precision, reliability, sensitivity, and selectivity which would simultaneously be cheap in operation to permit their use not only for research purposes, but also for wide-scale series control. The large sets of results should then be handled by modern computing techniques and incorporated in appropriate standards.

As a natural result of these requirements, a new analytical field has recently appeared that can be called the "analytical chemistry of the environment," which also comprises the "electrochemistry of the environment" as one of its specific branches. It can be assumed that environmental electrochemistry and related fields will contribute to the formation of "ecoanalytical chemistry," where the chemist-analyst, educated in various branches of analytical chemistry, with good orientation in instrumental techniques, in the principles of biology, and in ecological problems, will be able to broadly formulate a given problem and provide supporting analytical data to justify the necessity for solving the problem and to outline the means of its solution.

Analytical chemistry is one of the oldest chemical disciplines. The principal findings in stoichiometry and various chemical laws were obtained on the basis of analytical chemistry. The importance of analytical chemistry is also illustrated by Nobel prize awards for discoveries and development of new analytical methods, including polarography. At present, analytical chemistry is not only a chemical discipline, but also an independent scientific field that indicates ways for complex utilization of general chemistry, as well as physics, biology, mathematics, metrology, various technical branches, etc., for the determination of the qualitative and quantitative chemical compositions of substances and their structure, and points out optimal conditions and combinations for the data obtained.

However, if contemporary analytical chemistry is considered in a narrower sense, then the questions "what" and "how much" are still most important (in contrast to a broader concept which also requires answers to the questions "how" and "why," especially when the course of chemical reactions is to be clarified), and the main task is the development of new methods that are more sensitive, precise, and selective and are not too expensive.

Laboratories for subnanogram analyses are now common and some methods are so sensitive that a few hundred atoms can be determined. Figure 2.1 illustrates the relationship between the size of some particles, their mass, and the number of atoms contained in them.

The instrumentation required for these analyses (including, e.g., the analyses of minute particles deposited on the surface of satellites) is, of course, sophisticated and expensive (e.g., stereo-electron and polarization microscopes, x-ray and electron diffraction, electron and ion microprobes coupled with mass-spectrometers, activation analysis, spectral laser methods, etc.) and specially trained teams of workers are needed. Nevertheless, far less sophisticated methods can often be used when tackling everyday problems of environmental monitoring (e.g., pH-monitoring of soils and waters, monitoring of the most important pollutants in soils, waters, the atmosphere, and living organisms).

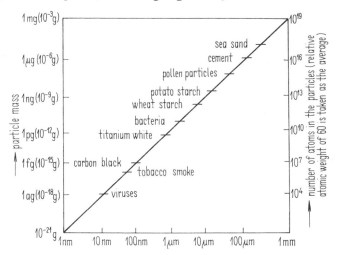

Fig. 2.1 The relationship between the size of some particles, their mass, and the number of atoms contained in them.

The analytical methods used at present in laboratories dealing with environmental control involve a great variety of optical methods (e.g., visible, uv, and ir spectrophotometry and atomic absorption spectrometry), separation methods, chiefly gas, liquid, and thin-layer chromatography, radiometric methods (limited to some specially equipped laboratories), and especially electroanalytical methods, such as voltammetry and poten- . . tiometry with ion-selective electrodes, which are advantageous from the point of view of low investment and operational costs. None of these methods is universal; some are suitable for determination of organic materials, others for analyses of inorganic substances. Most common analytical methods are summarized in Table 2.1. Gas chromatographic methods are not given, as their sensitivity depends on the detection technique; the sensitivity and reliability of measurements with membrane electrodes sometimes depend on the manufacturer of the instrumentation.

Electroanalytical methods have recently enjoyed a "renaissance" not only because of their low cost, but also because of their ability to elucidate chemical reactions, in addition to determination of substances (e.g., studies of redox reactions, even in living organisms in relation to carcinogenicity, investigation of "pollutant–environment–source–man" cycles, etc.). The wide range of methods based on electrochemical principles,

The applicability range of the most common analytical methods[a]   *Table 2.1*

Method	Sensitivity (mol liter^{-1})
Titration methods	10^{-2}–10^{-6}
Molecular absorption spectrophotometry	10^{-5}–10^{-6}
Molecular fluorescence spectrophotometry	10^{-7}–10^{-8}
Atomic absorption spectrometry	10^{-6}–10^{-7}
Atomic fluorescence spectrometry	10^{-7}–10^{-8}
Optical and emission spectroscopy	10^{-5}–10^{-6}
Neutron activation analysis	10^{-9}–10^{-10}
Ion-selective electrode potentiometry	10^{-4}–10^{-5}
Classical polarography	10^{-5}–10^{-6}
Square-wave and differential pulse polarography	10^{-7}–10^{-8}
Anodic stripping voltammetry with a hanging mercury drop electrode	10^{-7}–10^{-9}
Anodic stripping voltammetry with solid electrodes or with mercury film electrodes	10^{-9}–10^{-10}

[a] Gas chromatography is not given because its sensitivity depends on the detection technique used.

from chemical potentiometry through polarography, amperometry, coulometry, and conductometry to ion-selective electrodes, represents a very promising field, employing microprocessor techniques for automation. It is interesting to note that ion-selective membrane electrodes have been the only original contribution to instrumental analytical techniques during recent years; all the other methods are older and are simply being perfected as a result of instrumental development. The application of electroanalytical methods to environmental problems will not be discussed here in greater detail, as they are treated in the following chapters of the book.

Optical methods constitute a very varied field in environmental control. Classical colorimetry and spectrophotometry, based on the formation of colored compounds of the test substances with reagents, are still widely used. The rapid development of atomic absorption and emission (fluorescence) spectrometry should be emphasized; these methods can at present determine most elements in inorganic samples with a high sensitivity (down to amounts of about 10^{-14} g). Automated spectrometers, quantometers, are also quite common, in spite of their high cost.

Atomic absorption spectrometry is becoming common even in control laboratories, especially for analyses for pollutants in the atmosphere and in waters, when simple pretreatment of the sample or preconcentration (e.g., by extraction, by evaporation of water, or by trapping atmospheric pollutants on a suitable filter) leads to a further improvement in the sensitivity of the measurement.

Among these methods can further be classified, e.g., the electron microprobe method, x-ray fluorescence and diffraction analysis, molecular spectroscopy (uv, ir, NMR, EPR, mass, and Raman spectrometry), whose application involves the elucidation of the structure of substances and the investigation of processes, rather than the mere determination of substances.

Radiometric methods occupy a rather special position in contemporary analytical chemistry, because they cannot yet be applied widely in small control laboratories monitoring environmental pollution, as they require special equipment and maintenance of a variety of necessary safety conditions. Nevertheless, there are centers using radioisotopes for the solution of problems that cannot be solved by other methods, mainly because of extreme demands on the measuring sensitivity (e.g., the methods of radionuclide substoichiometric analysis or neutron activation analysis in the determination of trace elements in biological materials).

There has also been rapid progress in separation methods, especially gas chromatography (combined with mass, ir, and NMR spectrometry) and liquid chromatography (partition, ion-exchange, and adsorption), as well as electrophoretic methods (e.g., isotachophoresis). For identification and determination of organic substances with similar structures, chromatographic methods are often indispensable.

Great caution must always be exercised when evaluating the suitability of a method for a given purpose. In introducing a method for environmental applications, the economic aspect must always be considered, as well as the requirements on the sensitivity, resolution, rapidity, possibilities of automation, the sample type, the purpose of the analysis, etc. Here the analyst must possess a wide theoretical background, especially in considering the purpose of the measurement.

All chemical analysis methods are undergoing development and thus it can be expected that both common and new methods will further be perfected, using analog and digital techniques, automation and miniaturization of the instruments, and handling of large data sets by modern methods. The present stage, in which analytical chemistry plays a very important role in the protection and modification of the environment, is still chiefly based on obtaining data, which in turn creates problems with data evaluation, search for relationships, and formulation of approaches to a solution of the problems. Here the analyst with wide knowledge, whether he is called an analyst or analytical chemist or bio- or ecoanalyst, who has sufficiently mastered not only chemistry as a whole, but also physics (and biophysics), biology, computing techniques, and ecology and who thinks logically can become a key specialist in the maintenance of the required ecological equilibrium.

CHAPTER 3

ELECTROCHEMICAL METHODS
IN ENVIRONMENTAL PROTECTION

Robert Kalvoda

3.1   Introduction

As emphasized in Chapter 2, analytical chemistry plays an important role in environmental analysis and protection. Its applications include monitoring of emissions into the biosphere, both qualitatively and quantitatively, tracing of pathways of emissions from the source to man (or other object of interest) and following transformations into other substances, e.g., as a result of interactions among the emitted substances. Another task of analytical chemistry is to evaluate various processes employed to prevent emissions or liquidate substances already emitted.

Most applications of chemical analysis to environmental protection involve trace determinations, often at a ppb level or lower. High sensitivity of the methods must further be accompanied by sufficient selectivity, precision, and accuracy. Easy sample pretreatment and rapidity of the analytical procedure are also desirable. Because series analyses are often required, methods that are easy to automate are advantageous, especially if the whole procedure, i.e., from the sample introduction to the displaying of the results, is automated, as such series analyses are usually performed by technical staff. In the selection of the method, the cost of instrumentation that must be available in a great number of laboratories is also important. Measurements must often be carried out in the field and thus large apparatuses are *a priori* excluded, even if they meet all the other above criteria. It need not be emphasized that microanalytical instruments should be applicable to a wide range of substances (except for single-purpose analyzers and monitors) and that it is advantageous if several components can be determined simultaneously. Of course, successful application of a method requires a sound theoretical background.

Electrochemical methods fully satisfy the above criteria and were widely used for analyses of waters, the atmosphere, soils, and foodstuffs and for hygienic purposes even when the term "environmental protection" was not yet used. There are many physical methods that are much more sensitive (x-ray and electron diffraction, neutron activation analysis, gas and liquid chromatography, mass spectrometry, etc.), but these methods cannot be used in the field and are expensive. A similar situation is encountered with various spectral methods, among which atomic absorption spectrometry is used most extensively.

Virtually all common electroanalytical methods can be used for the purposes of environmental protection, depending on the nature of the determinands and the character of the sample matrix, as well as on the requirements on the sensitivity and selectivity; the principal methods are polarography, potentiometry, coulometry, and conductometry. In the following sections, these methods are briefly characterized in relation to environmental protection.

3.2   Polarography and Voltammetry

In these methods, based on the electrolytic principle, substances present in solution are reduced or oxidized on a polarizable electrode. The qualitative and quantitative composition of the sample can then be found from the recording of the dependence of the current on the voltage applied to the electrode system.

Among various polarographic methods, differential pulse polarography (DPP) is most useful in environmental chemistry, chiefly because of its high sensitivity; this method permits analysis of all substances determinable by classical polarography. Another polarographic trace analytical method, square-wave polarography, has the same sensitivity as DPP, but only for reversible electrode processes and thus it is used mainly for determinations of heavy metal ions. The DPP method can also be used for determining surface-active substances that alter the capacity of the electrode double-layer through adsorption on the electrode; polarographic investigation of capacity phenomena is sometimes called tensammetry.

Especially for determinations of heavy metal ions at concentrations below the ppm level, electrochemical stripping analysis (ESA) can be used, where the test metal is electrolytically deposited on the electrode and then is stripped under polarographic control. This method combined with DPP is one of the most sensitive analytical methods. The instrumentation is relatively simple and permits field analyses to be carried out. Entirely automated monitoring stations can also operate on this principle. The ESA method can be used to determine Cu, Pb, Bi, Sb, As, Sn, In, Ga, Ag, Tl, Cd, Zn, Hg, Au, Ge, Te, Ni, and Co ions and many anions. An important advantage of the method is its ability (in contrast to other analytical methods, such as atomic absorption spectrometry) to differentiate between free and bound (complexed) forms of metal ions, which is important both for evaluation of the physicochemical properties of the test substances and from a biological point of view (e.g., in analyses of waters). Also many organic compounds can be determined by ESA after their adsorptive accumulation onto the electrode surface.

Polarographic methods also make it possible to determine aerosols of various metals in the atmosphere, after trapping the aerosols on a suitable filter and then transferring them into solution. Organic compounds in the form of gases and vapors in the atmosphere can be determined after absorption in a suitable solution or solvent. Determinations of metals and various compounds in biological materials after an extraction procedure are common. All polarographic measurements (and ESA) can be fully automated, which is advantageous in series analyses. One of the important applications of polarography is the determination of oxygen dissolved in potable water. Amperometric sensors of the Clark type are most often used, with a current proportional to the oxygen concentration. By placing an immobilized enzyme over the membrane of the Clark sensor, various amperometric enzymic sensors can be prepared. They are mainly useful in biochemical and clinical analyses, but they may become applicable to environmental analyses as well. It can be expected that it will be possible to "tailor-make" such electrodes to solve particular problems, especially in the form of bioanalytical sensors with bacteria or thin layers of tissues immobilized on the electrode surface. Electrocatalytic electrodes are available for monitoring various gases (SO_2, H_2S, CO, NO_x) in the atmosphere; electrochemical reactions of these gases take place on the electrode surface, giving rise to a current in the electrode system. The application of polarography is not limited to analyses of individual samples, but it is being progressively more applied to continuous analyses of gases and liquids. New designs of such devices are constantly appearing in the manufacturers' literature. Many monitors (for O_2, CO, HCN, etc.) have been constructed for personal protection of workers, traffic wardens, etc. It can also be expected that personal monitors will be available for the general population (e.g., for SO_2, to indicate whether a building should be aired at a given moment, in relation to the SO_2 concentration in the outside air, etc.).

Polarographic and voltammetric detectors find important applications in high-performance liquid chromatography (HPLC). This combination of the highly selective separation method with sensitive polarography leads to considerable broadening of the range of substances determinable by this chromatographic method. It can be expected that this advantageous combination will become one of the most frequently used analytical methods, even in environmental chemistry, because the determination of traces of highly toxic substances in an excess of other substances (or highly toxic impurities in common products, such as herbicides, pharmaceuticals, growth stimulators, etc.) will be required more and more often.

The further development of polarography will involve the attainment of still higher sensitivity and automation of the measurement and the handling of the results. For special purposes, the polarographic instrumentation will be miniaturized (it can be made in the form of an integrated or hybrid circuit in a case the size of a transistor). For the chemistry of the environment, the research into new types of solid electrodes with excellent reproducibility and time stability is very important.

3.3 Potentiometry

The electromotive voltage of a cell consisting of a measuring and a reference electrode immersed in the test solution is measured at zero electrolytic current. The measured value can yield the test substance activity in the solution (pH, pION) or can be used to detect the end point in potentiometric titrations. From the point of view of environmental protection, one of the advantages of potentiometric sensors, especially ion-selective electrodes (ISE), is the possibility of monitoring the activity of ions in solution (e.g., in water).

One of the very important measurements for the characterization of the environment is the measurement of the pH, which is now mostly carried out using glass electrodes. For long-term measurements, special glass electrodes are available, with accessories ensuring that the electrode membrane is clean. A glass electrode covered with a semipermeable membrane with an electrolyte film is also the basis of various glass probes applicable to analyses of water and the atmosphere (NH_3, SO_2, NO_x, CO_2, H_2S, etc.). The development of ion-selective electrodes is one of the greatest successes of analytical chemistry in the sixties and seventies. Research in this field is still in rapid progress. ISEs for F^-, I^-, Br^-, Cl^-, CN^-, SCN^-, NO_3^-, NO_2^-, ClO_4^-, S^{2-}, Na^+, K^+, Ca^{2+}, Ag^+, Cu^{2+}, Cd^{2+}, and Pb^{2+} ions are common. These electrodes can be used for measurements at concentrations from ca. 10^0 or 10^{-2} down to 10^{-5} or even 10^{-6} or 10^{-7} M. Measurements with ISEs are rapid, simple, and easily applicable to continuous measurements. Electrodes selective for certain organic compounds, e.g., alkaloids, have also been described. It seems practicable to propose carriers selective for each type of organic compound; membranes differentiating between isomers can also be prepared. There also exist electrodes exhibiting selectivity towards detergents.

Potentiometry is also used to determine redox potentials in water analyses; this is usually a mixed potential, as more than one redox system is present. However, this nonspecific measurement is a good measure of the concentration of reducing and oxidizing components in water.

Good prospects for the use of sensors based on MOSFET semiconductors (ISFET, CSFET) should be pointed out here. Selectivity is again attained by choice of a suitable membrane or a layer deposited on the transistor gate. The system is immersed in the test solution and the potential difference between the reference electrode and the transistor gate affects the current passing between the source and drain of the transistor. In view of the selectivity of the membrane or the deposited layer, the current is then a function of the activity of the given component in the solution. Semiconductor sensors are also the basis of monitors of various gases and vapors. Miniaturization of the sensors makes it possible to concentrate several different sensors on a single support, in the form of a "mosaic," thus yielding an analyzer for a whole range of substances; the signals of the individual sensors can then be gradually and repeatedly recorded in a measuring center.

The development of electronics enables the construction of compact probe analyzers even with contemporary ISEs, where the handle of the probe contains the signal-handling circuitry and a display.

Research on selective probes is very intense. Selective detectors of organic substances using the "lock-key" principle seem to be very promising. A matrix proposed on the basis of quantum chemical calculations will be able to bind a single substance, matching the matrix stereochemically. These specific sensors might replace separation processes, which would accelerate field analyses.

3.4   Coulometry

Coulometry is among the oldest electroanalytical methods. The test substance is deposited (or converted to another oxidation state) on an electrode at a constant potential and the charge passed is proportional to the amount of substance converted. In another modification of this method, coulometric titrations, the titrant is generated in the test solution electrolytically, at a constant current. The titrant consumption is found from the charge passed through the solution in the generation of the titrant, until the equivalence point is attained.

One of the advantages of coulometric methods is the fact that standardization can often be omitted, as the Faraday constant is a standard. In view of the absolute character of this method, the amount and not the concentration of the substance is determined. A disadvantage of constant-potential coulometry is the long time required for complete electrolysis. Computing techniques make it possible to shorten this time, by predicting the end of the electrolysis from the time-course of the electrolytic current after the beginning of the electrolysis and by calculating the charge or the substance concentration in the solution. In analyses of multicomponent samples, scanning coulometry can be used, in which the electrolysis potential is varied continuously or stepwise. In analyses of complex samples, coulometric titration is more advantageous and simpler, because 100% current efficiency of titrant generation is easier to attain by judicious choice of the reagent and the solvent. Coulometric titrations are suitable for determining substances in amounts of 0.01 to 100 mg (sometimes even below 1 μg). Sample volumes from 10 to 50 ml are common. The method is characterized by high precision, as the error does not exceed a few tenths of a percent even in coulometric titrations of microgram amounts. Under optimum conditions, these titrations can be performed with a precision and accuracy of 0.01%. Various acid-base, redox, precipitation, and complexometric titrations of inorganic and organic substances can be carried out coulometrically. For field work, automated coulometric analyzers for gaseous pollutants (SO_2, O_3, H_2S, NO_x, etc.) in the atmosphere have yielded good results. Many coulometric analyzers are available commercially. Coulometry is also used in electrochemical detection in liquid chromatography.

3.5   Conductometry

This method is based on measurement of the conductance of conductors of the second kind (i.e., electrolyte solutions). In spite of its nonspecificity, it is the most commonly used electrochemical method in the control of industrial chemical processes. From the point of view of pollution of, e.g., waters and the atmosphere, the outlet controls of industrial processes are important. Owing to the very low conductance of pure water, the total content of pollutants can thus be monitored, which is often sufficient. Typical examples of the use of conductometric methods for environmental protection are analyzers of detergents in waste waters, of concentrations of synthetic fertilizers in irrigation waters, of the quality of potable waters, etc. Conductometric analyzers are also used for monitoring atmospheric pollutants, e.g., SO_2 and H_2SO_4. In addition to direct conductometry, indirect methods can also be employed to determine pollutants and the measurement can thus be very selective. The substance to be determined reacts with a suitable reagent prior to the measurement and the resultant change in the conductance is caused solely by the test substance. In this way it is possible to determine nitrogen oxides after catalytic reduction to ammonia, HCl, HBr, and CO_2 after reaction with barium or sodium hydroxide, mercaptans, and hydrogen sulfide after reaction with cadmium nitrate. This principle of the determination of CO_2 can also be applied to an indirect determination of organic substances in water.

In addition to classical conductometric methods, high-frequency methods (oscillometry) are also used, in which the electrode system is not in direct contact with the sample; this principle is often employed in the construction of continuous analyzers.

3.6   Other Electrochemical Methods

Trace analysis can be carried out using modified chronopotentiometric methods, in which the time dependence of the working electrode potential is measured. Cyclic derivative chronopotentiometry (such as Heyrovský's oscillographic polarography with alternating current) can be used to monitor a great variety of organic substances, even those that cannot be determined polarographically, e.g., detergents; in this method, adsorption of organic substances on the electrode surface leading to a change in the differential electrode capacity, which produces characteristic changes on the chronopotentiometric curves, can be used. Measurement of the electrode impedance is another method suitable for monitoring surfactants, e.g., traces of organic substances in water. Combination of polarography with nuclear methods − radiopolarography − is suitable for monitoring radioactive pollutants.

3.7   Conclusion

The potentialities of various electrochemical methods in monitoring pollution of the biosphere are discussed in greater detail in the following chapters.

Electrochemistry is important not only in its analytical aspects, i.e., in determining the extent of environmental pollution, but also in removal of toxic substances already formed or other kinds of waste polluting the environment as a result of industrial activity or occurring in the field of communal hygiene. Electrochemistry also plays a significant role in designing new technologies that should not be harmful to the environment. Moreover, electrochemistry can, and probably will, strongly affect energy production and transport, which are additional fields that are increasingly more harmful to the environment. These aspects of electrochemistry are treated in detail in Refs. 1–6. For a detailed study of the principles of various electroanalytical methods, Refs. 7–11 can be recommended.

REFERENCES

1. Bockris J. O'M. (Ed.): Electrochemistry of Cleaner Environment. Plenum Press, New York 1971.
2. Bockris J. O'M. (Ed.): Environmental Chemistry. Plenum Press, New York 1979.
3. Balej J. (Ed.): Electrochemistry and the Environment (in Czech). Academia, Prague 1982.
4. Blažej A. et al.: Chemical Aspects of the Environment (in Slovak). Alfa, Bratislava and SNTL, Prague 1981.
5. Smythe W. F.: Electroanalysis in Hygiene, Environmental, Clinical and Pharmaceutical Chemistry. Elsevier, Amsterdam 1980.
6. Kalvoda R. and Parsons R. (Eds.): Electrochemistry in Research and Development. Plenum Press, New York 1986.
7. Sawyer D. T. and Roberts J. Z.: Experimental Electrochemistry for Chemists. J. Wiley, New York 1974.
8. Bond A. M.: Modern Polarographic Methods in Analytical Chemistry. Marcel Dekker, New York and Basle 1980.
9. Vydra F., Štulík K., and Juláková E.: Electrochemical Stripping Analysis. Ellis Horwood, Chichester 1976.
10. Wang J.: Stripping Analysis. VCH Publishers, Deerfield Beach, USA 1984.
11. Veselý J., Weiss D., and Štulík K.: Analysis with Ion-Selective Electrodes. Ellis Horwood, Chichester 1978.

CHAPTER 4

POLAROGRAPHIC
AND VOLTAMMETRIC METHODS

Robert Kalvoda

4.1   Introduction — Classical Polarography

Polarography, discovered by Jaroslav Heyrovský in 1922, is, in its original
classical form, still one of the principal methods of analytical chemistry.
Classical dc polarography is still used in chemical analysis, because of its
simplicity and very cheap instrumentation, provided that its sensitivity
is sufficient, permitting determinations to be carried out down to concentrations of $10^{-5}\,M$ and with a final optimal concentration range
between 10^{-3} and $10^{-4}\,M$. Any polarographic research or development
of a new method of polarographic determination starts, or should start,
with recording of classical polarographic curves with a dropping mercury
electrode, and, for this reason, even multifunctional modern instruments
permit recording of these curves. As the method and the apparatus are
simple, "apparatus artefacts" that would simulate, affect, or mask electrode
processes are virtually nonexistent. The operational parameters are
easily adjusted on the instrument and thus information on the electrochemical behavior of the test substance, required for the development
of an analytical method, is obtained rapidly.

Classical polarography has also been applied using electrodes other
than mercury, i.e., solid electrodes of various materials; the method is
then termed voltammetry, in accordance with the international electrochemical nomenclature. According to the recent proposal of the Electroanalytical Commission of the International Union for Pure and Applied
Chemistry (IUPAC), the term polarography should be reserved for the
voltammetric method using an electrode with a renewed surface. Even
this definition should be modified according to the proposal of the workers
of the J. Heyrovský Institute of Physical Chemistry and Electrochemistry
of the Czechoslovak Academy of Sciences, namely: "Polarography is
a method employing potentiostatic control of polarization of the working
electrode under conditions of convective diffusion, in which the current
response (i) of the electrode system is measured in dependence on the

polarizing voltage (E), the time change of which (dE/dt) is negligible with respect to the time constant of the mass transport to the electrode surface, i.e., the measurable i does not depend on the rate of the time change of the polarizing voltage." Further, we will generally use the term polarography for the method in which the sensor is a mercury electrode formed by mercury flowing out of a capillary.

Voltammetry is one of the most common methods in electrochemical research because it yields valuable information on electrode processes in a relatively simple manner. This is especially true of cyclic voltammetry in which the potential is linearly cycled. The relative positions of the anodic and cathodic peaks characterize the degree of reversibility of the electrode process and the peak size may yield information on chemical reactions coupled with the electrode process. This method is not as advantageous analytically as polarography. Voltammetric peaks do not have the shape of derivative peaks, as the current slowly decreases after attainment of the maximum and the currents are not additive for subsequent electrode processes.

4.2   Improved Polarographic Methods

An important improvement that has been made in classical polarography by introducing the sampled dc method (tast polarography) involves measuring the polarographic current before the end of the mercury drop lifetime with a dropping mercury electrode. If a stationary mercury electrode is used, then it is advantageous to integrate the current signal at the end of the drop time when the electrode surface area is constant. Integration eliminates the noise and thus the measuring sensitivity improves (detection limit of 10^{-6} to 10^{-7} M) [1].

A limiting factor in the use of classical polarography in trace analysis is the charging current (i.e., the current required to charge the electrode to a given potential), which is comparable with the electrolytic current at depolarizer concentrations of around 10^{-5} M. Therefore, if the sensitivity of polarography is to be improved, then the effect of the charging current must be eliminated. The first approach was the linear compensation of the charging current, introduced by Ilkovič and Semerano in 1932 and used in classical polarographs. A substantial step forward was achieved only after the development of polarography with a superimposed ac component. Various polarographic methods based on superposition of sine-wave or rectangular voltage impulses on a linearly increasing polarization voltage are described in detail elsewhere [2]. Of these methods in which the charging current is eliminated, pulse polarography is used most often and a great majority of commercial instruments permit use of this method.

There are two basic techniques of pulse polarography, normal pulse polarography (NPP) and differential pulse polarography (DPP). In the former method, where the polarization voltage changes from E to $E + \Delta E$ at the end of the dropping mercury electrode (DME) lifetime* (where the ΔE value continually increases), polarographic curves exhibit waves in the presence of depolarizers, similarly to classical polarography. Owing to elimination of the charging current, this method is about one order of magnitude more sensitive than classical polarography. The method is especially useful in work with solid electrodes that are often passivated by adsorption of an electrode reaction product or other solution components. As the electrode mostly remains at the initial, constant potential, these effects are suppressed because the value of E can be chosen in such a way that adsorption does not occur or the electrode surface can even be regenerated.

In the DPP method, which is most suitable for analytical purposes, an impulse with a constant amplitude of 10 to 100 mV and with a width of around 100 ms is superimposed over a polarizing voltage increasing linearly with time at the end of the DME drop lifetime. In contrast to classical polarography and NPP, peaks are formed on the polarographic curves instead of waves. The peak potential is virtually equal to the half-wave potential in classical polarography and the peak height is proportional to the depolarizer concentration (for details, see Ref. 2). The DPP method permits routine work at concentrations of 10^{-6} M and concentrations of 10^{-7} or even 10^{-8} M can be attained when observing all purity conditions required for ultramicroanalysis. A good signal-to-noise ratio must be attained to be able to measure the peak height with sufficient accuracy and precision.

The peak potential and height may be affected by instrumental parameters, especially the time constant of the memory circuits. If the time constant equals 100 ms, as in most commercial instruments, then the peak potential shifts to more negative or more positive values when polarizing toward negative or positive potentials, respectively. This is caused by the inability of the memory circuit to record the full applied

---

* The drop time is controlled by a mechanical device operated by impulses from the polarograph with a frequency determined by the operator. The polarization impulses are applied to the electrode at the same intervals. A suitable pulse frequency (denoted on the instrument as the drop time) must also be selected when working with solid electrodes.

voltage during the sampling period (20 ms) and thus the correct value is only obtained with a delay of several pulses (i.e., several DME drops). The long time constant also has an adverse effect on the peak height with changing of the DME potential scan rate; the greater the scan rate, the lower the peak, although it should remain constant (with a DME). However, this effect is unimportant in quantitative analysis, provided that constant conditions are maintained. These difficulties can effectively be suppressed when decreasing the time constant of the memory circuits below 10 ms. However, the signal-to-noise ratio deteriorates with low time constants and the curves are distorted by irregular noise at depolarizer concentrations as high as 10^{-6} M. Therefore, larger time constants lead to effective filtering, but at the cost of a certain signal distortion. However, instruments have been proposed containing electronic circuits that permit curve recording at virtually zero time constant of the analog memories, with preservation of an effective filtering effect [3].

In the development of polarography, efforts to improve the sensitivity of the method have been accompanied by attempts to accelerate the polarographic recording. After application of various methods of oscillographic polarography, the advent of rapid recorders permitted an increase in the polarizing voltage scan rate to values around 100 mV/s. For example, in the fast-scan DPP method, the polarographic curve is recorded within a few seconds at the end of the growth of the mercury drop [4]. The polarographic analysis is not only accelerated, but its sensitivity also increases, by about a half an order of magnitude. A special spindle-shaped polarographic capillary is used as the DME, which has an advantage in that the mercury flow rate and the drop time can be varied independently over a wide range [5].

An important part of modern polarographs is the potentiostat that maintains the working electrode potential at the value of the pre-programmed voltage regardless of the electric resistance of the test solution in the electrolysis cell and is especially advantageous in analyses of organic compounds that are often poorly soluble in water. Moreover, in nonaqueous, aprotic media, many electrode processes are substantially simpler than in aqueous solutions, yielding simple peaks on polarographic curves. In work with potentiostatic circuits, a two-electrode circuit should be used during all manipulations with the capillary and the solution, during passage of nitrogen, etc., and only for the actual measurement should a three-electrode circuit be employed. In work with three-electrode circuits, it is preferable to employ a silver/silver chloride reference electrode, because it is much more resistant to current passage that might change its potential than the calomel electrode (interruption of the liquid junction between the electrodes may cause the potentiostat to oscillate or to bring its control amplifier to saturation, leading to passage of current through the reference electrode with consequent change in the potential or even damage to the reference electrode).

4.3   Electrochemical Stripping Analysis (ESA)

In this method, the test metal is first electrolytically accumulated on an electrode at a constant applied voltage. The accumulated metal is then stripped by polarizing with a voltage scanned from the preelectrolysis value to more positive values. The method has a number of modifications, is extremely simple, and can thus be performed even with the simplest polarographs (for details, see Chapter 5). The stripping process can be monitored not only by classical polarographic methods, but also by using methods involving elimination of the charging current, resulting in improved sensitivity (i.e., shorter preelectrolysis times can be employed) [6].

4.4   Electrodes Used in Polarography
and Voltammetry

The dropping mercury electrode (DME) remains the most frequently used electrode. This ideally reproducible electrode has been perfected especially from the point of view of the precision of the drop time. The static mercury electrode (SME) also finds practical applications, either as a hanging drop electrode (e.g., for ESA) or as a dropping electrode. The mercury flow through the capillary is controlled by a needle valve or a similar device [5, 7], controlled electronically by the polarograph. When a dropping electrode is used, the drops are periodically disconnected by a hammer, according to the preset drop time; the time of drop growth, i.e., the drop size, can be selected. The SME is usually contained in a special stand containing electronic circuits to automate the operation. For the purposes of ESA, circuits are included that control the time of nitrogen passage, solution stirring, etc.

A number of solid electrodes are used for voltammetric study of the oxidation of organic compounds, especially gold, carbon, and platinum electrodes that can be polarized to substantially more positive potentials than mercury electrodes. These electrodes are often used in the form of rotating disk electrodes. Their reproducibility and stability in time are still unsolved problems. There is no universal regeneration procedure for these electrodes, because the passivation is caused both by the initial and the final components of the electrode reaction. Every analytical method using solid electrodes also describes (or should describe) a procedure for regeneration of the electrode surface (mechanical polishing, polarization to negative and positive potentials, etc.).

Manufacturers of polarographs supply manuals for introduction to polarographic practice that also contain tests for instrument function. The excellent textbook by Heyrovský and Zuman [8] was written when polarographs still employed photographic recording, but most of the procedures can be modified for modern polarographs. The theoretical principles of polarography are dealt with in the monograph by Heyrovský and Kůta [9]. Advances in polarographic analysis are discussed in Ref. 10.

4.5 Polarography and Its Importance in Environmental Protection (EP)

Polarography, especially the most sensitive variants, DPP and square-wave polarography, finds use in all fields of chemical analysis. In addition to purely analytical applications, the methods can also be used to solve theoretical problems that, in turn, affect the solution of analytical problems (e.g., the investigation of electrode kinetics, reaction mechanisms, reaction kinetics, composition of complexes, etc.). Thus polarography is also useful in solving problems of environmental protection, as it fulfills all the criteria placed on methods for the analytical chemistry of the environment, primarily:

    (1) high sensitivity
    (2) good selectivity and resolution
    (3) sufficient accuracy and precision
    (4) rapidity
    (5) wide application range
    (6) possibility of determining several substances simultaneously
    (7) simple preparation of samples for the analysis
    (8) easy operation of the instrument
    (9) possibility of extensive automation
  (10) possibility of work in the field
  (11) acceptable cost

   The above criteria can be used to evaluate the polarographic methods:
(1) The DPP method permits routine analyses of solutions at a 10^{-6} M
concentration level (ca. 100 ppb),* and under optimal conditions, maintaining a high purity of the chemicals and water and cleanliness of the
polarographic cell, in the absence of impurities in the samples, even substances at a concentration of 10^{-8} M (1 ppb) can be determined. The
quality of the instrument is also important for obtaining a good signal-to-
noise ratio even at the lowest concentrations. The noise (appearing as
fluctuations on the curve) may cause substantial difficulties in determining
the peak height and consists of insufficiently compensated charging
current, the noise from the line frequency, and random noise. The noise
stems from the apparatus itself and the latter two kinds of noise also from
external conductors and instruments close to the electrode circuit and
the leads to the instrument. The voltage can also be affected by the noise
from various electrical appliances (motors, etc.). The combination of
DPP or square-wave polarography with ESA increases the sensitivity
by up to three orders of magnitude (detection limits of 10^{-3} ppb).
(2) The selectivity of the method is determined by the various deposition
potentials of inorganic and organic substances, depending on the base
electrolyte composition. With inorganic depolarizers, complexing media
can often be used to eliminate the interference from some substances.
If two substances are to be differentiated, their half-wave potentials must
differ by at least 100 mV to avoid distortion of the wave heights measured.
The DPP method can differentiate peaks with substantially smaller differences in their peak potentials (using as small a pulse amplitude as
possible), but the information can only be used for qualitative purposes
or for rough estimation of the concentrations. The situation can be improved by judicious choice of the base electrolyte or a preliminary separation (in ESA, e.g., by exchanging the base electrolyte after preelectrolysis). Computer methods have also been developed to deconvolute overlapping peaks. Special evaluation methods can also be used, involving
subtraction of the base electrode curve from the sample curve, thus also
improving the precision of the peak height measurement and hence also
the precision of the determination.
(3) Polarography (and DPP) has a highest precision of about 0.5 %.
At very low concentrations (and in ESA), the precision is poorer, about
5 % to 10 %, but in these cases, especially in analyses of complex samples,
even errors of 20 % are tolerable.

---

   * 1 ppm = 10^3 ppb = 10^{-4} % = 10^{-5} mol liter^{-1} (assuming the molecular weight of
the substance to be 100).

The accuracy of the determination may be affected by many factors, even if the errors committed in the sample preparation are not considered. The main sources of error are impurities in the chemicals used and adsorbable, especially high-molecular-weight, substances that must often be separated, e.g., on Sephadex (Pharmacia, Uppsala) [11] or Extrelut (Merck, Darmstadt) [12].

(4) The common time of polarographic curve recording, 5 to 10 min, can be shortened to several seconds by using a faster polarization voltage scan rate. The time of polarographic analysis further involves the time for solution deaeration (5 to 10 min) and preparatory operations; the latter are often substantially more time consuming than the polarographic analysis itself.

(5) Polarographic methods are applicable to both inorganic and organic compounds. In the inorganic field, most elements can be determined, but heavy metals are of the greatest interest in the environment. Among organic substances, those with reducible groups can be polarographed (aldehydes, ketones, quinones, nitro and nitroso compounds, compounds containing a hydroxylamine group, azo and azoxy compounds, azomethines, unsaturated compounds, halogeno compounds, etc.). The development of relatively stable and reproducible solid electrodes (glassy carbon, graphite paste, Au, Pt, etc.) has permitted determinations based on the oxidation of organic compounds on the electrode surface (e.g., aromatic hydrocarbons, aromatic amines, phenols, aliphatic acids, alcohols, amines, heterocyclic compounds, alkaloids, such as morphine, papaverine, etc., sulfur-containing substances, etc.). Adsorption effects are used more and more often and yield typical tensammetric peaks on polarograms obtained by methods with an ac component of the polarizing voltage, due to adsorption and desorption of the surfactant.

Polarography can be used for determinations of substances dissolved in solution, as well as for substances suitably trapped from the gaseous phase. There are many types of amperometric detectors for monitoring gases (O_2, CO, NH_3, SO_2, H_2S, HCN, etc.) in the atmosphere. In addition to analyzers and monitors (see Chapter 8) based on the polarographic principle, electrochemical (amperometric) detectors for high-performance liquid chromatography (HPLC), combining the high sensitivity of polarographic measurements with a highly efficient separation method, are gaining importance. Analogous detectors are also commercially available for ion chromatography and flow-injection analysis (FIA).

(6) The possibility of simultaneous determination of several substances is based on differences in the half-wave potentials of inorganic and organic substances that can be affected by judicious choice of the base electrolyte.

(7) The sample preparation for the analysis with various matrices is similar to that in other instrumental methods. It is somewhat simplified with DPP, which is very sensitive, and thus small samples are often sufficient. Various separation procedures are frequently necessary (extraction, liquid chromatography, etc.), as well as sample pretreatment (combustion, etc.). Most published procedures for determinations of emissions contain (or should contain) a prescription for sampling and sample pretreatment.

(8) Every manufacturer tries to simplify the instrument operation by decreasing the number of controls and using combined controls, so that the instrument can be operated by technicians after a brief training period. Nevertheless, the interpretation of the results obtained must mostly be carried out by a specialist. The operation is further simplified by partial or full automation of the instrument. Especially computer-controlled polarographs (e.g., the PARC 384 instrument from Princeton Applied Research Corp., USA, the ECM 700 produced by the Academy of Sciences of GDR, etc.) can be programmed to perform series analyses with minimal human intervention, especially when combined with an automated sample feeding system.

(9) The polarograph designed by Heyrovský and Shikata in 1924 was actually the first automatic instrument for chemical analysis, with recording of results in analog form. Computing techniques have permitted full automation of the polarographic measurements. A microcomputer not only controls the measuring procedure, but also handles the results obtained. A polarographic curve stored in a computer memory can be subsequently treated mathematically (subtraction of the base electrolyte curve from that of the sample, averaging of repeated recordings, differentiation, etc.). The result of the measurement is printed out in the required units (e.g., ppm, ppb), as well as the half-wave and peak potentials. The analog recording of the curve still retains importance as a control method, because some anomalies may be obscured during digitization of the signal. It should be borne in mind that the computer cannot produce correct results from erroneous data; this is, of course, valid generally for all micro- computer-controlled instruments.

Cheaper polarographs (without microcomputers) are also partially automated, especially in the control of ESA; the length of the individual operation periods (removal of oxygen from the solution, preelectrolysis, rest-period, solution stirring, etc.) can be preprogrammed.

Special types of polarographic instruments are various single- purpose analyzers and monitors that should operate for a long time without operator intervention.

(10) Field work requires instruments with a low weight, insensitive to

mechanical shocks, and with a low electric input. All these requirements are met by a polarograph constructed of semiconductor and integrated circuit components. The problem of the sensor is more difficult, especially with DMEs, but can be solved. Therefore, polarographs can be found in the equipment of numerous mobile laboratories for analyses of the atmosphere and waters. A number of works have described the use of polarography in floating laboratories, especially for analyses of sea water for trace heavy metal ions. Automated ESA of water, performing measurements at required intervals and recording the results, after digitization, on a magnetic tape, has been described [13]. The operator replaces the tape cassette once a week. The results from the individual cassettes are then handled by a computer.

Various types of single-purpose analyzers have also been designed for field work.

(11) The cost of common polarographic instrumentation with accessories (i.e., not that of a microcomputer instrument) is not high — substantially lower than, e.g., that of AAS or gas chromatographic instrumentation. A simple apparatus for classical polarography can be assembled from common electronic components at negligible cost [14] (see also Chapter 13); the only somewhat expensive instrument required is a recorder.

4.6   Examples of the Use of Polarography in EP

The development of polarographic instrumentation, especially towards higher measuring sensitivity, has been partly stimulated by the requirements of environmental analysts, who need a sensitive and cheap method for the determination of traces of heavy metals. DPP (possibly combined with ESA) is undoubtedly such a method and permits the determination of all substances (inorganic and organic) determinable by classical polarography and, in addition, of substances that can be adsorbed on the electrode surface.

The following sections summarize some applications of polarography to the determination of inorganic and organic substances important for environmental protection. The bibliography of polarographic works up to 1967 [15] should be mentioned here; many application possibilities that could not be discussed in this book can be found in this publication.

4.6.1   Determination of Inorganic Substances

Methods of determination of various inorganic substances can be found in classical polarographic monographs published when the term "environmental protection" was not yet used. Many "new" methods of polarographic analysis have merely been "rediscovered" and older methods have been modified (often without reference to the original author).

A thorough discussion of the determinations of metals and other inorganic as well as organic substances in water, the atmosphere, soils, biological materials, etc., can be found in the monograph by Březina and Zuman [16], including determination of Cr, Fe, Co, Mn, Ni, and heavy and alkaline metals in a great variety of matrices and determination of anions, such as SO_4^{2-}, $S_2O_3^{2-}$, Cl^-, I^-, ClO^-, and NO_3^-, oxygen, and surfactants in water. Determinations of many substances in the air are described, e.g., Pb, Cd, Bi, Cu, Zn, ZnO, As, Sb, Cr, CrO_3, Mn, V, O_2, Cl_2, SO_2, H_2S, benzene and its homologues, formaldehyde, hydrocarbons, nitromethane, and phenol. Detailed procedures for the determination of these substances, as well as of the aerosol of sulfuric acid and alkali sulfates, nitrogen oxide-containing gases, thallium compounds, and iron are given in Ref. 17.

In inorganic analysis, pollution of river water with metal ions emitted from factories, mines, garbage dumps, etc., must often be monitored. Direct determination by DPP is advantageous for pollutant concentrations between 10 and 100 ppb. The analysis of potable water contaminated by As (III) and Pb (II) from a zinc smelter is described in Ref. 18. In a solution of $2\,M$ H_2SO_4 + NaCl, the peaks of arsenic and lead have potentials of -0.29 and -0.46 V (SCE), respectively. An extensive review [19] is devoted to the determination of selenium in environmental waters, in which various analytical methods are discussed, including polarography. The published data on the speciation and the concentration levels of selenium in water from various localities are critically evaluated. Uranium can be determined by DPP at a sub-ppb level [20], using the classical catalytic reaction between the uranium (III) ion formed at the electrode through the reduction of the uranyl ions and the nitrate anions. Uranium can also be determined by accumulating it on a mercury electrode through adsorption of its complex with pyrocatechol, down to a concentration of $5 \times 10^{-9}M$ [21]. The method has been successfully applied to the determination of uranium in sea water; copper is separated on Dowex A1 ion-exchanger from an acidic medium containing DCTA (uranium is not sorbed). To determine molybdenum in industrial waste waters, the catalytic polarographic wave formed in the presence of 8-hydroxyquinoline in a weakly acidic medium containing an excess of nitrate can be utilized [22]. The DPP calibration curve is linear from 1×10^{-8} to $5 \times 10^{-7} M$ and the detection limit is 0.4 ppb.

To determine anions in water, ion-selective electrodes are useful; nevertheless, polarography is also sometimes useful, especially in the determination of nitrite and cyanide, where ion-selective electrodes have limited use. A newer method for the determination of nitrite and nitrate at a ppm level in solids and water can be found in Ref. 23. A determination of nitrite in the presence of a hundredfold excess of nitrate in a citrate buffer with pH 2.5 (with a peak at -1.0 V, SCE) has been described in Ref. 24. The procedure given must be followed exactly, because of rapid sample decomposition. To determine traces of cyanide in water, down to a concentration of 1 µg/liter, the catalytic current produced from the interaction of CN^- with the Ni–ethanolamine complex can be utilized. Fast-scan DPP (on a single mercury drop) was used and the peak potential equals -1.70 V (SCE) [25]. Cyanuric chloride, formed by polymerization of chlorocyan, can be determined in the atmosphere after absorption in methanol. The substance yields a DPP peak at $+0.125$ V in a base electrolyte containing a mixture of an acetate buffer and methanol. The substance can be accumulated on a mercury electrode by adsorption. Hundredths of a milligram of the substance per cubic meter can be determined [26].

Atmospheric sulfur dioxide can be monitored down to a concentration of 0.1 ppm, after trapping in dimethylsulfoxide containing 0.1 M LiCl [27].

Determination of traces of heavy metals in water is the subject of most papers dealing with applications of ESA. An independent chapter describes this topic (see p. 67) and thus only a few examples are given here. These techniques are practically standard methods for the determination of Cu, Pb, Cd, Zn, and Hg, down to a concentration level of 1 ng/liter. The advantages and reliability of the method are demonstrated by the fact that the method for the simultaneous determination of Cu, Cd, Pb, and Zn and of Pb and Tl in potable water has been included in a West German standard [28]. A continuous analyzer of Zn, Cd, Pb, and Cu in water down to a concentration of 0.1 µg/liter is described in Ref. 29. Combined ESA and DPP are used. The sample is uv irradiated, acidified, and heated to 95°C to liberate the metals, and is then cooled, filtered, and transferred to the electrolysis vessel. The pH is adjusted to 5 and oxygen is expelled. The instrument can operate automatically for 1 month. An automated, microcomputer-controlled analyzer for the determination of trace heavy metals in water has been described [30]. Polarography has been combined with coulometry [31] for monitoring Ag, Pb, Cu, Co, Fe, Cd, Zn, and Ni down to a concentration of 1 mg/liter in waste waters. The analyzer can operate for several months without control.

Many papers have been devoted to the monitoring of trace heavy metals in natural water [32–34] and in ecological chemistry in general [35]. An important contribution of the polarographic method is the possibility of studying various forms of bonding of metals in complexes with various ligands occurring in natural water (carbonate, phosphate, etc.) [36, 37]. Polarography is generally the most sensitive analytical method available for monitoring heavy metals in sea water directly at the sampling site.

To liberate various metals, e.g., Sn, Cu, Pb, and others, from biological materials, such as plants, plankton, sediments, fish tissues, and fruit juices, digestive solvent Lumatom (H. Kürner, Neuberg, FRG) containing quaternary ammonium hydroxide in toluene has been recommended. Lumatom can be present directly in the polarographed solution. The calibration curves are linear from 10^{-8} to 10^{-6} M concentration. ESA with a 60-s preelectrolysis permits the determination at a ppb level [38].

Trace metal aerosols in the air can also be analyzed by ESA. The test air is passed through a membrane filter which is then mineralized and the collected metals are dissolved. The determination is then carried out in a suitable base electrolyte. Part of this work has been undertaken in order to correlate the content of aerosols of heavy metals (especially Pb) with the density of traffic in various places [39, 40] and at various times of the day. A number of investigations have been devoted to the determination of traces of Cu, Pb, Zn, Mn, and Fe in the atmosphere of industrial areas [41]. The analysis of aerosols of some other metals (Ni, Cr, Se, Fe) and their mixtures around industrial enterprises has been reviewed [42]. About 2 000 liters of air are passed through an acetylcellulose filter at a rate of 130–150 liter/min. Various organic compounds of lead in vapors and aerosols can be determined analogously. As about 90% of toxic metals are contained in rain water and thus are readily absorbed by vegetation, atmospheric precipitation should be analyzed [43]. The toxic metal concentration is highest during the initial two hours of rainfall, owing to intense washout, and then it decreases substantially.

In the determination of silver aerosol after dispersion of AgI in clouds [44], the sensitivity of polarography (detection limit, 10^{-11} M) surpasses that of neutron activation analysis.

An example of the use of ESA for complex investigation of the environment at a certain locality (here Staten Island, New York) has been given [45]. This approach is based on the assumption that the presence of heavy metals is an indicator of environmental pollution, similar to the presence of *E. coli* indicating microbial pollution of water. In these measurements, the content of heavy metals was monitored in waters around the island and the island surface waters, in the atmosphere, and in samples of mussels that represented a stage in the effect of polluted water on the circulation of nutrients. Human hair and nails were also analyzed as the final objects affected by the pollutants. The high level of heavy metals found in the biosphere was correlated with the high occurrence of cancer and cardiovascular diseases in this area. In this work, ESA was also compared with AAS. An advantage of ESA is the possibility of using very small samples (20 mg), e.g., in the analysis of hair and nails. Drawbacks of ESA are the necessity to modify the procedure depending on the type of matrix and the longer time required for the analysis.

4.6.2   Determination of Organic Substances

The polarographic behavior of organic compounds is surveyed in the still useful monograph of Březina and Zuman [16]. This book provides basic information on the reducibility of various functional groups, on reaction mechanisms, and on isolation of test substances from biological materials. Many nonreducible organic compounds can be converted into electro-active substances (e.g., by conversion into nitroso or, more often, nitro compounds). Therefore, polarographic reducibility can be predicted from the structure of substances and the conditions for the determination can be selected by analogy with substances for which procedures for the determination have been found. The electrochemical behavior of organic compounds has been summarized in tabular form [46].

The question of whether a substance is voltammetrically oxidizable (i.e., using solid electrodes instead of mercury electrodes to which a positive voltage range cannot be applied) is more difficult to answer. Some types of compounds are readily electrooxidized on the electrode but an analytical determination also requires that the waves or peaks obtained be readily measurable and reproducible. The oxidizability further depends on the electrode material and the character of its surface. These problems are discussed in detail in Ref. 47, especially with regard to electrochemical detection in liquid chromatography.

The problem of application of polarography to the determination of organic compounds important in environmental protection lies in the method of isolation of the test substance from the matrix rather than in the polarographic analysis itself; however, isolation procedures are required for all measuring techniques. An advantage of polarography, especially DPP, is the possibility of using small samples, because of the high sensitivity of the method (in the μg/kg range). The isolation procedures are often substantially simpler than, e.g., in gas chromatography. The sample pretreatment may also be substantially affected by the possibility of polarographic measurements in nonaqueous solutions, in addition to aqueous media.

A polarographic determination usually consists of a number of steps, most of them related to the sample pretreatment. First, a method for sampling and sample storage must be found, followed by isolation of the test substance. The isolated substance must often be further purified but, in view of the high sensitivity of polarography, preconcentration is usually unnecessary (other than the ESA preconcentration). If the substance cannot be reduced, it is sometimes possible to form a derivative that is electro-active (e.g., phenol–nitrophenol). Only then does the polarographic determination follow.

Even before sampling, it must be considered whether the test substance can react with another compound in the sampled material to form some other compound, whether it can be degraded, metabolized, etc. An example of the complexity of the problem is the analysis of atmospheric emissions that react together and also interact with the atmosphere. Toxic substances to be determined can be converted into substances with an even higher toxicity during operations performed to liquidate them (e.g., phenol–chloro-phenol) and the products formed must also be determined. The isolation and other operations must be carried out under mild conditions, as some substances are readily decomposed (e.g., pesticides).

Therefore, the procedure for the determination will strongly depend on the sample composition. For example, the determination of a pollutant in waste water is affected by the origin of the water, i.e., by the contents of the accompanying substances. These, often specific, conditions must be respected during the sample treatment, even if the polarographic analysis itself is often performed under identical conditions. Therefore, most analytical papers dealing with environmental protection contain an exact procedure for sample treatment.

In view of the above facts, the summary of the published works below is purely informative and cannot replace a study of the original literature.

Pesticides are widely monitored from the point of view of environmental protection, as positive economical effects in agricultural application of pesticides are accompanied by negative consequences that are sometimes caused by careless manipulation with substances that are sometimes highly toxic. The monitoring of these substances in the environment is of great interest, not only to hygienists, but also to workers in agriculture, in the food industry, to physicians, and to veterinarians. Especially residues of pesticides in various, mostly biological, materials, water, etc., should be determined.

The determination itself is preceded by a number of operations in the sample treatment, as mentioned above, which are often difficult, because some of the test substances are highly labile. Some pesticides are poorly soluble in water and the base electrolyte must contain a nonaqueous solvent. Polarographic determination of pesticides is discussed in several monographs [48, 49] and reviews [50–53].

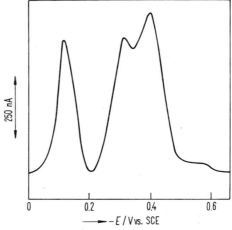

Fig. 4.1 The DPP curve of 2,4-dichlorophenoxyacetic acid at pH 6 after nitration with a 1 : 2 mixture of HNO_3 and H_2SO_4. Sample concentration, 100 µg/liter, DME, drop time, 1 s, pulse amplitude, 25 mV, scan rate, 2 mV/s. The first three peaks correspond to the reduction of the nitro groups (from Ref. 55).

Pesticides containing a nitro group, such as parathion and methyl-parathion, can be determined at any base electrolyte pH; the same holds for paraoxone and *p*-nitrophenol, which is a degradation or metabolic product of parathion [54]. After nitration, herbicides derived from 2,4-dichlorophenoxyacetic acid (which, in a mixture with the 2,4,5-trichloro-derivative, was used as a defoliant in the Vietnam War under the name Agent Orange) can be determined. Reference 55 describes a detailed study of the nitration process, i.e., the effects of the composition of the nitration mixture, the time of nitration, and the temperature (Fig. 4.1). A method has been described for the extraction of residues of these substances in irrigation water leaving fields, minimizing the extraction of accompanying compounds (humic acids, other pesticides). Phenitrothion has been determined in drainage water leaving fields [56]. DDT can be determined after extraction from aqueous solutions into tetrachloromethane, followed by evaporation and nitration of the residue [57]. The tetranitro derivative formed is determined after dilution of the mixture with water and methanol. Hexachlorocyclopentadiene and analogous compounds are determined in an acetone solution of tetraethylammonium iodide after extraction into benzene [58]. Organic tin compounds used as pesticides can be very sensitively determined by using ESA [59]. Organometallic compounds of dithiocarbamate pesticides (zineb, ziram) yield a polarographic wave for the reduction of zinc ions in 0.2 M NaOH at -1.42 V. The oxidation of diethyldithiocarbamate produces an anodic wave in 0.5 M NH$_4$OH + NH$_4$Cl at -0.41 V [60]. Zineb, maneb, and ziram are adsorbed on mercury electrodes and can be preconcentrated on the electrode in this way, down to amounts of 0.1 μg (e.g., after isolation from fruit, leaves, etc.) [61].

ESA can also be used to monitor pesticides that form insoluble compounds with mercury ions; they are preconcentrated at the mercury dissolution potential and the stripping of the compound formed is recorded polarographically. In this way, pesticides derived from thiourea can be determined [51]. The detection limit for the DPP determination of the strongly toxic herbicides paraquat and diquat, derived from bipyridyl salts, is 50 ppb [62]. In analyses of aqueous solutions (contaminated water, plant tissue extract), paraquat and diquat were sorbed on a strongly acidic cation-exchanger (Amberlite IR 120), eluted by a saturated solution of ammonium chloride, and determined by DPP in the effluent [63]. The peak potentials are -1.12 V for paraquat and -1.02 V for diquat (vs. SCE). The detection limit equals a concentration of 5×10^{-8} M. Polarography of organophosphates is discussed in a monograph [48]. They can also be determined by a method based on inhibition of carboxyl esterase: the unreacted enzyme liberates 2-naphthol from 2-naphthol acetate and this product is polarographed after nitrosation. Down to 10^{-8} M inhibitor can thus be determined [64]. Polarographic methods have been compared with spectroscopy and gas chromatography in the analysis of pesticides [65]. An advantage of polarography is its sensitivity and the fact that the isolation procedures required are simpler than with gas chromatography [66]. Some insecticides derived from methylcarbamate (methocarb, butacarb, carbaryl, etc.) are electroinactive, but yield tensammetric peaks (see p. 51), permitting the determination to be carried out down to a 5×10^{-6} M concentration [67] in a phosphate buffer of pH 8.4; the detection limit for aldecarb is 5×10^{-5} M. DPP with a DME was used to determine triazine herbicides (aziprotryn, desmetryn, methoprotryn, prometryn, terbutryn, atrazine, cyanazine, simazine, terbuthylazine, metribuzine), with a detection limit of about 5×10^{-8} M [68]. The base electrolyte is 0.05 M H$_2$SO$_4$ or acidic buffered solutions with a pH down to 2.5. In analyses of waters, the substances can be extracted into chloroform from the sample alkalized to a pH of 8 to 9. The polarographic determination of herbicides containing sulfur (methonyl, aldicarb) is based on catalytic effects [69].

Graphite paste electrodes have been used for the oxidation of many toxic and carcinogenic substances at concentration levels from ppm to ppb, involving amines and phenols, such as benzidine, *o*-dianisidine, naphthylamine, diphenylamine, *p*-chlorophenol, etc. [70–72]. Attention has been paid mainly to utilization of the electrooxidation of various polycyclic substances, such as benzpyrene (down to a concentration of 10^{-8} M in anhydrous acetonitrile containing $NaClO_4$). These substances have also been determined using a rotating platinum or glassy carbon electrode in sulfolan solvent containing 0.1 M tetrabutylammonium hexafluorophosphate.

From the point of view of chemical carcinogenicity, polarographic determination of nitroso compounds is gaining importance. These substances are often contained in foodstuffs, drinks, and the atmosphere and may be formed in the body by nitrosation of amines, especially in the stomach where sufficient acidity is ensured. In neutral and slightly alkaline media, *N*-nitrosation is possible in the presence of aldehydes, phenolic compounds, etc. However, *N*-nitrosamines can also be formed in the presence of nitrates and bacteria that are capable of reducing nitrate to nitrite. This is also one of the reasons why it is important to monitor nitrates that are often present in potable waters, vegetables, etc.

The polarographic behavior of nitrosamines was studied even before their carcinogenic effects were known [73]. These substances, e.g., derivatives of proline, pyrrolidine, and piperidine, aliphatic and aromatic derivatives, are reduced at the DME at pH 0 with exchange of four electrons in two steps, first to hydroxylamine and then to hydrazine, with half-wave potentials of between -0.6 and -0.7 V (SCE). Similar results have been obtained [74] for aliphatic nitrosamines. Traces of nitrosamines could, however, be determined only after introduction of modern polarographic methods with elimination of the charging current, such as DPP. The detection limit is 7×10^{-8} M for various carcinogenic nitrosamines (*N*-nitrosopyrrolidine, *N*-nitrosoproline, *N*-nitroso-4-hydroxypyroline, etc.) in 0.1 M HCl or H_2SO_4 [75]. *N*-Dimethylnitrosamine and *N*-nitrosoproline in water and serum (at concentrations from 0.2 to 20 ppm) [76] and *N*-nitroso-*N*-methylaniline and other nitrosamines in blood and urine [77] have been determined using DPP. A DPP determination of *N*-nitroso-*N*-methylaniline in biological materials [78] has a detection limit of 10^{-7} M. Volatile nitrosamines have been determined [79] after photochemical conversion into nitriles. The photochemical conversion takes place on sample injection into a stream of 5×10^{-3} M NaOH and the amperometric determination itself is carried out in an acidic medium, using a platinum electrode. The detection limit roughly corresponds to 1 ng of N-nitrosodipropylamine. The method was applied to analysis of cured smoked meat and bacon. The DPP method was also used to determine nitrosamines in liquids employed for grinding in the metallurgical and machine industries [80]. These liquids contain diethanol-amine and nitrite at high concentrations, so that N-nitrosodiethanolamine may be formed. Electroanalysis of carcinogenic substances has been re-viewed [70] and the work contains an extensive list of references.

The DPP determination of nitroso compounds at concentrations of 1 to 10 µg/kg is relatively simple, but the isolation of the test compound from actual samples is difficult. These problems have been examined thoroughly in a paper [81] dealing with the determination of dimethyl-nitrosamine in beer and malt. The method is evaluated and applications to other substrates are discussed.

High-performance liquid chromatography with polarographic de-tection is suitable for separation of nitrosamines [75, 82–84]. With regard to this combined technique, which is gaining importance, the polaro-graphic behavior of many N-nitrosamines (N-nitrosodiethanolamine, N-nitrosoproline, N-nitrosodiphenylamine, and N-nitrosodibutylamine) has been studied in base electrolytes containing methanol or acetonitrile [85]. Nitrosamines can be determined by much more sensitive methods than polarography, but the instrumentation for these methods is much more expensive and polarography can be useful in building extensive networks of analytical laboratories.

Polarography has also found use in the determination of other carcinogens, such as aflatoxins [86]. Trichocene toxins have also been determined polarographically [87].

In addition to purely analytical applications of polarography, the method can also be used for the screening of various compounds in $vitro$ for the mutagenic and carcinogenic effects resulting from their interaction with nucleic acids. For example, 7-methylguanine was determined in acidic hydrolyzates of nucleic acids after alkylation [88]. The alkylating agents, considered to be mutagenic and carcinogenic, are capable of affecting DNA and RNA through alkylation of the guanine N(7) atom. The con-version of guanine into 7-alkylguanine can thus be used as a measure of the alkylation of the nucleic acid. The DPP method is based on the electro-oxidation of 7-methylguanine at a glassy carbon electrode in a medium buffered at pH 1.2. To determine the degree of methylation of DNA, the ratio of 7-methylguanine to guanine is measured; both the substances yield well-developed peaks, that of 7-methylguanine being about 150 mV more positive. This method enables direct determination of the concen-tration ratio of the two substances after acid hydrolysis, without separation.

The DPP method can be used to demonstrate structural changes in natural DNA [89] caused by physical effects (temperature, irradiation) or by the action of chemical reagents (e.g., N-methyl-N-nitrosourea [90], N-acetoxy-N-2-acetylaminofluorenone, organophosphates). These interactions are manifested by a change in the height of the peak of natural DNA and possibly by the appearance of another peak corresponding to denatured DNA (natural and denatured DNA can be differentiated polarographically).

The "Priority Pollutant List" [91] also contains substances used in the manufacture of polymers, especially acrolein, acrylonitrile, and vinyl chloride. They can also be determined polarographically. The polarographic behavior of methacrylic acid and of its ethyl, propyl, and butyl esters and acrylonitrile have been studied [92]. The esters are reduced at a potential of about -2.0 V (SCE) in a solution of 0.1 M tetramethylammonium bromide in 50% ethanol and acrylonitrile is reduced at -1.85 V (SCE) in an aqueous solution. Acrylonitrile yields a polarographic wave with a half-wave potential of -2.2 V (SCE) in 0.1 M tetraethylammonium perchlorate [93]. Formaldehyde and glycolic acid nitriles do not interfere in the determination of acrylonitrile in 0.02 M tetramethylammonium iodide [94]. The DPP determination can be carried out from 1 to 100 ppm; the peak potential equals -2.0 V (SCE) [95]. Acrylonitrile (and methyl bromide, ethylene dibromide, and carbon tetrachloride) have been determined in the air by polarographic analysis in tetramethylammonium hydroxide [96]. Acrolein can be determined in water using the medium of phosphate buffer of pH 7.2; the peak potential is -1.22 V (SCE) and the detection limit is 50 µg/liter [97]. Some vinyl alcohol esters have been determined in the air and in water [98]. To determine unsaturated compounds of the type of vinyl chloride or 1,1-dichloroethylene, their easy bromination can be utilized, yielding polarographically active derivatives [99]. Vinyl chloride, 1,2-dichloroethylene, 1,1,2-trichloroethylene, and acetylene are brominated using a 5 M solution of bromine in glacial acetic acid [100]. For the determination of vinyl chloride, a 1 M bromine solution suffices, with a reaction time of 10 to 20 min. The half-wave potentials of the waves of the individual bromo compounds differ sufficiently for analyses of mixtures of these substances.

Various indirect determinations preceded by nitration include, e.g., the determination of alkylbenzene sulfonate-type detergents in waters [101]. Humic substances and lignine sulfonic acids have also been determined [102]. A substantially more sensitive determination of humic substances in potable water (a detection limit of 50 µg/liter) is based on a catalytic effect on the reduction of cupric ions on the surface of a DME blocked by surfactants [103].

Nitration has also been used in the determination of benzene, toluene, phenol, and other substances [16]. Benzene can be determined in water after displacing it by passage of nitrogen into the nitration mixture; phenol is determined analogously after extraction into ether [104]. Traces of various nitro compounds used as explosives (nitroglycerine, 1,2-propyleneglycol-dinitrate, TNT, etc.) in waters can be determined without extraction [105].

Lactic acid has been determined in contaminated waters [106], especially in waste waters from silage containers, by oxidizing it with potassium permanganate to pyruvic acid which yields a wave at -1.10 V (SCE) in a buffer of pH 4.

An important task of analytical chemistry has recently appeared in the determination of growth stimulators used in animal breeding, such as various biofactors and feed mixtures, and in the determination of residues and metabolites in organs and muscles of animals bred for meat production. The nitrovin growth stimulant can be determined by DPP down to a content of 1 mg/kg in the feed mixture [107]. Residues of carbadox and its metabolites have been determined in the liver, kidneys, and muscle tissues of calves after 14-day application of carbadox at doses of 60 ppm. Down to 50 ppb of the substance can be determined in the lyophilized biological material [108]. The determination of traces of estrogenic growth-promoting hormones (estriol, estrone, estradiol, diethylstilbestrol, dienestrol, hexestrol, zeranol) in meat can be performed by using HPLC with voltammetric detection [128].

Abuse of pharmaceuticals is becoming a progressively more serious problem. DPP has been useful as a very sensitive analytical method for pharmaceuticals and toxic substances, as well as their metabolites in biological matrices. A number of works have dealt with the determination of

Determination of organic substances in the air (DME)   *Table 4.1*

Substance	Base electrolyte	$E_{1/2}$ (SCE)	Detection limit (mg/m^3)
Hexachlorobutadiene	0.2 M tetrabutylammonium iodide in 80% ethanol	-1.12 V	1.6
Formaldehyde	0.1 M LiOH	-1.60 V	0.02
Acrolein	0.01 M HCl	-1.20 V	0.008
Acetaldehyde	0.1 M LiOH	-1.90 V	0.008
Furfural	0.1 M K-oxalate + 0.125 M hydroxylamine HCl	-0.75 V	0.02
Nitrocyclohexane	0.1 M NaOH + 5% formaldehyde	-1.05 V	0.33

substituted nitroimidazoles, benzodiazepine (Diazepam), chlorodiazepoxide (Radepur, Librium), nitrazepam (Megadon), etc. [109]. Diazepam can be determined in urine directly by DPP (urine with an acetate buffer of pH 4.6, 1:9). Diazepam and its metabolites can be monitored after thin-layer chromatographic separation.

Polarography can be utilized to determine many organic compounds present in the air, e.g., in the vicinity of factories, in the industrial atmosphere, etc. For example, a book [42] describes the determination of the substances listed in Table 4.1 and of their mixtures. The detection limit is usually 0.1 μg/liter of the substance in the base electrolyte and depends on the polarographic method used (it would be substantially lower with DPP). This determines the sample size, i.e., the amount of air passed at a rate of 0.5 to 2.0 liter/min through the trap containing the absorption solution, which takes 10 to 30 min. For the sake of illustration, the determination of furfural in the air will be outlined:

The tested air is passed through a trap containing 3 ml of 0.1 M potassium oxalate with 0.125 M hydroxylamine at a rate of 0.5 liter/min. The volume passed is 15 liters. For the determination, a calibration curve is used. The stock solution of furfural is prepared by weighing 3 to 4 drops into 10 ml ethanol and diluting with ethanol to 50 ml. By diluting the stock solution with water, a solution containing 100 μg furfurol/ml is prepared, from which standards containing 0, 2.5, 5.0, 10.0, 15.0, and 20.0 μg furfurol/ml in the base electrolyte are diluted (corresponding to ca. 2×10^{-5} to 2×10^{-6} M).

Similar problems are discussed in a book [17] that contains an extensive list of organic substances that can be determined polarographically. Detailed procedures are given for the determination of pollutants, such as phthalic acid, benzene and its homologues, trinitrotoluene, nitrobenzene, ethyleneglycol, formaldehyde, methacrylic acid methyl ester, and styrene, in the atmosphere.

4.6.3   Use of Adsorption Phenomena

A separate chapter involves the use of adsorption phenomena for the detection or determination of electroinactive substances. Immediately after the discovery of polarography, a method was developed for monitoring the purity of water after purification by coagulation, based on suppression of the oxygen maximum by colloids. An automatic instrument, the Coagulograph [110], operated on this principle and was used in the Prague Water Supply before World War II. Analytical methods based on the effects on polarographic maxima of various kinds are nonselective, but are still used, e.g., for the determination of traces of surfactants in sea water [111]. Suppression of a polarographic maximum has also been used to determine detergents in industrial and waste waters [112]. Tensides can also be determined by ESA, after accumulation on the DME through adsorption [113]. In addition to the effects on the height of polarographic maxima, changes in the magnitude of the charging current were also used for the determination of surfactants, using the Kalousek commutator switching the working electrode potential periodically from -0.6 V (SCE) to potentials continuously increasing towards more negative values [the Kalousek commutator is part of the polarographs manufactured by Metrohm (Switzerland) and Bruker (FRG)]. These two methods have been used to determine anionic and uncharged tensides in waste water from hospitals [114]. Similarly, oil hydrocarbons have been determined in sea water at concentrations from 0.02 to 100 mg/liter [115]. These substances can also be determined with a high sensitivity by measuring the electro-capillary curves [116].

Rapid periodic changes in the DME potential (i.e., basically ac polarography) are employed in an analyzer of surfactants in water [117], monitoring tenside concentrations from 1 to 20 mg/liter. The ac capacity current is measured, which is a function of the capacity of the DME electric double layer and varies in the presence of surfactants (tensammetry: see below). The difference in the electric double-layer capacity is measured for adsorption equilibrium at the electrode surface and for the absence of the adsorbable substance (i.e., for the blank).

For monitoring dextrane, quaternary bases, and analogous substances in waters at concentrations of the order of 10^{-7} M, depression of the electrocapillary curve can be measured with a DME in stirred solution [118].

The most common method of polarographic monitoring of surface-active substances is tensammetry. Basically, the polarographic curve is recorded in a solution containing a surfactant, using a method with an ac voltage or pulses superimposed on a linearly increasing voltage. Surfactants are adsorbed on the mercury electrode and alter (usually decrease) the electric double-layer capacity, which is manifested on the tensammetric curve by peaks at the adsorption and desorption potentials. These peaks are concentration-dependent within about two concentration orders; the dependence is not linear and resembles an adsorption isotherm. The peak potentials also vary with changing surfactant concentration. In the presence of two or more surfactants, competitive adsorption occurs; therefore, the test substances must be thoroughly isolated. Details on this method can be found in the books on the subject [119, 120]. Adsorption can also be used for accumulation of the surfactant on the electrode, followed by recording of the tensammetric curve from the accumulation potential to more negative values [113, 121, 122]. In addition to the above determination of tensides, oil components in water [116] and polychlorinated biphenyls (used as plasticizers, dielectrics, hydraulic liquids, heat-exchanging media, and fire-extinguishing substances) in water [123] can be determined. Adsorptive stripping voltammetry can also be employed to determine various pharmaceuticals [122, 129, 130] and pesticides [131], down to concentrations of 10^{-10} mol liter^{-1}, as well as many metals, e.g., uranium [21], nickel [132], cobalt [133], the alkaline earth metals [130], and aluminum [130], bound in adsorbable complexes. This technique, which considerably extends the applicability of voltammetric analysis, is generally treated in Refs. 129 and 130.

Here it should be mentioned that the use of adsorption phenomena for monitoring organic compounds from hydrocarbons to alkaloids is one of the main applications of oscillographic polarography with alternating current [124]. The method has been used, e.g., for the preparation of an identification system for toxicological analysis [125].

4.7 Combination of Polarography with High-Performance Liquid Chromatography

In the first chapter (see p. 15) it is emphasized that no method is universally applicable, but only most suitable for a certain application. It is sometimes useful to combine different methods, e.g., high-performance liquid chromatography (HPLC) with electrochemical detection, usually voltammetric or polarographic. For details, see Chapter 9 and Ref. 126. The applications of HPLC with electrochemical detection to environmental analyses (organometallic compounds, agrochemicals, etc.) are reviewed in Ref. 127. Applications of electrochemical detection in liquid chromatography (LCEC) are listed in the bibliography published by Bioanalytical Systems, Inc., West Lafayette, Indiana, USA.

REFERENCES

1. Kalvoda R.: *Chem. Listy* **71**, 530 (1977).
2. Bond A. M.: Modern Polarographic Methods in Analytical Chemistry, M. Dekker, New York and Basle 1980.
3. Horák K. and Gajda V.: *Chem. Listy* **76**, 561 (1982).

4. Gajda V. and Horák K.: *Anal. Chim. Acta* **134**, 219 (1982).
5. Novotný L.: Proc. J. Heyrovský Memorial Congress on Polarography, Prague 1980, Vol. II, p. 129.
6. Vydra, F., Štulík K., and Juláková E.: Electrochemical Stripping Analysis, J. Wiley and E. Horwood, Chichester 1976.
7. Gokhshtein A. Y. and Gokhshtein Y. P.: *Zh. Fiz. Khim.* **36**, 651 (1962).
8. Heyrovský J. and Zuman P.: Practical Polarography, Academic Press, London, New York 1968.
9. Heyrovský J. and Kůta J.: Principles of Polarography, Academic Press, London, New York 1965.
10. Kalvoda R.: in Thomas J., Belcher R., and West T. S. (eds.), Recent Advances in Analytical Chemistry, The Royal Society, London 1982, p. 151, and in *Phil.Trans.R.Soc. Lond.* A 305, 621 (1982).
11. Kalvoda R.: *Abh.Dtsch.Akad.Wiss.Berlin (Kl.Chem.)* 1964, No. 1, 285.
12. Breiter J.: *Erzneimittel-Forschung* **28**, 1941 (1978).
13. Turner D. R., Robinson S. G., and Whitfield M.: Proc.J.Heyrovský Memorial Congress on Polarography, Prague 1980, Vol. II, p. 179.
14. Kalvoda R.: Operational Amplifiers in Chemical Instrumentation, J. Wiley and E. Horwood, Chichester 1975.
15. Bibliography of Polarographic Literature 1922−1967, Sargent-Welch Scientific Co., Skokie, Ill., USA 1969.
16. Březina M. and Zuman P.: Polarography in Medicine, Biochemistry and Pharmacy, Interscience Publ., New York 1958.
17. Križan V.: Analysis of the Atmosphere, Alfa-SNTL, Bratislava and Prague 1981 (in Slovak).
18. Heckner H. N.: *Z.Anal.Chem.* **261**, 29 (1972).
19. Robberecht R. and Van Grieken R.: *Talanta* **29**, 823 (1982).
20. Keil R.: *Z.Anal.Chem.* **292**, 13 (1978).
21. Lam N. K., Kalvoda R., and Kopanica M.: *Anal.Chim.Acta* **154**, 79 (1983).
22. Navrátilová Z. and Kopanica M.: *Anal.Chim.Acta* − in press.
23. Boese S. W., Archer V. S., and O'Laughlin J. W.: *Anal.Chem.* **49**, 479 (1977).
24. Amino R. and McDonald J. A.: *Anal.Chem.* **33**, 475 (1961).
25. Stará V. and Kopanica M.: *Coll.Czech.Chem.Commun.* **47**, 2214 (1982).
26. Stará V., Jeník J., and Kopanica M.: *Anal.Chim.Acta* **147**, 371 (1983).
27. Garner R. W. and Wilson C. E.: *Anal.Chem.* **44**, 1357 (1972).
28. Klahre R., Valenta P., and Nürnberg H. W.: *Vom Wasser* **51**, 199 (1978).
29. van Duin P. J. and de Krenk C. W.: Proceedings of International Conference on Management and Control of Heavy Metals in the Environment, London 1979, CEP Consultants, Edinburgh 1979, p. 412.
30. Valenta P., Sipos L., Kramer I., Krumpen P., and Rützel H.: *Z.Anal.Chem.* **312**, 101 (1982).
31. Cnobloch H., Kellermann W., Kuhl D., Nischik H., Pantel K., and Poppa H.: *Anal. Chim.Acta* **114**, 303 (1980).
32. Nürnberg H. W., Mart L., and Valenta P.: *Rapp.Comm.Int.Mer.Médit.* **24**, 8 (1977).
33. Nürnberg H. W.: *The Science of the Total Environment* **13**, 35 (1979).
34. Roitman L. I., Pavlovich Yu. A., and Brainina Kh. Z.: *Zh.Anal.Khim.* **36**, 1009 (1981).
35. Nürnberg H. W.: *Pure Appl.Chem.* **54**, 853 (1982).
36. Nürnberg H. W., Valenta P., Mart L., Raspor B., and Sipos L.: *Z.Anal.Chem.* **282**, 357 (1976).

37. Piotrowicz S. R., Springer-Young M., Puig J. A., and JoSpencer M.: *Anal.Chem.* **54**, 1367 (1982).
38. Membrini P. G., Dogan S., and Haerdi W.: *Anal.Lett.* **13**, 947 (1980).
39. Grebenovský E., Mutínská T., Zvozníková Z., and Štulík K.: *Čs. Hygiena* **22**, 320 (1977).
40. Sturrock P. E. and Mendez-Merced R.: Proc. 4th Joint Conf. on Sensing of Environmental Pollutants, New Orleans, 1977, p. 189.
41. Peterka J.: Ph.D. Thesis, Mining Institute, Ostrava 1978.
42. Manita M. D., Salikhzhdanova R. M., and Yavorskaya C. F.: Sovremennyie Metody Opredeleniya Atmosfernykh Zagryaznenii Naselenykh Mest, Medicina, Moscow 1980.
43. Nürnberg H. W., Valenta P., and Nguyen V. D.: in Georgi H. W. and Pankrath J. (eds.), Deposition of Atmospheric Pollutants, D. Reidel, Dordrecht 1982, p. 143.
44. Eisner V. and Mark H. B., Jr.: *J.Electroanal.Chem.* **24**, 345 (1970).
45. Ferren W. P.: *Int.Lab.*, Sept./Oct. 1978, 55.
46. Meites L. and Zuman P.: Organic Electrochemistry, Vols. I – IV, CRC Press, Boca Raton, USA 1977 – 79.
47. Volke J.: in Ryan T. H. (ed.), Electrochemical Detectors, Plenum Press, New York and London 1984, p. 105.
48. Nangniot P.: La Polarographie en Agronomie et en Biologie, J. Duculot, Gembloux 1970.
49. Volke J. and Slamnik M.: in Dask G. (ed.), Pesticide Analysis, M. Dekker, New York and Basle 1981, p. 175.
50. Davídek J.: in Smyth W. F. (ed.), Electroanalysis in Hygiene, Environmental, Clinical and Pharmaceutical Chemistry, Elsevier, Amsterdam 1980, p. 399.
51. Osteryoung J., Whittaker J. W., and Smyth M. R.: in Smyth W. F. (ed.), Electroanalysis in Hygiene, Environmental, Clinical and Pharmaceutical Chemistry, Elsevier, Amsterdam 1980, p. 413.
52. Hance L. J.: *Pestic.Sci.* **1**, 120 (1970).
53. Gajan R. J.: *Res.Rev.* **5**, 80 (1964).
54. Smyth M. R. and Osteryoung J. G.: *Anal.Chim.Acta* **96**, 335 (1978).
55. Lechien A., Valenta P., Nürnberg H. W., and Patriarche G. J.: *Z.Anal.Chem.* **306**, 150, 156 (1981).
56. Sobina N. A., Kheifest L. Ya., Bondarenko L. M., and Glyadyaeva L. A.: *Zh.Anal.Khim.* **31**, 941 (1976).
57. Davídek J. and Janíček G.: *Z.Anal.Chem.* **194**, 431 (1963).
58. Lyalikov Yu. S.: *Zh.Anal.Khim.* **22**, 1579 (1967).
59. Woggon H., Säuberlich H., and Uhde W.: *Z.Anal.Chem.* **B 260**, 269 (1972).
60. Nangniot P.: *Bull.Inst.Agron.Res.Gembloux* **28**, 365, 373, 381 (1960).
61. Supin G. S. and Budnikov G. K.: *Zh.Anal.Khim.* **28**, 1459 (1973).
62. Franke G., Pietrula W., and Preussner K.: *Z.Anal.Chem.* **298**, 38 (1979).
63. Polák J. and Volke J.: *Chem.Listy* **77**, (1983).
64. Davídek J. and Seifert J.: *Die Nahrung* **15**, 691 (1971).
65. McKone C. E., Byast T. H., and Hance R. J.: *Analyst* **97**, 653 (1972).
66. Bronstad J. O. and Friestad H. O.: *Analyst* **101**, 820 (1976).
67. Booth M. D. and Fleet B.: *Talanta* **17**, 498 (1970).
68. Polák J. and Volke J.: *Čs.Farmacie* **32**, 282 (1983).
69. Stará V. and Kopanica M.: *Coll.Czech.Chem.Commun.* **49**, 1282 (1984).
70. Smyth M. R. and Osteryoung J. G.: in Smyth W. F. (ed.), Electroanalysis in Hygiene, Environmental, Clinical and Pharmaceutical Chemistry, Elsevier, Amsterdam 1980, p. 423.

71. Chey W. E., Adams R. N., and Yllo M. S.: *J.Electroanal.Chem.* **75**, 731 (1977).
72. Coetzee J. F., Kazi G. H., and Spurgeon G. M.: *Anal.Chem.* **48**, 2170 (1976).
73. Zahradník R., Svátek E., and Chvapil M.: *Coll.Czech.Chem.Commun.* **24**, 347 (1959).
74. Borghesani G., Pulidori F., Pedriali R., and Bighi C.: *J.Electroanal.Chem.* **32**, 303 (1971).
75. Hasebe K. and Osteryoung J.: *Anal.Chem.* **47**, 2412 (1975).
76. Chang S. K. and Harrington G. W.: *Anal.Chem.* **47**, 1857 (1975).
77. Pylypiv H. M. and Harrington G. W.: *Anal.Chem.* **53**, 2365 (1981).
78. Matrka M., Mejstřík V., and Ságner V.: *Chem.Průmysl* **54**, 466 (1979).
79. Snider B. G. and Johnson D. C.: *Anal.Chim.Acta* **106**, 1 (1979).
80. Samuelson R. and Rydström T.: in Smyth W. F. (ed.), Electrolysis in Hygiene, Environmental, Clinical and Pharmaceutical Chemistry, Elsevier, Amsterdam 1980, p. 435.
81. Pečenka V., Ságner L., and Mestřík V.: *Kvasný Průmysl* **28**, 536 (1982).
82. Samuelson R.: *Anal.Chim.Acta* **102**, 133 (1978).
83. Vohra S. and Harrington G.: *J.Chromatogr.Sci.* **18**, 379 (1980).
84. Samuelson R. and Osteryoung J. G.: *Anal.Chim.Acta* **123**, 97 (1981).
85. Samuelson R. and Sunström O.: *Anal.Chim.Acta* **138**, 375 (1982).
86. Smyth M. R., Lawellin D. W., and Osteryoung J. G.: *Analyst* **104**, 73 (1979).
87. Palmisano F., Visconti A., Bottalico A., Lerario P., and Zambonin P. G.: *Analyst* **106**, 992 (1981).
88. Séquaris J. M., Valenta P., and Nürnberg H. W.: *J.Electroanal.Chem.* **122**, 263 (1981).
89. Paleček E.: in Smyth W. F. (ed.), Electroanalysis in Hygiene, Environmental, Clinical and Pharmaceutical Chemistry, Elsevier, Amsterdam 1980, p. 79.
90. Lukášová E., Paleček E., Kruglyakova K. E., Zhizhina G. P., and Smotryaeva M. A.: *Rad.and Environ.Biophys.* **14**, 231 (1977).
91. Keith L. H. and Telliard W. A.: *Environmental Science and Technology* **13**, 416 (1979).
92. Matyska B. and Klier K.: *Coll.Czech.Chem.Commun.* **21**, 1592 (1956).
93. Coetzee J. F., Cunningham G. P., McGuire D. K., and Pandmanabhan G. R.: *Anal. Chem.* **34**, 1139 (1962).
94. Strause S. F. and Dyer E.: *Anal.Chem.* **27**, 1906 (1955).
95. Betso S. R. and McLean J. D.: *Anal.Chem.* **48**, 766 (1976).
96. Berck B.: *J.Agr.Food Chem.* **10**, 158 (1962).
97. Howe H. L.: *Anal.Chem.* **38**, 2167 (1976).
98. Filov V. A.: *Gigiena Truda i Professional.Zabolevanya* **4**, 54 (1960); C.A. 55, 7170 (1961).
99. Ryabov A. V. and Panova G. D.: *Dokl.Akad.Nauk USSR* **99**, 547 (1954).
100. Medonos V.: *Coll.Czech.Chem.Commun.* **23**, 1465 (1958).
101. Hart J. P., Smyth W. P., and Birch B. J.: *Analyst* **104**, 853 (1979).
102. Eberle S. H., Hoesle C., and Krückeberg C. H.: Report KFK 1969 UF, Kernforschungszentrum, Karlsruhe, BRD, June 1974, p. 44.
103. Sohr H. and Wienbold K.: *Anal.Chim.Acta* **121**, 309 (1980).
104. Adamovský M.: *Vodní Hospodářství* **16**, 102 (1966).
105. Whitnack G. C.: *Anal.Chem.* **47**, 618 (1975).
106. Kopanica M., Stará V., and Jeník J.: *Vodní Hospodářství* B **33**, 49 (1983).
107. Škarka P. and Šestáková J.: *Toxicological Aspects of Food Safety Arch.Toxicol.,* Suppl. 1, 207 (1978).
108. Šestáková J., Škarka P., and Manoušek O.: *Biol.Chem.Vet.* (Prague), 29 (1980).
109. Ellaithy M. M., Volke J., and Manoušek O.: *Talanta* **27**, 137 (1979).
110. Procházka R.: *Chem.et Ind.* **29**, 281 (1933).
111. Zvonarič T., Žutič V., and Branica M.: *Thalassia Jugoslavica* **9** (1/2), 65 (1973).
112. Linhart K.: *Tenside Detergents* **9**, 241 (1972).

113. Kalvoda R.: *Anal.Chim.Acta* **138**, 11 (1982).

114. Kozarac Z., Žutič V., and Čosovič B.: *Tenside Detergents* **13**, 260 (1976).

115. Žutič V.. Čosovič B.. and Kozarac Z.: *J.Electroanal.Chem.* **78**, 113 (1977).

116. Kalvoda R. and Novotný L.: *Vodní Hospodářství* **11**, 291 (1984).

117. Jehring H., Lohse H., and Horn E.: Internat.Soc.of Electrochemistry, 29th Meeting, Budapest 1978, Extended Abstracts, Part I, p. 188 (also WP 121187 (GDR), 12. 7. 1976).

118. Novotný L. and Smoler I.: Proc.J.Heyrovský Memorial Congress on Polarography, Prague 1980, Vol. II, p. 128.

119. Breyer B. and Bauer H. H.: Alternating Current Polarography and Tensammetry, Interscience Publ., New York 1963.

120. Jehring H.: Elektrosorptionsanalyse mit der Wechselstrompolarographies, Akademie Verlag, Berlin 1974.

121. Brainina Kh. Z.: *Z.Anal.Chem.* **312**, 428 (1982).

122. Kalvoda R.: *Anal.Chim.Acta* **162**, 197 (1984).

123. Kopanica M.: *The Science of the Total Environment* **37**, 83 (1984).

124. Kalvoda R.: Techniques of Oscillographic Polarography, Elsevier, Amsterdam 1964.

125. Faith L. and Dušinský G.: Identification of Pharmaceuticals and Poisons by Oscillographic Polarography, Osveta, Martin 1975 (in Slovak).

126. Štulík K. and Pacáková V.: *CRC Critical Reviews in Analytical Chemistry* **14**, 297 (1984).

127. Smyth W. F., Goold L., Dadgar D., Jan M. R., and Smyth M. R.: *Int.Lab.*, Sept. 1983, 40.

128. Smyth M. R. and Frischkorn C. G. B.: *Z.Anal.Chem.* **301**, 220 (1980).

129. Wang J.: Stripping Analysis. VCH Publishers, Deerfield Beach, USA 1984.

130. Wang J.: *Am.Lab.*, May 1985, 41.

131. Beňadiková H. and Kalvoda R.: *Anal.Letters* **17 (A13)**, 1519 (1984).

132. Pihlar B., Valenta P., and Nürnberg H. W.: *Z.Anal.Chem.* **307**, 337 (1981).

133. Gemercolos V., Scollary G., and Neeb R.: *Z.Anal.Chem.* **313**, 412 (1982).

134. Wang J., Farias P. A. M., and Mahmoud J. S.: *J.Electroanal.Chem.* **195**, 165 (1985).

CHAPTER 5

ELECTROCHEMICAL STRIPPING ANALYSIS

Miloslav Kopanica and František Opekar

5.1   Introduction

The sensitivity of determination by any analytical method is limited by the ratio of the signal from the test component to the background which involves the noise generated in the solution and in the apparatus. One of the ways of improving the signal-to-noise ratio is preconcentration of the component, which can be conveniently carried out by accumulating the test substance electrolytically on the working electrode, while selectivity can be controlled to a certain extent by judiciously choosing the solution composition and the preelectrolysis potential. The preconcentrated component is then electrolytically stripped into the solution. As the test component concentration is increased by several orders of magnitude during the preelectrolysis, the analytical signal also increases.

Another way of increasing the signal-to-noise ratio is the use of some special techniques, such as nonstationary electroanalytical methods of ac and pulse polarography. The combination of preconcentration and a nonstationary measuring technique enables the attainment of detection limits of 10^{-3} µg test substance in 1000 ml of solution. The method employing electrolytic preconcentration and subsequent electrochemical stripping of the preconcentrated substances is termed electrochemical stripping analysis (ESA).

ESA has been treated in many reviews and books [1–6] and in hundreds of original papers describing the determination of almost all elements of the periodic table and of many anions. However, the main practical importance is connected with the determination of zinc, cadmium, lead, copper, and mercury. As the problems of ESA are discussed in detail in the above publications, only the principles of the method are given below.

5.2   ESA Methods and Conditions

5.2.1   Electrolytic Preconcentration

Substances are generally preconcentrated potentiostatically under conditions at which the reaction is sufficiently rapid. Therefore, provided that a special choice of the potential is not dictated by selectivity requirements, the preelectrolysis potential is selected in the region of the polarographic limiting current of the test substance. The solution is stirred during the electrolysis to maintain constant transport of the electroactive substance from the bulk of the solution towards the electrode. The convection in the solution is sometimes caused by a defined motion of the working electrode (vibration, rotation). After a certain time (minutes to tens of minutes) the stirring is stopped and, after a rest period (30 to 60 s), the stripping step is begun. The rest period is not necessary when using solid electrodes, with which even the stripping is sometimes carried out in stirred solutions.

There exist a number of possibilities for preconcentration of the test substance on the electrode, of which the most important and most often used are the following:

(a) A metal cation is reduced to the metal, forming an amalgam with a mercury electrode or forming a film on the electrode surface. The deposited metal is then anodically stripped (oxidized) and the method is called anodic ESA.

(b) The substance is preconcentrated on the electrode at a potential at which the electrode material is oxidized in the given medium and the metal ions formed react with the test component, which produces a film on the electrode. The insoluble compound is then cathodically stripped and the method is called cathodic ESA.

(c) The test substance can be accumulated in the form of a sparingly soluble compound produced by its reaction in the particular valence state formed by electrode oxidation or reduction with a solution component. The film of the sparingly soluble compound is then stripped by the opposite electrochemical process.

5.2.2   Electrolytic Stripping

Electrolytic stripping is commonly carried out at linearly scanned potential, using common polarographic or voltammetric methods monitoring the current vs. potential dependence. The curve exhibits peaks whose position, characterized by the half-peak potential, specifies the nature of the substance and whose area (or, more often, height) is a measure of the concentration of the substance in the solution.

The peak height depends on the amount of substance deposited on the electrode and is a function of many parameters. In order to be able to correlate the peak height with the original concentration of the test substance in the solution, the deposited fraction of the substance must be constant in all measurements that are to be compared. This can be ensured by maintaining constant the preelectrolysis time and potential, the convection conditions (cell geometry, stirring), the electrode properties (surface area, volume, material, pretreatment), temperature, solution composition, and the stripping conditions (scan rate, pulse amplitude, etc.).

In addition to potentiostatic stripping, the deposited substance can be stripped at a constant current (galvanostatic ESA; see Refs. 7 and 10) or chemically (potentiometric ESA [8]).

In galvanostatic stripping, the potential of the working electrode is monitored as a function of time during the stripping step performed at a constant current. The stripping curve has the shape of a chronopotentiometric curve whose transition times are proportional to the amount of the substance deposited on the electrode. An advantage of galvanostatic stripping is improved selectivity of the determination compared with potentiostatic ESA, especially when mercury film or solid electrodes are employed. Breaks on the $E - t$ curve, determining the transition times, usually occur only after complete stripping of given components and thus all of the current passing is consumed in the charging of the electrode to the potential determined by the further redox system, so that the potential change on the $E - t$ curve is well pronounced. In potentiostatic stripping, the residues of the metal stripped at a more negative potential are still stripped during the stripping of the following metal, which leads to overlapping of the peaks of systems with close half-wave potentials. This can be illustrated by the determination of lead by galvanostatic ESA in the presence of a thousandfold excess of cadmium, whereas even a tenfold excess of cadmium interferes in the potentiostatic stripping [9].

In the potentiometric modification of ESA [8, 11], metals deposited on a mercury electrode are stripped by chemical oxidation, usually using Hg^{2+} ions as the oxidant; these are added to the test solution at a concentration several orders of magnitude higher than that of the test metals, to eliminate oxidation by the test metal ions themselves. The Hg^{2+} ions simultaneously provide the mercury for the film formation (see below). After completion of the preelectrolysis, the working electrode is disconnected from the polarizing circuit and its potential is measured against a reference electrode in a stirred solution. The $E - t$ curve has a shape similar to the galvanostatic curve, the method is instrumentally simple, and the results obtained are satisfactory. Certain disadvantages lie in the fact that the oxidation reaction is affected by a number of factors of a fundamental and experimental nature (the equilibrium constant of the redox reaction, the necessity of maintaining a constant concentration of the oxidant at the electrode surface, the character of the solution stirring during the stripping, etc.).

5.3   Selection of the Experimental Conditions

5.3.1   Sample Preparation

For ESA, similarly to classical polarography and voltammetry, it is required that the sample be converted into a solution containing a sufficient amount of a base electrolyte and usually be also purged of oxygen, as metals deposited on the electrode in the form of an amalgam or a film are usually very sensitive to oxidation and oxygen also interferes in the determination by its reduction double-wave.

The base electrolytes are usually solutions of simple inorganic acids and salts or buffers. In the presence of complexing agents some components of the sample can be masked and thus the selectivity of the determination can be improved. The base electrolyte composition is generally dependent on the nature of the test substance and the character of the sample matrix.

A serious problem of trace analysis is the loss of the test component during the sample pretreatment, storage, and determination. In transferring insoluble samples into solution, direct or pressure decomposition with acids is preferred over procedures involving fusion or sintering. Extraction and other separations should also be limited to an acceptable minimal number; in general, the number of operations with the samples must be as low as possible.

In the sample preparation and the determination itself it must be borne in mind that dilute solutions are unstable owing to various hydrolytic, adsorption, and redox reactions and thus the samples must be analyzed as soon after the decomposition as possible. Surfaces of glass and plastic vessels exhibit strong adsorption of solution components and thus it is recommended to use quartz and PTFE electrolysis vessels, or to make the surface of glass vessels hydrophobic by siliconization. For these reasons, dilute standard solutions must be prepared immediately before use by dilution of more concentrated (above 10^{-2} M) stock solutions.

5.3.2 Electrodes

Electrodes can, in principle, be classified as mercury or solid (noble metals, various forms of carbon). The choice of the electrode, similarly to the base electrolyte composition, depends on the nature of the test substance. Mercury electrodes are generally used to determine the metals that form amalgams and the substances (anions) that form insoluble compounds with mercury ions (cathodic ESA). Solid electrodes are used to determine mercury and other metals forming films or insoluble compounds with a solution component.

There are two principal types of mercury electrodes, the hanging mercury drop electrode (HMDE) and the mercury film electrode (MFE). The latter has some advantages over the HMDE. It is formed by a mercury film on an inert support, usually carbon (actually, the mercury "film" on carbon consists of an array of mercury droplets [12]). The film is generally deposited electrolytically [13], often simultaneously with the test substance during the preelectrolysis, when a mercuric salt is added to the sample solution at a concentration of about 10^{-5} M [14]. The films thus formed are very thin (1–100 μm) and the stripping peaks are consequently very sharp (the peak half-width is up to 60% smaller than that with the HMDE), which is important from the point of view of the resolution. The ratio of the film surface area to the film volume is several orders of magnitude higher than with the HMDE and thus a higher metal concentration in the amalgam is attained under identical experimental conditions, which leads to a higher measuring sensitivity. It also follows from the theory [15] that the peak height is proportional to the potential scan rate for the MFE, whereas for the HMDE it is proportional to the square root of the scan rate; therefore an increase in the stripping potential scan rate leads to a greater increase in the determination sensitivity when the MFE is used. Moreover, a uniform distribution of the metal concentration in the amalgam is attained more rapidly with the MFE (within as little as 4 s of the electrolysis for a film thickness of 1 μm [16]), so that the rest period is unnecessary, in contrast to the HMDE.

A certain problem in the use of the MFE is easy oxidation of the film and the dependence of the film reproducibility on the condition of the support surface. The value of the hydrogen overvoltage is also lower than on the HMDE, which causes problems in the determination of metals that are deposited at very negative potentials (Zn). The HMDE is also usually preferred when pulse methods are used to monitor the stripping step. On the other hand, an MFE can be prepared in the form of a rotating disk electrode, enabling defined material transport toward the electrode surface during the preelectrolysis, which improves the reproducibility of the results. The MFE is also easy to prepare and is very cheap compared with the HMDE.

As solid electrodes, various forms of carbon are mostly used. Metal electrodes have a number of drawbacks for use in ESA (high residual currents caused by the oxidation of the electrode material and reduction of the oxides formed, adsorption and desorption of various substances, etc.). Attempts have been made to overcome these difficulties by using special electrode constructions, e.g., by using a rotated ring–disk electrode [17, 18] where the deposited metals are stripped from the disk at a linearly varying potential and the ring current is recorded at a constant potential corresponding to the reduction of the metal ions transported from the disk by convective diffusion, so that the residual current is low.

It is also possible to use split-disk electrodes [19]. Two halves of the disk are potentiostatted independently, first at the same potential, and then, immediately before the stripping step, the potential of one half is changed so that the deposit is stripped off, with recording of the difference in the currents passing through the two halves. These types of electrodes are difficult to manufacture and thus are not widely used in practice. The problem of the residual current elimination is more often solved by electronic signal handling, mostly using a digital computer.

Carbon electrodes are not as prone to oxidation as metal electrodes, the determination is not complicated by the formation of intermetallic compounds with the electrode material, their accessible potential range is sufficiently wide, and they can be prepared from cheap materials. Detailed information on the properties and preparation of various types of carbon electrode can be found in Refs. 5 and 20.

The most common electrodes are made from glassy carbon, graphite, or spectral carbon impregnated with paraffin or a polymer, or carbon paste prepared by mixing a pulverized carbon material with a suitable nonvolatile, electrochemically inactive, and sufficiently viscous diluent (nujol, silicone oil, bromonaphthalene, etc.). Carbon electrodes are also the most suitable supports for the MFE. They are generally prepared in the form of disks whose surface can be pretreated relatively reproducibly by grinding and polishing, retaining the electrode geometry and permitting utilization of the favorable convective-diffusional properties of the rotating disk.

5.3.3 Electrolysis Vessel

Common polarographic or voltammetric vessels, e.g., as described in Ref. 21, can be used for ESA, but they must be made of a material suitable for trace analysis (see Section 5.3.1) and must enable reproducible stirring of the electrolyte and fixing of other parts of the apparatus (electrodes, inlet tube for an inert gas). Common reference and auxiliary electrodes are employed in ESA. From the point of view of sample contamination it is desirable that the reference and auxiliary electrodes be suitably separated from the working space of the vessel. The solution temperature must be maintained constant within at least $\pm 1°C$, as the voltammetric currents depend on temperature.

For some ESA determinations, special vessels have been proposed: vessels that permit solution exchange after the preelectrolysis, so that the substance is stripped into a pure base electrolyte [22]; vessels for determinations in small volumes (down to 1 μl) [23–25]; and cells permitting separation on ion-exchangers simultaneously with the preelectrolysis [26, 27]. Although ESA is basically a discontinuous method, voltammetric cell systems have been proposed for semicontinuous ESA measurements in flowing liquids (see, e.g., Refs. 28, 29), and special electrodes of reticulated glassy carbon with which up to a 100% substance conversion can be attained during the electrolysis have been reported [30], whereas in normal ESA the depletion of the test substance during the preelectrolysis is negligible.

5.4 ESA Instrumentation and Automation

As mentioned above, the ESA measuring principle is analogous to polarographic or voltammetric measurements, with the difference that a series of operations must be carried out before measuring the signal, i.e., the current. This series is virtually the same for all determinations, only with differences in the duration of the individual operations. This constant series of operations, which are, moreover, time-consuming and must be timed precisely, is more or less automated in modern instruments permitting ESA measurements. The best instruments contain a microcomputer and thus acquire a certain degree of "intelligence" permitting minimization of manual work and automated handling of the experimental results.

Automation in ESA can be divided into three principal groups: (a) automation of the procedure (i.e., control of the time of solution deaeration by passage of an inert gas, electrochemical pretreatment of the electrode, preelectrolysis, rest period, starting the stripping curve recording, etc.); (b) automation of the result handling (i.e., the calculation and display of information on the concentration or amount of the test substances in the sample on the basis of a preprogrammed calibration method); (c) automation of sample manipulation (automatic sample feeding).

5.4.1 Automation of the Procedure

The procedure is most often automated in ESA analyzers. Of well-known instruments with this degree of automation, the AS-01 analyzer (Mitsubishi Chemical Industries, Japan) can be mentioned; this system automates the working procedure from the aspiration of the sample into the vessel, through the recording of the polarization curve, to expulsion of the test solution from the vessel. The well-known Polarographic Analyzer 741 (Princeton Applied Research, USA) forms an automated system with Automated Electrolysis Controller 315A from the same company that also enables automated electrochemical pretreatment of the electrode and repeated measurements on the same sample with an automated shift of the recorder pen. Among other instruments, Stripping Voltammeter E 200 (Bruker, FRG) and Polarographic Analyzers PA 3 and PA 4 (Laboratorní Přístroje, Czechoslovakia) can be mentioned. All these analyzers employ the dc and DPP techniques. The possibilites of automation of the procedure have been improved by the introduction of automatically operated HMDEs (e.g., 303 SMDE from Princeton Applied Research or SMDE from Laboratorní Přístroje).

Automated operation, mainly control of the preelectrolysis time, is also possible with instruments employing other ESA modifications. For potentiometric ESA, simple analyzers are manufactured, e.g., the Striptec System (Tecator, Sweden) or Ion Scanning System ISS 820 (Radiometer, Denmark).

The automation of the procedure is so simple that the programming unit can be designed in the laboratory, as an accessory for common polarographs (see, e.g., Refs. 31, 32). To control the ESA operations (and in a higher degree of ESA automation), laboratory computers can be used, which also permit the use of special techniques of working electrode polarization [33, 34].

5.4.2 Automated Handling of the Experimental Data

The foremost requirement in automated quantitative evaluation of voltammetric experimental data is the subtraction of the background (noise) and averaging of repeated measurements [35], which is best carried out by using a computer. Classical examples of commercial analyzers with built-in computers are models 374 and 384 from Princeton Applied Research. These instruments fully automate the ESA procedure and further make it possible to store the experimental curve in the memory, subtract the background, and calculate the concentrations of up to ten test substances, provided that the necessary calibration data are fed into the instrument, either in the form of a calibration curve stored in the memory or as the result of a measurement with standard addition. The result of the measurement is a graphical recording of the stripping curve, supplemented with alphanumerical data on the quality and amount of the test substances in preselected units, possibly with other data, e.g., on the experimental conditions of the determination.

A great problem in the automated (and not only automated) evaluation of the experimental curves in the analysis of multicomponent samples is the overlapping of the stripping peaks. It is usually solved by a computer on the basis of a suitable program (see the literature in Ref. 36 and Refs. 37–40). More detailed information on the use of computers in practice, automation, and optimization of ESA can be found in Ref. 41.

In some special analyses, the evaluation of the experiments is simpler. For example, in analyses of samples containing a single test component it is possible to integrate the stripping current, with the integrator constants set at values such that the output signal corresponds numerically to the test substance concentration in the solution. The 3010 automaton from Environmental Sciences Associates, USA, operates on this principle. The presence of only a single metal in the sample is ensured by using special ion-exchangers. (This company manufactures many other special ESA analyzers, e.g., models 7010, 2011, and 2014.)

Futher possibilities for simplification of the automated data handling are provided by other ESA methods. In automated measurements of the transition times in galvanostatic ESA, the E–t curve can be differentiated and the derivative peaks be used to start and stop a digital timer complemented with electronic circuits, so that the result printout is numerically equal to the concentrations of the test substances in the solution. The instrument employs the standard addition method [42, 43]. An analogous method has also been applied to potentiometric ESA, but with the use of a microcomputer for the evaluation of the transition times [44, 45].

5.4.3   Automation of the Sample Manipulation

The principle of this aspect of ESA automation (in contrast to the above stages) lies in the mechanical arrangement and construction of this part of the apparatus. Two approaches can be taken.

In the flow-through arrangement, sample exchage is ensured by the flow of the test liquid. Several systems studied for use in flow-through ESA have been mentioned in Section 5.3.3 and are incorporated in laboratory instruments.

In commercial analyzers, discrete samples are analyzed and are transported toward the electrode system, e.g., by aspiration from vessels (the AS 01 analyzer), or are brought to the electrode systems in the vessel, such as in the Automatic Cell Sequencer 316 or the Sample Changer 319 from Princeton Applied Research.

5.5   Application of the ESA Methods

The determination of trace concentrations of metals in samples from the biosphere is very important ecologically, as it yields information on the quality, quantity, and movement of metals in various parts of the biosphere. ESA is advantageous for these analyses because (a) it is highly precise in the determination of trace concentrations of biologically important metals, such as copper, lead, mercury, cadmium, and zinc; (b) it enables simultaneous determination of several metals; (c) it is economical in analyses of large sets of samples.

5.6   Determination of Trace Metals in Water

ESA has been applied to the determination of metals in all types of water, including natural water (oceans, seas, rivers, springs, lakes), potable water, and waste water contaminated by communal and industrial wastes. In addition to the advantages mentioned in the preceding section, there is an additional advantage in the electrochemical approach to analyses of natural water, namely, the possibility of determining the chemical form in which the metal is present in the water, i.e., speciation.

5.6.1   Sampling and Sample Pretreatment

Because of the high sensitivity of ESA methods, contamination of the sample, of the chemicals used, of vessels, and of the whole apparatus that comes in contact with the sample must be prevented. It has been found that the precision of the results suffers most during the sample preparation and thus great attention must be paid to this operation.

For example, when samples are taken in the open sea from a rubber boat, it is located upwind from the expedition ship and, to eliminate contamination from the rubber boat, the sample is taken from the prow of the boat, in the windward direction, into a 2-liter polyethylene bottle fixed at the end of a telescopic pole about 3 m long [46]. The polyethylene bottles must be pretreated because perfectly clean bottle walls exhibit a tendency to adsorb dissolved metals; they are rinsed with sea water, which leads to the occupation of the adsorption sites by calcium and magnesium ions and a substantial decrease in the adsorption of heavy metals.

Samples of coast water and water from rivers and lakes are taken analogously, using a rubber or wooden boat. Samples of sea water at various depths are collected using a special apparatus. Samples of rain water must be protected against contamination by dust and thus for prolonged sampling special vessels are used that are automatically closed during periods without rain [47]. Samples of potable water are collected directly from the pipe.

Among economically acceptable materials, polyethylene bottles are best suited for sampling and sample storage. Mercury ions are strongly adsorbed on polyethylene and thus part of the samples to be analyzed for mercury must be transferred into glass or, better, quartz vessels within 3 h after the sampling.

5.6.2   Filtration of Natural Water Samples

Natural water, especially sea coast, river, and lake water, contains certain amounts of insoluble substances. According to traditional convention, samples of these waters are filtered through filters with a pore size of 0.45 μm. Of course, the insoluble particles are not completely removed, only the fraction with a particle size larger than the filter pore diameter.

Filtration must also be performed carefully, to avoid contamination and adsorption of the dissolved metals on the filter. Therefore, the filtration is carried out in a closed system under a nitrogen atmosphere [48]. A simple but efficient filtration apparatus that can also be used on a ship or in the field has been described by Mart [48]. Some authors recommend immediate acidification of the filtrate, using 1 ml of concentrated hydrochloric acid (suprapure, Merck) per liter of the sample; others prefer acidification only in the laboratory, to prevent contamination.

The filter with the insoluble residue is placed in a plastic case, protected by a polyethylene foil. To analyze this residue, the amount of water passed through the filter must, of course, be recorded.

5.6.3   Storage of Filtered Samples

It is best to determine trace metals within the shortest possible time after sampling. However, this requirement cannot be satisfied in most cases. A measurable decrease in the concentration of dissolved metals (cadmium and lead at a level of 10 to 30 ng/liter) in natural water with suspended particles occurs within as little as five hours after sampling [48]. Hence the necessity of sample filtration.

Analyses of filtered samples of sea water have shown that no loss in the contents of cadmium and lead occurs during one week after sampling.

Prolonged storage requires more complex measures. It has been found from analyses of acidified (pH 2) samples of sea and lake water stored in polyethylene or Teflon bottles at a temperature of $+5°C$ that the content of cadmium decreases by about 10% and that of lead increases by about 10% during 75 days [49]. It is still not certain whether the changes stem from adsorption of cadmium on the vessel walls and desorption of lead from them or from interactions among the trace components of the solution. For prolonged storage it is best to rapidly freeze the filtered sample, either acidified or untreated, to $-20°C$; the trace concentrations of metals such as cadmium, lead, zinc, copper, and bismuth remain constant for one year.

5.7   Laboratory for Trace Analyses

Laboratories for common chemical analyses are mainly contaminated by dust, cigarette smoke, corrosion products, etc., and are unsuitable for trace analysis. For electrochemical trace analysis, the cost of laboratory modification is low, because the space that must be effectively protected against contamination is small.

First of all, all metallic objects must be removed and the plumbing must be of plastic pipes or metallic pipes protected by a nonabrasive paint whose content of trace metals has been checked. Wooden furniture is satisfactory and the floor should be covered by a material that readily collects dust particles from footwear and is easy to clean.

It is further suitable to provide filtered air that is fed into the laboratory at a small overpressure, thus suppressing the penetration of dust through windows and doors. Dust sedimentation is then almost completely eliminated, but still insufficiently for trace analysis, as the blank determinations still yield relatively high contents of trace metals. Therefore, benches for manipulations with samples are provided with plastic boxes into which laboratory air is aspirated through two filters. The air enters the box through a plastic membrane covering the whole separated area and containing many small holes that ensure constant laminar flow in the working space. For easy manipulation of the samples inside the box, the front part of the box is only covered by a thin plastic curtain. All manipulations inside the box are carried out using protective gloves free of all impurities.

5.7.1   Cleaning of Vessels

Thorough cleanliness of all vessels with which the sample comes into contact is another prerequisite for the attainment of accurate results in trace analyses. Polyethylene bottles (250 ml) for sampling and sample storage are normally washed and then treated with diluted hydrochloric acid p.a. (1 : 5) for four days at a temperature of about 75°C. The cleaning is continued using 1 : 10 hydrochloric acid, followed by 2% hydrochloric acid "suprapure" and finally by 1% hydrochloric acid "suprapure", each stage taking again four days at a temperature of 75°C. The cleaning solutions are stirred by a stream of pure nitrogen. The bottles are then emptied in a clean box and filled with distilled water acidified with concentrated hydrochloric acid "suprapure" (1 ml/liter). The bottles are finally wrapped in polyethylene bags, sealed, and stored for future use. The polyethylene bags are cleaned with water, dilute HCl p.a. (1 : 10) for one day, dilute HCl "suprapure" (1 : 10) for one day, rinsed with distilled water, and dried in a clean box. Other polyethylene objects, such as protective gloves, exchangeable pipette tips, etc., are cleaned analogously.

Teflon and quartz vessels can be cleaned using the same procedure as above, with the exception of the first step, which involves prolonged (1 week) treatment with nitric acid p.a. (1 : 2) at a temperature of 70 to

80°C. The filters used for filtration of waters (mostly cellulose acetate) are first cleaned with HCl p.a. (1 : 2) and then with 1% HCl "suprapure" for at least one week. They are stored in 1% HCl "suprapure". Before use, the filters are rinsed with distilled water and pretreated with a solution containing the alkaline earth ions at concentrations identical with those in sea water, thus substantially decreasing the loss of dissolved metals through adsorption on the filter.

5.7.2   Distilled Water

Water is distilled three times in a quartz apparatus for use in electrochemical trace analysis. The greatest attention must be paid to the water inlet and outlet; silicone rubber tubes are usually used and are replaced after each distillation batch. Excellent results have recently been obtained in water pretreatment using adsorption columns, ion-exchangers, and microfilters, e.g., with the Milli-Q-System (Millipore, Bedford, USA).

5.8   Removal of Sources of Contamination in ESA

To prevent contamination of the test solution during the electrochemical measurement, the electrolysis vessel with the electrodes must be placed in a clean box. Rotating electrodes with the driving motor above the electrode are unsuitable, because dust and corrosion products from the motor can cause serious contamination. Therefore, the driving apparatus is separated and the electrode is driven using a belt transmission. The electromotor is placed in a closed plexiglass box, together with the belt cover and the bearing of the rotating electrode. A small underpressure is maintained in this space during the analysis.

The reference and auxiliary electrodes are connected with the sample solution by a liquid bridge filled with a solution of potassium chloride, "suprapure".

Silicone rubber tubes are most suitable for transport of nitrogen. With long tubing, oxygen penetrates through the tube wall; the silicone rubber is then placed in a larger polyethylene tube through which nitrogen is also passed.

5.9   Sample Treatment Prior to Analysis

5.9.1   Preparation of Filtered Water Samples

The sample preparation depends on the type of water analyzed, mainly on the amount of dissolved organic matter (DOM). DOM is a mixture of many substances of organic and biological origin, whose composition is not completely known, including, e.g., proteins, protein degradation products, carbohydrates, nucleic acids, and other substances, most of which are strongly surface active. DOM components may form soluble inert chelates with the test metal ions, thus mostly preventing the electro-chemical determination, and surfactants may adversely affect the electrode reactions of the test cations. The equilibrium in which test ions (Me^{n+}) participate at trace concentrations in the medium of sea water can be schematically written as

$$ MeL \; \rightleftarrows \; L + Me^{n+} + iX^- \; \rightleftarrows \; MeX_i $$

where X^- are anions present in the water at high concentrations (Cl^-, SO_4^{2-}, CO_3^{2-}), L are the chelate-forming components of DOM, MeX_i are labile inorganic complexes, and MeL are inert chelates. Therefore, DOM can make the electrochemical determination of metals difficult or even impossible and thus must be removed before the measurement.

The amount of DOM depends on the origin of the sample and the removal is simple with samples containing small amounts of DOM (potable and rain water, most ocean water). Then it is sufficient to acidify the sample to pH 2 and the labile chelates are decomposed within 2 to 4 h. With higher DOM contents, more effective methods must be used. Decomposition with acids is not particularly suitable, because the danger of contamination is high. Good results have been obtained by using photolysis with uv light at an elevated temperature [50].

The photolysis is carried out directly in the Teflon or quartz electro-lysis vessel on an acidified solution (pH 2) for 1 to 4 h; hydrogen peroxide is added to samples with a high DOM content (1 ml concentrated H_2O_2 per 50 ml of the sample). The solution must be protected against contamina-tion during this stage because the irradiation cannot be carried out in a clean box. A 150-W uv lamp is used and the sample temperature increases to about 100°C, so that part of the sample is evaporated. It is thus necessary to shorten the photolysis time in the determination of mercury or to use a special apparatus. The irradiation is terminated when the development of hydrogen ceases and the sample is clear. A special apparatus has been proposed for sample irradiation [51], permitting simultaneous photolysis of six samples in closed quartz vessels that are placed in a closed, air-cooled space. Such an apparatus is suitable for the determinations of mercury, but the increase of number of necessary sample treatment operations is disadvantageous. The removal of organic substances by photolysis has also yielded good results in analyses of wines [52] and communal waste waters [50].

5.9.2   Filters

Nitric acid is unsuitable for dissolution of the collected particles and decomposition of the filter, because the nitrogen oxides formed may interfere in the voltammetric determination of trace metals. It is preferable to decompose the sample with a mixture of sulfuric acid and hydrogen peroxide or a mixture of perchloric and sulfuric acids; however, both procedures lead to an increase in the blank value. So far it seems that the best procedure involves combustion of the filter in an oxygen plasma at a decreased temperature of 150°C [52].

In this procedure, the filter does not come into contact with the atmosphere — the process takes place in a closed space at an oxygen partial pressure of 0.1 to 1.0 torr and the oxygen used is of the highest purity (99.999%). As combustion occurs at a relatively low temperature, in contrast to common combustion techniques, loss of the metals as a result of volatilization of their chlorides is negligible, especially with copper, cadmium, lead, and zinc. Combustion in oxygen plasma cannot, of course, be applied to samples for the determination of mercury, selenium, or arsenic and these samples must be mineralized with acids.

Oxygen plasma combustion can be carried out in the apparatus manufactured by the International Plasma Corp., Hayward, USA, in which the oxygen plasma is excited by microwaves.

5.10   Selection of the ESA Method

As the main requirement in the analysis of natural water is a high sensitivity, potentiostatic ESA is used and the stripping step is monitored by DPP, as the technique with the most efficient elimination of the charging current. To further improve the sensitivity of the DPP measurements, a fast-scan technique is useful, e.g., the FSDPP technique that can be carried out using the PA-3 polarographic analyzer (Laboratorní Přístroje, Czecho-slovakia) or the DPP modification described by Valenta *et al.* [53].

Differential pulse cathodic stripping voltammetry (DPCSV) has a smaller importance in analyses of water and is used, e.g., for the determination of selenium in rain water [54]. Among other ESA methods, subtractive differential pulse stripping voltammetry (SDPSV) with the rotating split-disk electrode should be mentioned. This technique is suitable for the determination of mercury in the presence of excess copper [55]. In general, DPASV is most suitable for the determination of trace metals in natural water, in optimal cases with the above modification of the anodic polarization regime.

5.11   Selection of the Electrode

The maximum sensitivity of DPASV can be attained by accumulating the test metals in the MFE. The film electrode is prepared *in situ,* simultaneously with the preconcentration of the test metals. For the determination of traces of mercury and copper in natural water, the rotating gold disk electrode is important. The HMDE is used when the expected content of the test metals is above ca. 200 ng liter^{-1}.

5.12   The Determination Itself

In the DPASV determination of trace metals, the electrolysis vessel is placed in a clean box. The sample solution has a pH of 2 and its volume is 50 to 80 ml. Before the determination, 50 μl of a 2×10^{-2} M Hg(NO$_3$)$_2$ solution are added ("suprapure") to yield a final concentration of ca. 10^{-5} M, the vessel is connected to the electrode head, and the solution is deaerated for 10 to 20 min. Then the preelectrolysis is carried out for about 5 min at a potential of -1.0 V with the rotating electrode (2 000 rpm). After the rest period (30 s) the stripping curve is recorded from -0.9 to -0.10 V. The electrode potential is then maintained at -0.1 V for 3 min with the electrode rotating to completely strip all the metals. On the basis of the size of the stripping peaks, the sensitivity and the preelectrolysis time (3 to 15 min) are selected. The procedure is repeated under the selected conditions. If the amounts of the test metal differ considerably, the stripping peak heights are also very different; then it is suitable to change the measuring sensitivity during the recording, stopping the potential scan between the peaks. Provided that this operation does not take more than 10 s, the effect of the reduction limiting current in quiescent solution is negligible. After the stripping of the last metal, after which the mercury would dissolve in the more positive potential region (mostly copper at about -0.1 V), the electrode rotation is started and all the metals are oxidized within 3 min at this potential; during this period, a standard solution of one of the test metals is added to the sample. The analysis is repeated under the same conditions, but with a preelectrolysis time, t_{s1}, equal to one half of the original preelectrolysis time ($t_{s1} = t_{e1}/2$). This is repeated with the second standard addition, with a preelectrolysis time (t_{s2}) equal to one third of the original preelectrolysis time (t_{e1}).

The standard addition should lead roughly to doubling of the metal concentration in the sample. As with the MFE, the amount of the amalgam formed is proportional to the metal concentration in the sample (c) and the preelectrolysis time; shortening of the preelectrolysis time will cause the amount of the amalgam to be virtually constant in all the measurements, thus eliminating possible interference caused by the formation of inter-metallic compounds, yielding approximately the same peak heights in the three measurements, and substantially shortening the analysis time.

To evaluate the analysis results the measured peak heights must be normalized to one common preelectrolysis time, e.g., t_{s2}. The original concentration of the metal in the sample, c (μg kg^{-1}), is calculated from the relationship

$$c = mi_s/w\Delta i$$

where m is the standard addition in ng, w is the mass of the water sample in the vessel in g, i_s is the height of the sample peak normalized to pre-electrolysis time t_{s2}, and Δi is the normalized average value of the increase in the sample peak height after two subsequent additions of the standard solution.

Table 5.1

Precision of the ESA determination of lead and cadmium in sea water

Pb and Cd (ng/liter)	Rel. stand. deviation (%)
0.5	± 20
1.0	± 10
10	± 4.0
20–2 000	± 2.5
> 2 000	± 4.0

The precision of the ESA determination using the above procedure is very good even for trace metal concentrations in the sample. Table 5.1 lists the relative standard deviations for various contents of lead and cadmium. The accuracy of the results is also very good, as shown by the comparison of the DPASV, mass spectrometric, and AAS results for sea water samples [48].

5.13   Determination of Individual Metals

Cadmium, lead, and copper are mostly determined in acidic solutions (HCl, pH 2) simultaneously. Because of the low solubility of copper in mercury, the standard addition of copper is made least when the thickness of the mercury film is greatest. Because only a small amount of the deposited copper forms an amalgam during the preelectrolysis, it is necessary to shorten the preelectrolysis time after the individual additions of the standard (the amount of mercury increases after each measurement) and with small amounts of copper the preelectrolysis time should be as short as possible and the instrument sensitivity as high as possible (the deposited amount of copper is small and thus all the mercury forms an amalgam). On the other hand, with a high mercury content the preelectrolysis time should be long and the instrument sensitivity low (most of the mercury is not in the form of an amalgam). Under these conditions, a linear dependence between the peak height and the copper concentration in the sample is preserved. A rotating gold disk electrode [55] is also suitable for the determination of copper and is used for the simultaneous determination of mercury and copper.

Zinc is usually determined together with lead and cadmium in 0.01 M HCl or in an acetate buffer at pH 4.5. The determination in hydrochloric acid medium is more advantageous because the danger of sample contamination is smaller. Organic substances must be completely removed (photolysis in the presence of H_2O_2) to prevent interference from the catalytic evolution of hydrogen.

Thallium interferes in the determination of lead on mercury electrodes in acidic media (HCl, pH 2). Therefore, cadmium and copper are determined in an acidic solution and lead and thallium are determined in an alkaline medium after addition of NaOH [56].

Mercury is determined using gold or glassy carbon electrodes [55, 57]; a gold split-disk electrode is recommended for small amounts of mercury.

Selenium is determined by DPCSV in an acidic medium on the HMDE [19]; it is preconcentrated on the electrode in the form of HgSe which is then cathodically stripped [58]. Elemental selenium can be deposited on a gold electrode and stripped anodically or cathodically [59].

Arsenic is difficult to determine with mercury electrodes and gold or glassy carbon rotating disk electrodes are recommended [60]. Organic substances that are adsorbed on the electrode must be removed.

Nickel and cobalt can be determined with a high sensitivity after adsorptive accumulation of the dimethylglyoxime complexes of the two metals on the HMDE and using pulse stripping techniques [61].

Chromium can only be determined using ASV after accumulation on a glassy carbon electrode in a weakly acidic medium; nitrogen-containing organic compounds, copper, and other metals interfere.

The ESA determinations of trace metals contained in atmospheric aerosols are analogous to the above procedures. The test air is passed through a filter which is then decomposed and the collected material dissolved, followed by the determination in a suitable base electrolyte.

ESA also finds use in analyses of biological materials, as the contents of trace metals in them yield important information on the circulation of the metals in the biosphere. Heavy metals (Pb, Cd, Hg, Zn, and Cu) are determined in urine, after the sample decomposition in acids (HNO_3 + $HClO_4$), using an acetate buffer base electrolyte [62]. Similar procedures are used for analyses in milk, after sample combustion at $560°C$ [63], and in blood, after oxygen plasma combustion of the material at a low temperature [64]. The combustion in oxygen plasma has also yielded good results in DPASV determinations of heavy metals in marine organisms, fish organs, and canned foodstuffs [62].

5.14   Importance of ESA for Trace Metal Speciation in Natural Water

A first, rough speciation of metals follows from the results of ESA of the filtered natural water and of the filter. The filtrate can further be chemically treated, e.g., by digestion with acids, irradiation with uv light, or chromatography on complexing ion-exchangers (Dowex A-1, Chelex 100). The ESA of the filtrate fractions thus obtained gives more detailed data on the distribution of metals in labile and inert complexes, bound to colloid particles, etc. [65].

As the hydrogen ion concentration in natural water is mostly determined by the relative concentrations of H_2CO_3, HCO_3^-, and CO_3^{2-}, it is advantageous to carry out the ESA determinations for speciation purposes directly in untreated (unacidified) water samples to avoid changes in the chemical equilibria caused by a change in the pH value. Such a change might also occur during removal of the dissolved oxygen from the sample, as the passage of an inert gas leads to volatilization of CO_2 and a change in the pH by as much as 0.6. To prevent this effect, Lacomte et al. [68] proposed using a mixture of nitrogen and carbon dioxide for the sample deaeration. The CO_2 partial pressure is adjusted so that the solution pH remains constant. The sample is divided into three parts. During the deaeration of the first part, the nitrogen flow-rate is adjusted so that the solution pH does not change by more than 0.1. During the deaeration of the second part, the nitrogen flow-rate is adjusted more finely to attain a maximum pH change of 0.02. The ESA determination itself is carried out with the third part of the sample whose pH is not measured to prevent contamination of the sample by the glass electrode and the deaeration conditions are identical with those in the previous experiments.

To determine the stability of trace metal labile complexes with inorganic ligands present at high concentrations in natural waters, such as chloride and carbonate in sea water, the well-known relationships describing the dependence of the polarographic half-wave potential on the ligand concentration [66] can be employed. The experimental data for the application of these relationships cannot, of course, be obtained polarographically because the metal concentrations in natural waters are too low. However, it has been found [67] that on plotting the stripping peak currents, i_p, as functions of the preelectrolysis potentials that are varied within the interval $E_{1/2} + 200\,mV$ and $E_{1/2} - 200\,mV$, a curve analogous to the classical polarographic curve is obtained. If these series of measurements are performed with various ligand concentrations, the dependence of $E_{1/2}^x$ on the ligand concentration can be determined from the plots and consequently also the number of ligands bound in the complex and the stability constant. Of course, the electrode reaction must be reversible.

The determination of trace metals bound in stable chelates, usually with organic ligands, is different and is based on the assumption that a metal bound in a stable complex is reduced irreversibly, at a more negative potential than that for the reduction of the hydrated ion. Therefore, it is experimentally possible to find the dependence of the free metal ion concentration on the chelating agent concentration. DPASV with short preelectrolysis times is carried out to minimize the disturbance of the chelate dissociation equilibrium during the preelectrolysis. If the ionic strength, the pH, and the content of the major components (Ca^{2+}, Mg^{2+}, Cl^-) in the solution correspond to those in natural water, then the shape of the dependence of i_p (the free metal concentration) on the ligand concentration is the same as that for natural sea water. Nitrilotriacetic acid (NTA) or EDTA is used as a chelating agent because the chelating properties of these substances are well known and because NTA is contained in some kinds of water as a result of pollution by detergents.

Analysis of the i_p vs. the ligand concentration plots permits estimation of the conditional stability constant of the chelate. If the dependence of the free ion concentration (i_p values) on time is also measured under identical conditions, further data are obtained for the calculation of the rate constant for the chelate formation described schematically, e.g., for cadmium, as

$$Ca-NTA + Cd^{2+} \xrightarrow{k_f} Cd-NTA + Ca^{2+} + Mg^{2+}$$
$$\updownarrow \qquad\qquad \updownarrow$$
$$Mg-NTA \quad CdX$$
$$\updownarrow$$
$$CdX_j$$

where CdX are labile inorganic complexes of cadmium.

5.15   Use of ESA in the Analysis of Biological Materials

The increasing interest in the methods for trace metal determination in biological materials is also documented by an increasing use of ESA in this field, e.g., in clinical chemistry, toxicology, in the determination of the nutritional value and any danger to health for foodstuffs and animal fodder, and in environmental hygiene. ESA methods are advantageous here mainly because of their good accuracy and precision, high sensitivity, and the possibility of determining several metals simultaneously. The principal problem, whose solution determines the success of the analysis, is the sample preparation, i.e., a suitable method for obtaining a representative sample and chiefly the choice of a suitable procedure for sample mineralization. The mineralization is carried out by employing modified classical procedures, i.e., combustion or acid digestion. From the point of view of minimal contamination, oxygen plasma combustion seems to be optimal. In the application of electroanalytical methods it is necessary that trace metals be completely released and that the matrix be perfectly decomposed, as organic substances mostly exert unfavorable effects on the voltammetric determination of metal, either through interfering electrode reactions or through adsorption processes.

At present, the determination of lead and cadmium accounts for more than 70% of published papers dealing with the electroanalysis of biological materials. These methods are preferred to AAS because of better precision and accuracy. The interference from tin in ESA is suppressed by tin separation with hydrobromic acid [69] or, more simply, by exchange of the electrolyte after the preelectrolysis; the stripping is carried out in a citrate buffer [70]. It is also possible to use a methanolic solution of hydrochloric acid as the base electrolyte, where lead can be determined by the ACASV method in the presence of a 50-fold excess of tin [71]. A number of authors [52, 72, 73] have compared the ASV and AAS methods for the determination of lead and cadmium in biological materials and have recommended the use of electroanalytical methods.

Mercury can be determined in biological materials using a rotating gold [74] or an impregnated graphite [75] electrode. However, these methods have so far not found wide application and the AAS determination based on the evolution of cold mercury vapors has been preferred in practice. On the other hand, the superiority of electroanalytical methods is generally acknowledged in the determination of copper and zinc. The ESA methods are mostly based on the use of the HMDE, GCE, or MFE with the DPASV measuring technique [76 − 78]. In the determination of copper and zinc, the formation of intermetallic compounds ($CuZn$, $CuZn_2$, $CuZn_3$) may interfere, leading to an increase in the copper stripping peak and a decrease in that of zinc [79]. Then it is unsuitable to use the MFE, because the small volume of mercury involved causes the formation of concentrated amalgams and possible exceeding of the solubility product of the $CuZn_3$ compound. The formation of intermetallic compounds can be prevented, e.g., by a selective preelectrolysis with two working electrodes; either copper is selectively deposited on one electrode (at a relatively positive preelectrolysis potential) and the other metals on the other electrode [80], or copper, lead, and cadmium are deposited on one electrode, followed by the deposition of zinc on the other electrode [81]. Of course, these methods require complete deposition of the test metals and thus small volumes of the test solutions must be used. Trace amounts of arsenic can be determined in biological materials by using ASV with a gold electrode [82, 83] or by CSV with the HMDE [84].

Determination of tin is important in the control of the quality of canned foodstuffs. As lead is usually also present, the method should permit the simultaneous determination of both the metals (see the paragraph describing the determination of lead).

Selenium may be present in grain, plant oils, etc., and can be sensitively determined by CSV with the HMDE [85] or by ASV with a gold electrode [59, 86]. The determination of thallium is required in clinical analysis and two methods for its determination in urine have been proposed [87, 88], employing DPASV with the HMDE.

5.16   Determination of Organic Compounds

ESA is basically useful for the determination of trace metals, but even many organic compounds can be determined, first of all those that form insoluble compounds with the mercury ions on the surface of mercury electrodes polarized at potentials corresponding to mercury anodic dissolution; these compounds are then cathodically stripped from the electrode. Hence CSV can be applied to the determination of these substances, such as some herbicides and rodenticides derived from thiourea (phenylthiourea, naphthylthiourea) that have been determined using the DPP technique at concentrations down to units of ng ml^{-1} [89].

However, CSV is not limited to the above class of organic substances. Compounds with surface-active properties can be accumulated on a mercury electrode by adsorption and then be electrochemically reduced or oxidized. For example, cyanuric chloride can be determined in dye industry atmosphere. This substance yields a polarographic wave in a weakly acidic medium, with a half-wave potential of $+0.15$ V (SCE); after adsorptive accumulation of the substance at $+0.25$ V (SCE) the detection limit equals 0.2 µg ml^{-1} [90].

The principle of adsorptive accumulation can also be applied to the determination of some electroinactive substances, such as certain alkaloids and pharmaceuticals, codeine, papaverine, cocaine, various detergents, etc. [91]. Adsorptive accumulation has been used for the determination of surfactants in distilled water, at concentrations of 10 to 300 µg liter^{-1} [92]. To monitor oil products in sea water at concentrations of 0.02 to 100 mg liter^{-1}, polarographic maximum suppression and the effects of the substances on Kalousek commutator polarographic curves have been employed [93]. The substances are accumulated by adsorption on a HMDE. DPP has also been used for the study of adsorption phenomena caused by oil components present in water [94]. The water sample is added to the base electrolyte, 1 M NaOH or 5 M KF, and adsorption accumulation is carried out at -0.7 V (SCE) with 1 M NaOH (the desorption peak lies at a potential of ca. -1.2 V) or in the open electrode circuit with 5 M KF (the peak potential is ca. -1.4 V).

Adsorption accumulation can further be used to determine highly toxic biphenyl and its chlorinated derivatives. Biphenyl is polarographically reduced in 70% dioxan containing tetrabutylammonium iodide. In aqueous solutions containing 20% methanol of pH 4–6, biphenyl and its chlorinated derivatives yield a drawn out dc polarographic wave at a potential of -1.10 V (SCE); after adsorptive accumulation on the HMDE at a potential of -0.4 V, the substance can be determined from 3 µg liter^{-1} to 1 mg liter^{-1} using the DPP technique [95].

Adsorptive accumulation is also used in the determination of metals; the metal is complexed with a suitable reagent and the complex formed is adsorbed on the surface of the working electrode. A typical example is the determination of nickel based on the adsorption of the nickel complex with dimethylglyoxime on the HMDE surface (see Section 5.13). Solid electrodes have also been recommended [96] for this type of determination.

Adsorption effects often play a role in the determination of substances forming sparingly soluble compounds with the mercury ions. In the CSV determination of thiourea in a sodium hydroxide base electrolyte, the detection limit is 1 ng ml^{-1} using the DPP technique. When this determination is carried out in a neutral medium (0.5 M NaNO$_3$), the detection limit is 2 pg ml^{-1} using a 10-min adsorptive accumulation at a potential of $+0.20$ V (SCE) and the dc measuring technique [97]. In all the above procedures, the amount of the substance accumulated on the electrode surface depends on the accumulation time and the electrode potential, although in some cases substances can even be adsorptively accumulated with the electric circuit open.

REFERENCES

1. Vydra F., Štulík K., and Juláková E.: Electrochemical Stripping Analysis, E. Horwood, Chichester 1976.
2. Barendrecht E.: in Bard A. J. (ed.), Electroanalytical Chemistry, M. Dekker, New York 1972, Vol. 2, p. 53.
3. Neeb R.: Inverse Polarographie und Voltammetrie, Verlag Chemie, Weinheim 1969.
4. Brainina Kh. Z.: Inversionnaya Voltamperometriya Tverdykh Faz, Khimiya, Moscow 1972.
5. Vydra V.: Chem.Listy **70**, 337 (1976).
6. Hrabánková E. and Doležal J.: Chem.Listy **62**, 1164 (1968).
7. Opekar F.: Chem.Listy **75**, 132 (1981).
8. Jagner D. and Granéli A.: Anal.Chim.Acta **83**, 19 (1974).

9. Vydra F. and Luong L.: *J.Electroanal.Chem.* **54**, 447 (1974).

10. Barański A.: *Chem.Anal.* (Warsaw) **16**, 989 (1971).

11. Bruckenstein S. and Bixler J. W.: *Anal.Chem.* **37**, 786 (1965).

12. Štulíková M.: *J.Electroanal.Chem.* **48**, 33 (1973).

13. Matson W. R., Roe D. K., and Carritt D. E.: *Anal.Chem.* **37**, 1594 (1965).

14. Florence T. M.: *J.Electroanal.Chem.* **26**, 293 (1970); **27**, 273 (1970).

15. deVries W. T. and van Dalen E.: *J.Electroanal.Chem.* **8**, 366 (1964); **9**, 448 (1965).

16. Igolinskii V. A.: *Zavod.Lab.* **32**, 1310 (1966).

17. Tindall G. W. and Bruckenstein S.: *Anal.Chem.* **40**, 1637 (1968).

18. Johnson D. C. and Allen R. E.: *Talanta* **20**, 305 (1973).

19. Sipos L., Kozar S., Kontušič I., and Branica M.: *J.Electroanal.Chem.* **87**, 347 (1978).

20. Štulíková M. and Štulík K.: *Chem.Listy* **68**, 800 (1974).

21. Hanzlík J.: *Chem.Listy* **66**, 313 (1972).

22. Zieglerová L., Štulík K. and Doležal J.: *Talanta* **18**, 603 (1971).

23. Underkofler W. L. and Shain I.: *Anal.Chem.* **33**, 1966 (1961).

24. Huderová L. and Štulík K.: *Talanta* **19**, 1285 (1972).

25. Štulík K. and Štulíková M.: *Anal.Lett.* **6**, 441 (1973).

26. Koster G. and Ariel M.: *J.Electroanal.Chem.* **33**, 943 (1974).

27. Štulík K. and Bedroš P.: *Talanta* **23**, 563 (1976).

28. Wang J. and Ariel M.: *Anal.Chim.Acta* **99**, 98 (1978); **101**, 1 (1978).

29. Trojánek A. and Opekar F.: *Anal.Chim.Acta* **126**, 15 (1981).

30. Blaedel W. J. and Wang J.: *Anal.Chem.* **51**, 799, 1725 (1979).

31. Both M. D., Brand M. J. D., and Fleet B.: *Talanta* **18**, 603 (1971).

32. Opekar F. and Herout M.: *Chem.Listy* **71**, 867 (1977).

33. Jagner D. and Kryger L.: *Anal.Chim.Acta* **78**, 251 (1975); **80**, 255 (1975).

34. Perone S. P., Jones D. O. and Gutknecht W. F.: *Anal.Chem.* **41**, 1154 (1969).

35. Rifkin S. C. and Evans D. H.: *Anal.Chem.* **48**, 2174 (1976).

36. Perone S. P.: *Anal.Chem.* **43**, 1288 (1971).

37. Gutknecht W. F. and Perone S. P.: *Anal.Chem.* **42**, 906 (1970).

38. Sybrandt L. B. and Perone S. P.: *Anal. Chem.* **44**, 2331 (1972).

39. Čipak J., Ružič I. and Jeftič L.: *J.Electroanal.Chem.* **75**, 9 (1977).

40. Thomas Q. V., Kryger L. and Perone S. P.: *Anal.Chem.* **48**, 761 (1976).

41. Perone S. P. and Jones D. O.: Digital Computers in Scientific Instrumentation, McGraw — Hill, New York 1973.

42. Opekar F., Herout M. and Kalvoda R.: *Chem.Listy* **74**, 542 (1980).

43. Herout M. and Opekar F.: *Chem.Listy* **76**, 645 (1982).

44. Anfält T. and Starndberg M.: *Anal.Chim.Acta* **103**, 379 (1978).

45. Mortensen J., Ouziel E., Skov H. J., and Kryger L.: *Anal.Chim.Acta* **112**, 297 (1979).

46. Mart L.: *Z.Anal.Chem.* **299**, 97 (1979).

47. Nguyen N. D. and Valenta P.: *Ger.Pat.Appl.* P 28318403, June 20, 1978.

48. Mart L.: *Z.Anal.Chem.* **296**, 350 (1979).

49. Valenta P., Mart L., Nürnberg H. W. and Stoeppler M.: *Vom Wasser* **48**, 89 (1977).

50. Valenta P. and Nürnberg H. W.: in Böhnke B. (ed.), Gewässerschutz-Wasser-Abwasser, Ges. Siedlungswasserwirtschaft, Aachen 1980, p. 105.

51. Armstrong F. A. J. and Tibbits S.: *J. Mar.Biol.Assoc.UK* **48**, 143 (1968).

52. Golimowski J., Valenta P. and Nürnberg H. W.: *Z.Lebensmittelunters.Forsch.* **168**, 353 (1979).

53. Valenta P., Mart L. and Rützel H.: *J.Electroanal.Chem.* **82**, 327 (1977).

54. Nürnberg H. W., Valenta P. and Nguyen V. D.: KFA-Jahresbericht 1978/79, Jülich 1979, p. 47.

55. Sipos L., Golimowski J., Valenta P. and Nürnberg H. W.: *Z.Anal.Chem.* **298**, 1 (1979).
56. Klahre P., Valenta P. and Nürnberg H. W.: *Vom Wasser* **51**, 199 (1978).
57. Sonntag G., Kerschbaumer M. and Kainz G.: *Mikrochim.Acta* **1976**, II, 411.
58. Vajda R.: *Acta Chim.Acad.Sci.Hung.* **63**, 257 (1970).
59. Andrews R. W. and Johnson D. C.: *Anal.Chem.* **47**, 1056 (1975); **48**, 1611 (1976).
60. Forsberg G., O'Laughlin J. W., Megargle R. G. and Koirtyohann S. R.: *Anal.Chem.* **47**, 1586 (1975).
61. Hien H. M.: Thesis. Charles University, Prague 1981.
62. Nürnberg H. W.: in Smyth W. F. (ed.), Electroanalysis in Hygiene, Environmental, Clinical and Pharmaceutical Chemistry, Elsevier, Amsterdam 1980, p. 184.
63. Jöhnsson H.: *Z.Lebensmittelunters.Forsch.* **160**, 1 (1976).
64. Valenta P., Rützel H., Nürnberg H. W. and Stoeppler M.: *Z.Anal.Chem.* **282**, 25 (1976).
65. Batley G. E. and Florence T. M.: *Anal.Lett.* **9**, 379 (1976).
66. Heyrovský J. and Kůta J.: Principles of Polarography, Academia, Prague 1965, Chapters 8 and 14.
67. Bubic S. and Branica M.: *Thalassia Jugoslavica* **9**, 47 (1973).
68. Lacomte P., Mericam P., Astruc A. and Astruc M.: *Anal.Chem.* **53**, 2372 (1981).
69. Geissler M., Schiffel B. and Kuhnhardt C.: *Z.Chem.* **15**, 408 (1975).
70. Desimoni E., Palmisano F. and Sabbatini L.: *Anal.Chem.* **52**, 1889 (1980).
71. Metzger L., Willems G. G. and Neeb R.: *Z.Anal.Chem.* **293**, 16 (1978).
72. Sulek A. M., Elkins E. and Zink E. W.: *J.Ass.Offic.Anal.Chem.* **61**, 931 (1978).
73. Nürnberg H. W.: *Sci.Total Envir.* **12**, 35 (1979).
74. Ahmed R., Valenta P. and Nürnberg H. W.: *Mikrochim.Acta 1*, 171 (1981).
75. Ulrich L.: *Z.Anal.Chem.* **277**, 349 (1975).
76. Smart R. S. and Weber J. H.: *Anal.Chim.Acta* **115**, 331 (1980).
77. Neeb R. and Kiehnast I.: *Z.Anal.Chem.* **285**, 121 (1977).
78. Wandat F. and Neeb R.: *Z.Anal.Chem.* **288**, 32 (1977).
79. Shuman H. S. and Woodward G. P.: *Anal.Chem.* **48**, 1979 (1976).
80. Roston D. A., Brooks E. E. and Heineman W. R.: *Anal.Chem.* **51**, 1728 (1979).
81. De-Angelis T. P., Bond R. E., Brooks E. E. and Heineman W. R.: *Anal.Chem.* **49**, 1792 (1977).
82. Davis P. H., Dulude G. R., Griffin R. M., Matson W. R. and Zink E. W.: *Anal.Chem.* **50**, 137 (1978).
83. Marchidan S., Vatires I. and Totir N. D.: *Rev.Roum.Chim.* **25**, 33 (1980).
84. Holak W.: *J.Ass.Offic.Anal.Chem.* **59**, 650 (1976).
85. Blades M. W., Dalziel J. A. and Elson C. M.: *J.Ass.Offic.Anal.Chem.* **59**, 1234 (1976).
86. Posey R. S. and Andrews R. W.: *Anal.Chim.Acta* **124**, 107 (1981).
87. Kinard J. T.: *Anal.Lett.* **10**, 1147 (1977).
88. Franke J. P. and Zeeuw R. A.: *J.Anal.Toxicol.* **1**, 291 (1977).
89. Smyth M. R. and Osteryoung J. G.: *Anal.Chem.* **49**, 2310 (1977).
90. Stará V., Jeník J. and Kopanica M.: *Anal.Chim.Acta* **147**, 371 (1983).
91. Kalvoda R.: *Anal.Chim.Acta* **138**, 11 (1982).
92. Bednarkiewicz E., Donten M. and Kublik Z.: *J.Electroanal.Chem.* **127**, 241 (1981).
93. Žutič V., Čosovič B. and Kozarac Z.: *J.Electroanal.Chem.* **78**, 113 (1977).
94. Kalvoda R. and Novotný L.: *Vodní Hospodářství* **11**, 291 (1984).
95. Lam N. K., Kalvoda R. and Kopanica M.: *Anal.Chim.Acta* **154**, 79 (1983).
96. Brainina Kh. Z.: *Res.Z.Anal.Chem.* **312**, 428 (1982).
97. Stará V. and Kopanica M.: *Anal.Chim.Acta* **159**, 105 (1984).

CHAPTER 6

POTENTIOMETRY WITH ION-SELECTIVE ELECTRODES

Josef Veselý and Karel Štulík

6.1   Introduction

Following the rapid development at the end of the sixties and the beginning of the seventies, the field of potentiometry with ion-selective electrodes (ISEs) has stabilized. There are a number of more or less reliable electrodes, manufactured by many companies, whose applicability is well known. Recently, perhaps only two significant discoveries have been made that are important for analysis, the discovery of an electrode sensitive to calcium ions that is highly selective with respect to sodium ions, which is of prime importance in clinical analysis, and of an electrode sensitive to chloride with a membrane containing a mixture of Hg_2Cl_2 and HgS that permits the determination of substantially lower concentrations of chloride than the classical $AgCl$ electrode. A great deal of attention is now being paid to the development of electrodes based on biochemical processes, such as tissue and bacterial electrodes, and electrodes containing modern electronic elements combined with ion-selective membranes (e.g., CHEMFET [1]). At the time of writing this chapter, these sensors, however, have not left the research laboratories.

However, during recent years experimental techniques and instrumentation have undergone rapid progress. Flow techniques are especially important, either continuous monitoring of substances or rapid flow-injection analyses (see, e.g., Refs. 2 and 3). Flow measurements not only render the analysis more rapid, but also lead to a substantial decrease in the determination limit for some solid-membrane electrodes, to values of 10^{-8} to 10^{-9} mol liter^{-1}. This is primarily a result of suppression of the contribution of the membrane to the amount of the test substance (e.g., through membrane dissolution). This problem is more complicated with electrodes containing liquid ion-exchangers and neutral ion-carriers in polyvinyl chloride membranes and especially with gas sensors, with which many theoretical and practical problems are encountered during measurements in the vicinity of the determination limit.

While many reviews and monographs have been devoted to practical applications of ion-selective electrodes (recently, e.g., Refs. 4–7), there are only a few reviews dealing with the use of ISEs in monitoring environmental pollution (see, e.g., Refs. 8–11). The analysis of the environment has some specific features, the most important being that the toxicity of substances depends on their chemical form; from this point of view there is a basic difference between free ions and ions bound in complexes. Most instrumental analytical methods are only capable of determining the overall concentration of substances, and thus potentiometry with ISEs, yielding the activity, i.e., the free ion content, is of a great importance in the study of ion distribution among various chemical forms.

Another characteristic feature in environmental analyses is that it is important to compare the actual amount of a pollutant with a certain critical value that is considered to be the toxicity limit in the biosphere and is usually specified by a standard or a law, rather than to determine the absolute content of the substance. From this point of view it is important that measurement with ISEs is instrumentally and methodically simple, permitting field work, and the measurement is readily adapted to continuous monitoring of pollutants in the biosphere (the atmosphere, rivers, waste water, etc.).

ISEs find use chiefly in the determination of anions, for which far fewer methods are traditionally available than for the determination of cations. From the point of view of environmental analysis, electrodes capable of determining ionic tensides are important. Finally, gas probes, combining a membrane selectively permeable for gases with an ISE, are of great importance for monitoring toxic gases, such as sulfur dioxide, hydrogen sulfide, ammonia, and cyanogen.

ISEs are most often used for analysis of water, such as surface, sea, ground, potable, or waste water, and also for analysis of the atmosphere after absorption of the gases in a solution or deposition of the aerosol on a filter. Sedimented dust and soils are analyzed far less often, chiefly because the tedious sample preparation largely offsets the main advantages of the measurement itself, i.e., simplicity and rapidity. Applications to the analysis of foodstuffs are important, but this field is only loosely connected with the environment and will be treated only marginally here. An extensive and significant field is the application of ISEs in clinical analysis, but this is outside the scope of this chapter.

6.2   Survey of Modern Measuring Methods with ISEs

Several basic conditions must be met in measurements with ISEs (for a more detailed discussion see, e.g., Ref. 5): (1) A constant temperature; (2) a constant and relatively high ionic strength of the solution; (3) a pH value lying within the optimal interval from the point of view of both the electrode function and the chemical form of the test substance in the solution; (4) a suitable composition of the test solution, considering the selectivity of the measurement (masking of interferents), as well as maximal possible precision, accuracy, and sensitivity of measurement. Prior to the determination, the electrodes must usually be activated by repeated measurements in standard solutions with various concentrations of the test substance, until the electrode response is rapid and reproducible.

The simplest and most common measuring technique is direct potentiometry, quite analogous to pH measurement with the glass electrode, which is especially well suited for field measurements (either *in situ* or at a field base). For less precise measurement, calibration with one or two standards is often sufficient, but for precise measurement a calibration curve constructed from a great number of points is necessary, or single and multiple standard addition methods or titration can be used.

The enormous progress in electronics has brought about a great development in measurement automation. Modern instruments are commonly provided with microprocessors and are capable of storing the values obtained in standard solutions, comparing them with the values measured in sample solutions, calculating the test substance concentration by a preset method, and displaying it in the required units. These instruments can, to a certain extent, detect and eliminate gross errors committed by the operator, correct for curvature of the calibration dependence and for time dependence of the potential, and take into account the blank value. A number of specialized automatons are marketed chiefly for clinical analysis. In more complicated measurements requiring a high precision, minicomputers are often used that control the analytical procedure itself and handle the data obtained, e.g., by the least squares method using Gran functions, etc. (For a more detailed discussion and references to the literature see, e.g., Ref. 1.)

For environmental analyses, methods of automated continuous monitoring and periodical measurements in flowing liquids are of special importance. Potentiometry with ISEs is well suited for these purposes. A more detailed description of these techniques can be found, e.g., in Refs. 1 − 3 and 12.

6.3   Determination of Nitrogen Compounds

When nitrogen oxides, NO_x (NO, NO_2), are emitted, e.g., from automobile engines and other sources in which combustion takes place at a high temperature in the presence of air, highly toxic substances can be formed, such as peroxyacetyl nitrates (PAN). On sunny days, smog can be formed, of the Los Angeles (photochemical) type. Oxidation of nitrogen oxides leads to nitrates that are removed by atmospheric precipitation. The level of nitrates is increasing dangerously in virtually all kinds of water, but this is caused rather by pollution with artificial fertilizers. The adverse effect of nitrates lies in the eutrophization of rivers (excessive growth of aquatic flora) and in direct damage to human health (nitrate alimentary methoglobinemia). In the presence of ammoniacal nitrogen and SO_2 emissions, ammonium sulfate can be formed in the atmosphere in addition to sulfuric acid, which adversely affects not only human health, but also visibility and corrosion of materials. The determination of ammoniacal nitrogen in waste waters is also important. Although only a few papers in the literature deal with this subject, many monitors of ammoniacal nitrogen in various types of water operate in industrially developed countries and are based on the ammonia gas probe.

6.3.1 NH_4/NH_3 Determination

In determining nitrogen-containing compounds, several kinds of ISEs can be used. The ammonia gas probe is most selective; it consists of a glass pH-ISE on the surface of which a thin film of an unbuffered ammonium chloride solution is fixed by a gas-permeable, hydrophobic membrane. Ammonia diffuses into this film from an alkaline sample solution (pH $\geqq$ 11.2) and reacts with hydrogen ions according to the equation

$$NH_3 + H_2O \quad \rightleftarrows \quad NH_4^+ + OH^- \qquad (6.1)$$

causing measurable changes in the pH. The electrode response is linear, provided that the NH_4^+ concentration in the film is constant. Samples for measurement of ammoniacal nitrogen using this probe are made alkaline with sodium hydroxide to which EDTA is usually added to prevent binding of ammonia in complexes, e.g., with copper. At pH $\geqq$ 11.2 the $[NH_4^+]/[NH_3]$ ratio is negligibly small and acid gases (CO_2, SO_2, and H_2S) are converted into the corresponding anions and therefore cannot interfere by diffusion through the membrane. However, volatile amines, e.g., chloroamines, interfere and must be reduced prior to the measurement, e.g., by adding hydrosulfite. For the NH_3 probe, the Nernstian response is usually given as extending down to 2 to 5×10^{-6} M, with a detection limit of ca. 5×10^{-7} to 1×10^{-6} M NH_3; however, these values depend on many factors. In determinations close to the determination limit, errors stemming from impurities in the water used must be eliminated. It is also necessary to bear in mind that the potential of all gas probes depends markedly on the temperature (> 1 mV/°C).

The content of ammoniacal nitrogen in water varies widely, from less than 0.01 mg liter^{-1} in distilled water to 40 mg liter^{-1} in heavily polluted waste waters. Determination with a gas probe is rapid, but may suffer from poor precision at low concentrations. Evans and Partridge [13] determined the content of ammoniacal nitrogen in various water samples, including those from swimming pools. They prevented interference from chloramines by adding sodium thiosulfate and obtained results that were in a good agreement with those of the spectrophotometric determination using indophenol blue, although the potential stabilized somewhat slowly at NH_3 concentrations below 0.1 mg liter^{-1}. Růžička et al. [14] demonstrated the use of the air-gap electrode on the determination of ammoniacal nitrogen in waste waters. The construction of this electrode has certain advantages, especially if surfactants are contained in the sample. However, a somewhat higher determination limit has been found [15] for this electrode, compared with the membrane NH_3 electrode.

Concentrations of NH_4^+/NH_3 in the pure atmosphere attain values of units of μg m^{-3}; the highest permissible average daily concentration of NH_3 is 0.1 mg m^{-3} in Czechoslovakia. Ammonia adsorbed on solid particles (trapped in filters preceding the absorbers) and the residual NH_3, absorbed in a solution of hydrochloric or sulfuric acid, are determined. Eagan and Dubois [16] passed large volumes of air (2 000 m^3/24 h) through a glass fiber filter and were able to determine concentrations of ammoniacal nitrogen greater than 0.03 μg m^{-3} after extracting the ammonia from the filter into water. Ferm [17] passed air through a vertical glass tube whose inner walls were coated with a film of oxalic acid that absorbed the ammonia present. After 24 h the absorber was washed with sodium hydroxide and concentrations of ammonia greater than 5×10^{-10} M could be determined.

6.3.2   Determination of Nitrogen Oxides, NO_x

The NO_x content in the air varies from units of $\mu g\ m^{-3}$ to values above $100\ \mu g\ m^{-3}$; the latter values are usually considered dangerous. As the NO_2 gas probe suffers from interference, nitrate ISEs are used in the potentiometric determination of nitrogen oxides.

Dee *et al.* [18] described a method based on chemisorption of NO_x on lead dioxide heated to $190°C$; lead(II) nitrate is formed in the process. Hydrogen chloride that might be present is eliminated by additon of lead (II) fluoride and the nitrate content is measured after extraction into hot water, using a nitrate ISE. However, the presence of more than about a tenfold excess of SO_2 causes a positive error when about 200 ppm NO_x is determined; carbon dioxide, water vapor, and nitrogen are not sorbed. The kind of lead dioxide used is important.

Driscoll *et al.* [19] determined the NO_x content in the combustion products of oil and gases, also using a nitrate ISE after oxidative absorption. In determining 40 to 250 ppm NO_x, good agreement with standards was attained, even in the presence of sulfur dioxide, similarly to the work of Kneebone and Freiser [20]. The latter authors passed air through three solutions of 2% hydrogen peroxide for 24 h at a flow rate of 2 liters/min. Excess H_2O_2 was then removed from the absorption solution by adding manganese dioxide and NO_x contents of 100 to 200 $\mu g\ m^{-3}$ were measured using a coated-wire nitrate ISE; sulfur dioxide did not interfere if present in less than a 40-fold excess. Ferber *et al.* [21] described a personal NO_x dosimeter containing a nitrate ISE; NO_x was trapped in a glass filter impregnated with sodium dichromate and sulfuric acid (oxidation of NO to NO_2) and the NO_2 passed through a silicone rubber membrane into a solution of hydrogen peroxide and sulfuric acid, where it was absorbed with formation of nitrate.

6.3.3   Determination of Nitrates

The several known types of nitrate-ISE differ mainly in the detection limit and less in the not very high selectivity, which limits the application of these electrodes to low- and medium-mineralized waters. The oldest electrode is based on an ion-associate in which the complex of Ni^{2+} with a substituted *o*-phenanthroline is the cation and nitrate is the anion. This ion-exchanger is used, e.g., in the Orion 93-07 or Crytur 07-15 electrodes and the Nernstian response is usually attained in a concentration range down to 3×10^{-5} to $1 \times 10^{-4}\ M$. Electrodes containing substituted ammonium salts, e.g., tetradecylammonium nitrate [22], have lower detection limits and can be used above ca. 10^{-6} M NO_3^-. In Czechoslovakia, a nitrate electrode based on a basic dye, e.g., crystal violet, dissolved in nitrobenzene has been developed [23]. Its main advantage is easy renewal of the active surface, which is important because the properties of the electrode ion-exchanger change after prolonged exposure to high interferent concentrations (e.g., in strongly mineralized waters) due to ion exchange.

To suppress the effect of interfering ions, primarily HCO_3^-, Cl^-, and NO_2^-, addition of buffers containing compounds of silver, amidosulfonic acid, and aluminum sulfate (Al^{3+} ions complex the anions of organic acids) has been recommended. Interference from HCO_3^- can be prevented by carrying out the measurement at a pH of less than 4 or by removing gaseous carbon dioxide.

Water samples for the determination of nitrate are protected against reduction by addition of 1 ml of chloroform or a 0.1% solution of phenylmercuric acetate per liter of water. Weiss [24] found that the determination of nitrate at levels below 5 mg liter^{-1} in strongly mineralized waters is unreliable using the Orion 92-07 electrode, even if concentrations down to 0.6 mg liter^{-1} could be measured. However, this is unimportant for evaluation of potable water, for which the Czechoslovak standard permits concentrations of up to 50 mg liter^{-1}.

Simeonov et al. [25] compared various potentiometric techniques with the spectrophotometric determination of nitrate. The most precise results were obtained using the Gran technique of multiple addition and it was even possible to monitor small changes in the nitrate content in lake waters at a level of $>$ 1 mg liter^{-1}.

In strongly mineralized waters, where interferences in the determination are serious, it is preferable to reduce nitrate to ammonia using Devard's alloy (50 % Cu, 45 % Al, 5 % Zn) and to use the ammonia gas probe for the determination. The ammonia originally present in greater than a twofold excess must be removed, e.g., by passage of purified nitrogen at 80 to 90°C. When the original ammonia content is small, it can be measured before the reduction of nitrate and the appropriate correction can be made. Mertens et al. [26] used this method to measure nitrate contents of 1 to 50 mg liter^{-1} in a flow-through system, with a maximal frequency of 30 samples per hour. Under laboratory conditions, the determination limit was 0.05 mg liter^{-1}, even though the calibration curve was no longer linear at nitrate concentrations below 0.6 mg liter^{-1}.

A continuously operating apparatus was described by Forney and McCoy [27] for the determination of nitrates adsorbed on solid particles in the atmosphere. The sample is collected in an aerosol impact device and is dissolved in a small amount of water containing silver fluoride, to eliminate interference from halides and sulfide and to maintain a constant fluoride activity permitting the use of a fluoride-ISE reference electrode. The flow-through cell is constructed so that bubble formation is prevented and the solution flow rate is 6 to 10 ml min^{-1}.

6.3.4   Determination of Total Nitrogen

In evaluation of water quality, ammoniacal and nitrate nitrogen, i.e., the total inorganic nitrogen, and sometimes also organic nitrogen, must often be determined. The determination of inorganic nitrogen is described above in connection with the Mertens method [26]. Devard's alloy and an air-gap electrode were used by Hansen *et al.* [28] in the determination of inorganic nitrogen. Organic nitrogen does not interfere in this determination and this fact was utilized by McKenzie and Young [29] for determination of all three basic components of dissolved nitrogen in waste waters. Ammonia is distilled off from the water, the nitrate is reduced by Devard's alloy in an aliquot part, and the organic nitrogen is finally converted into ammonia by the Kjeldahl procedure. The measurement is carried out after making the solution alkaline, in the same way for all the three determinations and thus the total dissolved nitrogen can be determined relatively rapidly.

Lowry and Mancy [30] described an automatic apparatus for the determination of organic nitrogen in natural waters. The Kjeldahl procedure is replaced by 17-min irradiation with ultraviolet light and reduction of the decomposition products of the organic nitrogen compounds.

6.4   Determination of Sulfur Compounds

The chemical form of sulfur in solution depends on the oxidation-reduction conditions and the pH value. First of all, sulfides can be determined potentiometrically using the Ag_2S electrode. The gas probe for sulfur dioxide has so far found substantially narrower application and a satisfactorily selective electrode for sulfate has not yet been constructed. Therefore, sulfate is determined indirectly with ISEs, usually by titration. In determination of sulfur compounds using ISEs, the oxidation state can also be changed, but this is less common than with nitrogen compounds and it may sometimes be difficult if the conversion must be quantitative (e.g., the reduction of sulfate to sulfide).

6.4.1   Determination of Sulfides and of Hydrogen Sulfide

The ISE containing silver sulfide can be used to measure the sulfide activity over an unusually wide range, because of the very low solubility of the membrane material. For the same reason, other anions, e.g., chloride, cannot precipitate on the electrode active surface as a result of ion-exchange and the determination is thus highly selective. Only cyanide interferes at concentrations greater than 10^{-3} M. However, difficulties arise from redox changes, both in solution (oxidation of sulfide by dissolved oxygen) and at the electrode surface. To protect sulfide against oxidation, various reagents are added that are medium-strong reductants, because prolonged action of strong reductants can cause reduction of the silver ion in the membrane and accumulation of metallic silver on the electrode surface, leading to deterioration of the response. Gulens [31] thus recommends a pH of about 5 for the measurement of hydrogen sulfide in waters using continuous monitors, and reductants need not necessarily be included in the acidic reagent added to the sample. The electrode exhibits a Nernstian response to the total sulfidic sulfur, i.e., the sum $[H_2S] + [HS^-] + [S^{2-}]$, down to a concentration of 10^{-7} M and the electrode can work in the monitor for one year without loss of response to sulfide.

At a pH of about 5, most sulfidic sulfur is present as H_2S in the solution, but the ISE measures only the activity of free sulfide anions, which is very low but is maintained constant by the hydrolytic reactions,

$$S^{2-} + H_2O \ \rightleftarrows \ HS^- + OH^- \text{ and } HS^- + H_2O \ \rightleftarrows \ H_2S + OH^-$$
$$(6.2)$$

provided that oxidation is suppressed. Therefore the measurement of low sulfide activities is reproducible. The sulfide activity is related to the total sulfidic sulfur concentration, $[S^{2-}]_{tot}$, by the relationship

$$a_{S^{2-}} = \frac{\gamma_{S^2} [S^{2-}]_{tot}}{1 + \dfrac{a_{H^+}}{K_{a_2}} + \dfrac{(a_{H^+})^2}{K_{a_1} K_{a_2}}}$$
$$(6.3)$$

where K_{a_1} and K_{a_2} are the dissociation constants of hydrogen sulfide and

$\gamma_{S^{2-}}$ is the sulfide activity coefficient. The solution pH must either be accurately adjusted, or simultaneously measured, if correct values of $[S^{2-}]_{tot}$ are to be calculated.

Measurement of sulfide in discrete samples is still mostly carried out in strongly alkaline solutions, at pH greater than 13 where most sulfidic sulfur is in the form of free sulfide ions, S^{2-}, and thus $[S^{2-}]_{tot} = [S^{2-}]$. A composite buffer solution, usually containing sodium hydroxide, ascorbic acid, and sodium salicylate, is added to the sample. The lifetime of this buffer solution can be substantially prolonged by careful removal of atmospheric oxygen by passage of nitrogen. If this solution is to be successfully used for measurement of sulfide at concentrations around 10^{-7} M, heavy metals forming insoluble sulfides must be removed from it. Sekerka and Lechner [32] added an amount of sulfide, accurately determined by titration of an aliquot part of the buffer with sulfide with the end-point indication using an Ag$_2$S-ISE, to the stock solution. They were then able to determine very low concentrations of sulfide, down to 10^{-8} M. The authors recommend that a large amount of ascorbic acid be added (more than 40 g liter^{-1}), which, according to Gulens's results [31], should lead to formation of silver metal on the electrode surface; however, the electrode surface can readily be repolished in the laboratory, in contrast to the electrodes built into continuous monitors.

If the sample is protected against oxidation, trace metals are removed from the buffer, and the electrode active surface is occasionally cleaned mechanically, sulfide can readily be determined at concentrations down to 10^{-7} M. Therefore, the tedious sample enrichment by distillation of H$_2$S which was originally recommended, e.g., for natural waters, is mostly unnecessary. Similar to most other ISEs, determinations of low concentrations are more readily carried out in flowing systems.

Sulfides present at relatively high concentrations, e.g., in waste waters, can be determined by monitoring the change in the potential of an Ag$_2$S-ISE immersed in a silver nitrate solution on addition of the sample to the solution. Titration determinations have better precison. Solutions of lead(II) or cadmium ions are mostly used as titrants, as silver and mercury(II) ions cannot be used in reducing media. However, the precipitation of lead (II) sulfide is nonstoichiometric in alkaline reducing media and depends on the medium composition. Therefore, the lead(II) solution must be standardized using a poorly stable standard sulfide solution [33]. Ehman [34] titrated sulfide after absorption of hydrogen sulfide from the atmosphere in 10 ml of a solution of sodium hydroxide containing ascorbic acid, with a solution of 6×10^{-6} M CdSO$_4$ using an Ag$_2$S-ISE. He was able to determine 50 to 1 000 ppb of H$_2$S in the air after 20 min sampling, with excellent precision and without interference from nitrogen and sulfur dioxides and ozone at concentrations similar to that of the analyte. The author found that titrations with lead(II) acetate and perchlorate, or with mercury(II) chloride and nitrate, cannot be used for the determination of very small sulfide concentrations.

It is also possible to determine sulfides using electrodes other than the Ag_2S-ISE. For example, hydrogen sulfide catalyzes the redox reaction between iodine and azide [35]. Two platinum electrodes are used to measure the redox potential in a solution containing iodine and sodium azide; one electrode is separated by a frit through which the solution slowly flows. Hydrogen sulfide is introduced by a stream of a carrier gas into the space beyond the frit, catalyzes the reaction between iodine and azide, and thus causes a difference in the potentials of the two electrodes. This reaction is also catalyzed by other sulfur-containing ions, e.g., thiosulfate and thiocyanate, but these ions are not in the gaseous state at normal temperature; sulfur dioxide does not interfere if present in not more than a 100-fold excess.

6.4.2   Determination of Sulfur Dioxide and Sulfites

Emissions of sulfur dioxide are mainly produced by combustion of fossil fuels with high sulfur contents. Unfortunately, Czechoslovakia is near the top of the list of countries in the total amount of sulfur dioxide emitted, not only per capita, but in absolute terms [36, 37]. The highest permissible daily average concentration prescribed by the Czechoslovak standard, i.e., 0.15 mg m^{-3} SO_2, is relatively stringent, but is often exceeded, especially in the northwestern Bohemian coal mining area and in the center of Prague. In strongly polluted atmophere, the sulfur dioxide concentrations may reach a value of 1–5 mg m^{-3}, which leads to formation of London type smog during winter and in rainy weather.

In the potentiometric determination of $SO_2/HSO_3^-/SO_3^{2-}$, indirect methods using the SO_2 gas probe are employed. In the determination of SO_2 emissions the West–Gaeke sampling technique is often employed, based on trapping of sulfur dioxide as the $Hg(SO_3)_2^{2-}$ complex during passage of the gas through a solution containing tetrachloromercurate, according to the equation

$$HgCl_4^{2-} + 2SO_2 + 2H_2O \rightleftarrows Hg(SO_3)_2^{2-} + 4Cl^- + 4H^+$$

$$(6.4)$$

The $HgCl_4^{2-}$ complex is stabilized by adding EDTA to the absorption solution and an addition of glycerol perhaps inhibits the chain reactions in which free sulfur dioxide is oxidized. The completeness of the absorption depends on the pH, which is usually maintained at a value of 6.9. Before the analysis, amidosulfonic acid is added to the solution containing $Hg(SO_3)_2^{2-}$ (removal of nitrite) and the pH is adjusted to about 1 using a mixture of sodium sulfate and sulfuric acid. In this acidic solution, about 90% of the sulfite is present as free sulfur dioxide [38] and can be determined by the SO_2 gas probe. The potentiometric determination of this absorbed sulfite is substantially faster than the spectrophotometric determination, but is disturbed by hydrogen sulfide and acetic acid. The method has been applied, e.g., by Krueger [38], to the determination of sulfur dioxide produced by burning of oil with a high sulfur content.

When sulfur dioxide is to be monitored in the purest air, which requires the determination of $\mu g \ m^{-3}$ amounts of SO_2, the sampling technique described by Axelrod and Hansen [39] can be used. A filter is impregnated with a tetrachloromercurate solution, so that the sampling can be carried out at a higher gas flow rate and for longer time periods than with the absorption solution.

Indirect methods can be employed for the determination of higher sulfur dioxide contents. For example, sulfate formed on absorption of SO_2 in a 3% solution of hydrogen peroxide can be titrated with a lead(II) standard solution [40]. It is also possible to utilize the reduction of iodine by sulfur dioxide and measure the iodide formed by an iodide-ISE. As the content of iodine should not exceed that of iodide more than 10-fold during the determination, the residual iodine must sometimes be removed by extraction into tetrachloromethane. Mascini and Muratori [41] employed this principle and determined SO_2 at concentrations of 0.1 to 10 mg m^{-3}. A monitor of this type is suitable, e.g., for measurements in heavily polluted air or for checking the doses of SO_2 in dechlorination of waste waters.

In foodstuffs to which sulfite is added as a conserving agent, the "total" SO_2 is determined by an SO_2 gas probe after short exposure of the sample to an alkaline solution in which aldehyde and hydrosulfite admixtures are decomposed. This exposure is omitted when "free" SO_2 is to be determined; direct measurement is then performed after sample acidification to pH ≤ 0.7 [42]. The sulfite content in, e.g., wine, beer, fruit juices, and smoked meat products can thus be determined.

An interesting method for the determination of sulfide was proposed by Sekerka and Lechner [43]. The sulfite concentration is found from the rate of change of the Ag_2S-ISE potential during precipitation of silver bromide as a result of reaction of sulfite with bromate. The method permits the determination of sulfite down to a concentration of $5 \times 10^{-7} \ M$.

6.4.3   Determination of Sulfate

Rapid and selective determination of sulfate is still a serious problem in environmental control [44] and in analytical chemistry in general. Sulfate can be titrated by solutions of lead(II) or barium(II) ions in the presence of nonaqueous solvents, such as 1,4-dioxan, acetone, or alcohols, and the end point can be detected by electodes selective to Pb^{2+} or Ba^{2+}.

The titration of sulfate with a lead(II) standard solution suffers from oxidation of the active surface of the lead-ISE by oxygen, by components of the sample, and by the solvent added. The solubility of lead(II) sulfate is suppressed most in dioxan and thus the presence of the latter should permit the determination of the lowest concentrations of sulfate. However, traces of peroxides produced by the decomposition of dioxan rapidly oxidize the electrode surface. Therefore, Veselý [45] proposed purification of dioxan by the subsequent addition of solid sodium hydroxide and sodium metal, followed by distillation. The stability of the electrode potential is then substantially improved and it was possible to semiautomatically titrate solutions with a sulfate concentration greater than $10^{-4}\,M$, containing 60% v/v dioxan. However, when determining sulfates at a concentration of units of mg per liter, a long time is required for potential stabilization. Anions that are precipitated by lead(II) ions, especially phosphate, can cause a positive error in the determination.

The electrode sensitive to barium ions is based on a neutral ion carrier. In the titration determination of sulfate [46], some common cations sensed by the electrode, e.g., potassium, interfere. When the sample is acidified (pH ca. 2) by adding hydrochloric acid, interference from some anions, such as carbonate and phosphate, is suppressed. However, it still remains unclear how this titration is affected by the complicated processes occurring during the precipitation of barium sulfate that are known, e.g., from the gravimetric determination of sulfate.

Sulfate can be determined in waters with low contents of mineral substances and in snow by a differential titration, in which transient response of a sodium glass ISE, e.g., to barium ions is utilized [47]. The titrant is added at a constant rate to the sample containing dioxan and the rate of potential change is recorded as a function of the volume ($\Delta E/\Delta V$). The abrupt change in the Ba^{2+} concentration at the equivalence point causes a transient response of the Na^{+}-ISE which appears as a peak on the recording. The minimal titrable concentration is strongly dependent on the sample ionic strength and thus better results are often obtained in more dilute solutions.

Trojanowicz [48] has described an apparatus for continuous determination of higher sulfate contents (30 to 400 mg liter^{-1}). In a system with two Pb^{2+}-ISEs, the decrease in the concentration of Pb^{2+} in the standard solution is measured after mixing with the sample.

6.4.4   Determination of Total Sulfur

In environmental monitoring, the determination of sulfur in fuels and fly ash is often required, as up to 30% of the sulfur in these materials may be present in forms insoluble in water, in contrast to aerosols. The determination of sulfur in oil by the SO_2 gas probe was described above. Chakraborti and Adams [49] have developed a method for the determination of sulfur in various materials including fly ash after its reduction to hydrogen sulfide by a mixture of hydrogen iodide, sodium hypophosphite, and acetic acid. The small amount of sulfur dioxide formed (ca. 4%) is reduced in another reaction vessel connected in series. The sulfite is then titrated by a lead(II) salt solution using a cadmium sensitive ISE for the end-point detection, as it has been found that this electrode is more resistant than Ag_2S-ISE in the strongly alkaline medium of 1 M NaOH and the use of a lead(II) standard solution yields the most precise results. It is recommended that the Cd^{2+}-ISE be calibrated three or four times with a sulfate solution, until reproducible response is attained. The results obtained in titrations of sulfide are sometimes contradictory (see also Section 6.4.1) and comparative studies will be required to clarify this problem.

6.5   Determination of Residual Chlorine and of the Total Content of Oxidants

Potable and waste waters, as well as water in swimming pools, are commonly disinfected by chlorination. To determine the residual chlorine (i.e., not only Cl_2, but also $HOCl^-$, OCl^-, chloramines, and bromine compounds), its ability to oxidize iodide can be utilized. The decrease in the iodide activity is measured in an acid-base buffered potassium iodide solution. The residual chlorine electrode (Orion 97-70) is based on the same principle, but contains a platinum redox electrode and an iodide-ISE; this is actually a combined electrode that does not require an external reference electrode. The potential of the platinum electrode depends on the ratio, $[I_2]^{1/2}/[I^-]$, but that of the iodide-ISE only on $[I^-]$. If these two electrodes are connected in a cell, the measured change in the voltage is proportional only to the concentration of iodine, because the terms containing $[I^-]$ cancel out:

$$E = \underbrace{\left[E_{0(I_2/I^-)} + \frac{RT}{F} \ln \frac{[I_2]^{1/2}}{[I^-]} \right]}_{\text{Pt-electrode}} - \underbrace{\left[E'_0 - \frac{RT}{F} \ln [I^-] \right]}_{I^- \text{-ISE}} =$$

$$= \text{const} + \frac{RT}{2F} \ln [I_2] \qquad (6.5)$$

In measurements on discrete samples, a Nernstian (29.6 mV at 25°C) or slightly higher slope is obtained at concentrations down to 10^{-7} M Cl_2 [15]. Rigdon et al. [50], who reported a Nernstian response down to 3.5 ppb Cl_2, used a computer-controlled Gran addition technique. At the ppb level, atmospheric or absorbed oxygen may substantially increase the values. A similar effect is observed if the iodide employed contains traces of iodine. For chlorine at concentrations below 3×10^{-6} M, it is recommended to add 1 ml of 5×10^{-3} M KI to a 100-ml sample [15]. On the other hand, a negative error in the determination may be caused by loss of iodine from the solution, e.g., through volatilization. According to Jenkins and Baird [51], measurement is possible only in water with a BOD (biological oxygen demand) value less than 50 mg liter^{-1}, as at higher values the potential is subject to drift, perhaps due to reactions of the iodine liberated with organic substances in the sample. Of course, this is a problem encountered in all iodometric methods. The authors prefer potentiometric determination of Cl_2 over the PAO method (back-titration of phenylarsine oxide) at levels below 1 ppm; above 1 ppm the two methods yield comparable results. An advantage of potentiometry is easy applicability to field measurements, provided that the sample temperature is measured simultaneously; if a temperature difference between the sample and standards of 10°C is neglected, then the error incurred amounts to about 20%. The authors have proposed corrections for these differences employing an empirical graph.

In continuous measurements, in which iodine is removed from the iodide immediately before mixing with the sample on a graphite or cadmium column, Nernstian response can be obtained, as specified by the manufacturer (model 1770 Cl_2 monitor from Orion Research), down to a concentration of 3×10^{-8} M. The dosing of iodide is different in the Orion SLeD instrument, in which the sample passes through a lead(II) iodide pellet. Unfortunately, the accessible literature contains little information on the practical usefulness of these instruments.

Iodometric determination of the total content of oxidants in the atmosphere (O_3 + PAN) employing an iodide-ISE has been described by Mascini and Muratori [41]. The loss of iodide was measured by the ISE, permitting the determination of 1 to 100 µg O_3 per m^3 with a precision of 5%. Trachtenberg and Suffet [52] employed a monitor based on a similar principle and were able to follow changes in the total content of oxidants in the ambient atmosphere, because the instrument was sufficiently sensitive (0.01–0.16 ppm O_3).

6.6   Determination of Heavy Metals

Although several commercial ISEs are available for heavy metal cations (e.g., Ag^+, Cu^{2+}, Pb^{2+}, Cd^{2+}), they find little use in trace analysis of these metals, with the exception of the determination of copper. The lead and cadmium ISEs suffer from poor potential reproducibility and are mainly used for end-point detection in titrations; analyses for silver occur rarely and small concentrations of Ag^+ are rapidly and strongly sorbed on surfaces or are reduced to the metal. The determination of overall concentrations of heavy metals is much more often carried out using atomic absorption spectrometry or stripping voltammetry. Therefore, it can be expected that ISEs will chiefly be used for speciation of heavy metals, i.e., for studying the distribution of the metals among various chemical forms in solution, utilizing the fact that ISEs measure the activities of ions.

An example of such an application of ISEs in environmental chemistry is the work [53] based on the use of the copper-ISE. Toxic effects of copper on aquatic life depend on its chemical form and thus knowledge of the distribution of copper is of great interest. The authors separate colloidal particles from the solution, measure the pH and the cupric ion activity, and calculate the concentration of carbonate from the total alkalinity. On this basis, conclusions can be drawn concerning the distribution of copper among various anions.

The overall concentrations of some metals can also be determined indirectly. For example, the toxicity of mercury is utilized for the determination of mercury at concentrations of up to 7×10^{-10} M, on the basis of urease inhibition [54]. The ammonia formed by the enzymic reaction is monitored by the ammonia gas probe. If the enzymic reaction takes place in a preceding reactor, the decrease in the amount of ammonia formed is proportional to the content of mercury in samples injected into a stream of a standard urea solution. As mercury is strongly bound to urease, not only free mercury ions but also some mercury complexes are detected in this way. Between measurements on samples, the urea in the reactor is regenerated using thioacetamide and EDTA. This determination of mercury is not sufficiently selective, as copper and silver interfere, but this principle could perhaps be employed to determine pesticides that inhibit choline esterase; a choline ISE has been described. Another interesting example is an enzyme electrode that has been developed for monitoring oil in water that is pumped from ships in coastal areas [55].

6.7   Determination of Cyanides and Cyanogen

The extreme toxicity of free cyanides and of cyanogen is generally known. For monitoring the concentration of cyanide in waste waters (maximum permitted concentrations are usually from 0.1 to 0.05 mg liter^{-1}), analyzers containing ISEs, e.g., Orion 1206, have been constructed. Flue gases are also sometimes analyzed for the cyanogen content.

The determination of cyanides using ISEs has some special features. The membranes of ISEs do not contain cyanide, as iodide-ISEs, or preferably Ag_2S-ISEs, are used. The response mechanism for these electrodes is always connected with the $Ag(CN)_2^-$ complex, which is either added to the solution or is formed at the electrode surface during dissolution of the membrane. Furthermore, the electrodes respond not only to free CN^-, but also to some complexes (e.g., with zinc) that are less stable than $Ag(CN)_2^-$. The kinetics of the complex decomposition in the given medium are also important. If the overall cyanide content in solution is required, then very stable complexes of cyanide, e.g., with iron, nickel, gold, and cobalt, must be decomposed, either by irradiation with uv rays in a quartz tube [56] or by acidification of the solution and addition of EDTA (Ni, Cu).

In the presence of cyanide at a concentration of less than 10^{-2} M, the $Ag(CN)_2^-$ complex is formed at the surface of Ag_2S electrodes and it holds for the measured activity of silver ions that

$$\beta_{2,[Ag(CN)_2^-]} = \frac{[Ag(CN)_2^-]}{[Ag^+][CN^-]^2} \doteq 2.7 \times 10^{20} \qquad (6.6)$$

If it holds that $[Ag(CN)_2^-] \gg [Ag^+]$, then the complex concentration in Eq. (6.6) can be considered constant and the activity of Ag^+ depends on the CN^- activity in solution. A tenfold change in the concentration of cyanide causes a hundredfold change in the concentration of silver ions and thus the Nernstian slope is 118.3 mV. However, if the conditions in the Prandtl layer at the electrode are not rigorously defined, the measurement of cyanide using an Ag_2S-ISE suffers from poor precision [57].

Therefore, the $Ag(CN)_2^-$ complex is added to the sample solutions at a concentration of 10^{-3} to 10^{-5} M, as originally proposed by Frant et al. [58]. The determination then has a higher sensitivity than that using the I^--ISE and is more selective; Ag_2S-ISEs also exhibit longer lifetimes and perhaps a lower detection limit for cyanide [59]. The minimal determinable amount is affected by the amount of cyanide added to the sample together with the $Ag(CN)_2^-$ complex (to attain a sufficient stability of the complex, about a 1% excess of cyanide is used). In simple solutions it is common to determine cyanide concentrations down to 5×10^{-7} M (0.01 mg liter^{-1}) at pH ca. 11.5. Thiosulfate and iodide do not interfere even if present in a thousandfold excess. According to Clysters et al. [59], the serious interference from sulfide can be eliminated by adding lead(II) nitrate which does not interfere in excess. In the determination of 2×10^{-6} M CN^-, up to a hundredfold excess of sulfide can thus be precipitated without coprecipitation of the cyanide. A similar effect can be attained when using salts of bismuth(III) and cadmium(II).

An interesting apparatus for monitoring cyanides has been described by Durst [60] (see Fig. 6.1). The acidified sample is pumped through a gas dialyzer in which the solution containing cyanide is separated from the indicator solution of $KAg(CN)_2$ by a porous and hydrophobic polytetrafluoroethylene membrane. The conditions are selected so that the transfer of the cyanogen into this slowly flowing medium is not only quantitative, but the cyanide is also concentrated, because the acidified sample solution flows faster. The apparatus permits the determination of cyanide at levels of 30 to 400 μg liter^{-1} and the lowest detectable concentration (defined here as that causing a 5-mV deviation of the signal from the background value) is 0.5 μg liter^{-1}.

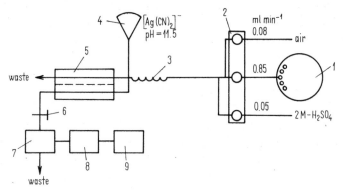

Fig. 6.1 A block scheme of a CN^- analyzer. 1, sampling unit; 2, proportioning pump; 3, mixing coil; 4, $Ag(CN)_2^-$ reservoir; 5, HCN gas diffuser; 6, flow-rate control; 7, ISE detector; 8, meter; 9, recorder.

The monitor described by Hofton [61] has been designed for river and waste waters and detects concentrations around the maximal permissible value, i.e., usually around 50 µg liter^{-1}. The determination limit is about five times lower when using an $Ag(CN)_2^-$ indicator at a concentration of $2.5 \times 10^{-3} M$. The analyzer is standardized at the monitored concentration every twelve hours.

In a determination of cyanogen in cigarette smoke, the gas is freed of solid particles and led through an ascarite filter. The cyanide content is measured after adding solutions of lead(II) nitrate and sodium hydroxide [62].

Sekerka and Lechner [63] described a determination of cyanide from the rate of the change in the potential of an Ag_2S-ISE provided with a cyanide-permeable membrane (zero-current chronopotentiometry). Cyanogen diffuses from the acidified sample solution into a silver nitrate solution of pH ca. 11.5 fixed on the electrode surface by the membrane. It is bound to $Ag(CN)_2^-$ and the electrode potential thus changes in time in proportion to the decreasing silver ion activity. The film of the internal electrolyte between the electrode and the membrane is replaced after each measurement.

6.8   Determination of Hydrogen Fluoride and Fluorides

The fluoride-ISE with a lanthanum(III) fluoride single crystal has made a great impact on the analytical chemistry of fluorine and probably is the second most widely used ISE (after the pH-ISE). Some authors consider the F^-- and pH-ISEs analogous to a certain extent, because both respond only to univalent hydrophilic ions, the glass electrode to H_3O^+, Li^+, and Na^+ and fluoride-ISE to F^- and OH^-. As the latter electrode does not respond to La^{3+} ions, it can actually be considered as an anion-sensitive analog of the pH-ISE [64].

The interference from OH^- ions can be simply suppressed, at fluoride contents above $10^{-7} M$, by adjusting the pH of the solution to a value less than 6 (usually the pH is between 5 and 6). Under optimal conditions, down to 2 ppb, i.e., $10^{-7} M$ concentration, can be determined in water; however, at concentrations below $10^{-6} M$ the potential stabilization is sluggish. The most frequent problem is liberation of fluoride from the relatively stable fluoro complexes of Fe^{3+}, Ti^{4+}, and especially Al^{3+}, which is successfully solved by adding buffer solutions containing agents that bind the metals into complexes more stable than the corresponding fluoro complexes. When the content of Al^{3+} exceeds 2 mg liter^{-1}, tiron is usually used for masking [5].

In environmental chemistry, the determination of hydrogen fluoride emissions and of the fluoride content in precipitation water is mostly required. Increased occurrence of fluorine compounds is found mainly in areas around factories manufacturing superphosphate (HF, H_2SiF_6, and SO_2), aluminum smelters, and plants for production of enameled ware, where anthropogenic emissions of fluorine may constitute a serious problem. The air usually contains 0.01 to 2 μg HF per cubic meter, rain-water tens of μg per liter, but the level may fluctuate between 1 and 1 000 μg liter^{-1}.

Great attention has been devoted to monitoring of fluorine emissions using the F$^-$-ISE. Filters impregnated with various reagents, e.g., sodium formate, have been used for sampling. More recently, particle-bound and gaseous fluorine compounds have been distinguished and at present are usually determined using automatic instruments. For example, Mascini [65] traps the particles in a filter impregnated with citric acid and heated to 80°C. The filter collects the particles, but the fluoride adsorbed on the particles passes through and is absorbed in a thin layer of sodium carbonate in a spiral absorber. Once an hour the absorber is rinsed with 5 ml of sodium citrate and the layer of sodium carbonate is replenished from a solution in a reservoir. At an air flow-rate of 25 liters/min, average 1-h levels of HF emissions could be measured down to 0.1 μg m^{-3}.

Substantially higher hydrogen fluoride concentrations (9 to 75 mg m^{-3}) were determined in the air sampled above an electrolyzer for production of aluminum [66]. The gas was passed through TISAB buffer. Particles with adsorbed fluorine compounds were collected on a paper filter with a pore diameter of 2 μm, placed before the absorber. The filter was extracted with water and with 1 M sodium hydroxide, so that cryolite was also determined. MacLeod and Christ [67] also monitored industrial emissions of hydrogen fluoride and compared the potentiometric determination with the spectrophotometric method SPADNS (the zirconium method). They found that the determination with the F$^-$-ISE is ten times faster.

An interesting semiquantitative procedure for measurement of very low fluoride contents below 1 μg liter^{-1} in rainwater, mist, and aerosols has been described by Warner and Bressan [68]. The rates of the electrode potential changes on immersion of the F$^-$-ISE into the sample and a standard are compared. The concentration range is thus determined, given by the two standards closest to the sample. Slanina et $al.$ [69] have described a computer-controlled system for determination of fluoride, chloride, and ammonia in rainwater, employing a Gran addition technique and applying the apparatus in an industrial region of the Netherlands. Satisfactory results were obtained even for the lowest fluoride contents investigated, i.e., 50 μg liter^{-1}.

From the environmental point of view, the determination of fluoride in other types of water is not as interesting. On the other hand, potable waters are often enriched in fluorine for stomatological reasons and the F^{-}-ISE is a very suitable sensor for control of the dosing pump [70].

6.9   Determination of Surfactants

Contamination of the environment, especially natural waters, by surfactants is important, in view of the extensive exploitation of detergents in industry and households. Reliable monitoring of these substances is difficult, because they occur in a wide concentration range, from traces (e.g., in waste waters from households) to concentrations of tens of percent in some industrial wastes, there are many types of these substances, and they are often contained in complex matrices.

The classical methods of determination of surfactants are titration and colorimetry, which are tedious and suffer from many interferences. The development of potentiometric electrodes selective for cationic and anionic tensides (see, e.g., Refs. 71−77) has led to faster and more reliable methods. The potentiometric determination of tensides is treated in detail in a review [78].

Electrodes selective for ionic tensides are ISEs with a liquid membrane containing a complex of a cationic and anionic tenside that is insoluble in water and behaves as a liquid ion-exchanger. These electrodes are not specific for individual kinds of tensides and are only capable of differentiating between cationic and anionic surfactants. The active substance is usually the complex of cetylbenzyldimethylammonium chloride with sodium dodecyl sulfate or an alkylbenzene sulfonate. The classical liquid-membrane electrodes with the liquid phase deposited on a porous support are not sufficiently strong for practical measurements and thus polyvinyl chloride membranes are mostly used, prepared by common technologies (for details see, e.g., Ref. 78). The electroactive substance need not necessarily be added to the plastic membrane, as after two or three titrations the sufficiently polar plasticizer is saturated with the surfactant ion-pair through extraction from the sample solution. Such electrodes are usually prepared as "coated wire" electrodes [79].

These electrodes are not suitable for direct potentiometric measurements, because the calibration curves exhibit sharp breaks on attaining the critical micelle concentration (c.m.c.) of the tenside, the absolute electrode potential values are poorly reproducible because of complexation of the tenside in more complicated matrices, and adsorption of the tenside on the membrane and the liquid-junction potential in the system are high and variable. The c.m.c. values can actually be determined from the measurements with these electrodes [80]. Therefore, electrodes selective for tensides are used to detect titration end points, where these difficulties are much less important. For the determination of cationic tensides, a suitable anionic tenside or sodium tetraphenylborate is used as the titrant and, vice versa, anionic tensides are titrated by a cationic tenside. The titrand and titrant form an insoluble complex. The titration end point is attained when an excess of the cationic (anionic) tenside is replaced by an excess of the anionic (cationic) tenside, which is accompanied by a substantial change in the electrode potential (around 200 mV). In this way, e.g., alkyl sulfates, alkyl sulfonates, alkyl sulfosuccinates, various soaps, and mixtures of tensides have been monitored (for details see, e.g., Refs. 81 – 85).

Electrodes provided with a PVC membrane containing tritolyl phosfate or dibutyl phthalate as a plasticizer yield good results in these titrations and their lifetime is at least two months. Typical values of the determination limit and the relative standard deviation of determination of macroscopic amounts are $7 \times 10^{-7} M$ and 1.5%, respectively. In the presence of proteins and fats at high concentrations, however, the membranes are rapidly destroyed due to irreversible adsorption of these substances.

From the point of view of monitoring tensides, the application of these electrodes to flow systems using gradient titration has also been studied. This technique has yielded good results in analyses of solutions, whereas in determination of tensides in solid samples, in which the most tedious operation is the preparation of a sample solution, discontinuous automated titrations are more suitable [78].

REFERENCES

1. Veselý J. and Štulík K.: Advances in Potentiometry with Ion-Selective Electrodes, in J. Zýka (ed.), New Trends in Analytical Chemistry, SNTL, Prague 1983 (in Czech).
2. *Anal.Chim.Acta* **114** (1980) – entire volume.
3. Růžička J. and Hansen E. H.: Flow-Injection Analysis, Wiley, New York, Chichester, Brisbane, Toronto 1981.

4. Bailey P. L.: Analysis with Ion-Selective Electrodes, Heyden, London 1976.

5. Veselý J., Weiss D. and Štulík K.: Analysis with Ion-Selective Electrodes, E. Horwood, Chichester 1978.

6. Baiulescu F. and Cosofret V. V.: Application of Ion-Selective Membrane Electrodes, E. Horwood, Chichester 1978.

7. Freiser H. (ed.): Ion-Selective Electrodes in Analytical Chemistry, Vols. 1 and 2, Plenum Press, New York 1978 and 1980.

8. Linch A. L.: Health Lab.Sci. 2, 182 (1977).

9. Thomas J. D. R.: Ion-Selective Electrodes in Environmental and Toxicological Analysis, in J. Albeiges (ed.), Anal. Techniques in Environmental Chemistry, Pergamon Press, Oxford 1980, p. 543.

10. Hesman H. B.: Environ.Sci.Res. 13, 103 (1979).

11. Eicken D.: Gewasserschutz, Wasser, Abwasser 39, 181 (1980).

12. Štulík K. and Pacáková V.: Continuous Measurement in Flowing Liquids, in: J. Zýka (ed.), New Trends in Analytical Chemistry, SNTL, Prague 1983.

13. Evans W. H. and Partridge B. F.: Analyst 99, 367 (1974).

14. Růžička J., Hansen E. H., Bisgaard P. and Reymann E.: Anal.Chim.Acta 72, 215 (1974).

15. Midgley D.: Ion-Selective Electrode Rev. 3, 43 (1981).

16. Eagan M. L. and Dubois L.: Anal.Chim.Acta 70, 157 (1974).

17. Ferm M.: Atmos.Environ. 13, 1385 (1979).

18. Dee L. A., Martens H. H., Merill C. I., Nakamura J. T. and Jaye F. C.: Anal.Chem. 45, 1477 (1973).

19. Driscoll J. N., Berger A. W., Becker J. H., Funkhouser J. T. and Valentine J. R.: J.APCA 22, 119 (1972).

20. Kneebone B. M. and Freiser H.: Anal.Chem. 45, 449 (1973).

21. Ferber B. I., Sharp F. A., and Freedman R. W.: J.Am.Ind.Hyg.Assoc. 37, 1 (1976).

22. Nielsen H. J. and Hansen E. H.: Anal.Chim.Acta 85, 1 (1976).

23. Šenkýř J. and Petr J.: Chem.Listy 73, 1097 (1979).

24. Weiss D.: Chem.Listy 69, 202 (1975).

25. Simeonov V., Andreev G. and Stoianov A.: Fres.Z.Anal.Chem. 297, 418 (1979).

26. Mertens J., Van den Winkel P. and Massart D. L.: Anal.Chem. 47. 522 (1975).

27. Forney L. J. and McCoy J. F.: Analyst 100, 157 (1975).

28. Hansen E. H., Růžička J. and Larsen N. R.: Anal.Chim.Acta 79, 1 (1975).

29. McKenzie J. R. and Young P. N. W.: Analyst 100, 620 (1975).

30. Lowry J. H. and Mancy K. H.: Water Res. 12, 471 (1978).

31. Gulens J.: Ion-Selective Electrode Rev. 2, 118 (1980).

32. Sekerka I. and Lechner J. P.: Anal.Chim.Acta 93, 139 (1977).

33. Florence T. M. and Farrar Y. J.: Anal.Chim.Acta 116, 175 (1980).

34. Ehman D. L.: Anal.Chem. 48, 918 (1976).

35. Kiba N. and Furusawa M.: Talanta 23, 637 (1976).

36. Moldan B.: Věst.Ústřed.Úst.Geol. 56, 237 (1981).

37. Moldan B.: Geochemistry of the Atmosphere, Academia, Prague 1977.

38. Krueger J. A.: Anal.Chem. 46, 1338 (1974).

39. Axelrod H. D. and Hansen S. G.: Anal.Chem. 47, 2460 (1975).

40. Young M., Driscoll J. N. and Mahoney K.: Anal.Chem. 45, 2283 (1973).

41. Mascini M. and Muratori T.: Anal.Chim. (Roma) 65, 287 (1975).

42. Bailey P. L.: J.Sci.Food Agr. 26, 558 (1975).

43. Sekerka I. and Lechner J. F.: Anal.Chim.Acta 99, 99 (1978).

44. Tanner R. L. and Newman L.: J. APCA 26, 737 (1976).

45. Veselý J.: *Coll.Czech.Chem.Commun.* **46**, 368 (1981).

46. Jones D. L., Moody G. J. and Thomas J. D. R.: *Analyst* **104**, 973 (1979).

47. Veselý J.: *Anal.Lett.* **13**, 543 (1980).

48. Trojanowicz M.: *Anal.Chim.Acta* **114**, 293 (1980).

49. Chakraborti D. and Adams F.: *Anal.Chim.Acta* **109**, 307 (1979).

50. Rigdon L. P., Moody G. J. and Frazer J. W.: *Anal.Chem.* **50**, 465 (1978).

51. Jenkins R. L. and Baird R. B.: *Anal.Lett.* **12**, 125 (1979).

52. Trachtenberg A. F. and Suffet I. H.: *J. APCA* **24**, 836 (1974).

53. Stella R. and Ganzerli-Valentini M. T.: *Anal.Chem.* **51**, 2148 (1979).

54. Ögren L. and Johansson G.: *Anal.Chim.Acta* **96**, 1 (1978).

55. Cundell A. M. and Findl E.: Report ER-013179 (1979), Bio Res.Inc., USA; *Chem.Abstr.* **91**, 216495s (1979).

56. Sekerka I. and Lechner J. F.: *Water Res.* **10**, 479 (1976).

57. Veselý J., Jensen O. J. and Nicolaisen B.: *Anal.Chim.Acta* **61**, 1 (1978).

58. Frant M. S., Ross J. W. and Riseman J. H.: *Anal.Chem.* **44**, 2227 (1972).

59. Clysters H., Adams F. and Verbeek F.: *Anal.Chim.Acta* **83**, 27 (1976).

60. Durst R. A.: *Anal.Lett.* **10**, 961 (1977).

61. Hofton M.: *Environ.Sci.Technol.* **10**, 277 (1976).

62. Vickroy D. G. and Gunt G. L.: *Tobacco* **174**, 50 (1972).

63. Sekerka I. and Lechner J. F.: *Anal.Chim.Acta* **93**, 129 (1977).

64. Veselý J.: *J.Electroanal.Chem.* **41**, 134 (1973).

65. Mascini M.: *Anal.Chim.Acta* **85**, 287 (1976).

66. Hrabéczy-Páll A., Valló F., Tóth K. and Pungor E.: *Hung.Sci.Instr.* **41**, 55 (1977).

67. MacLeod K. E. and Christ H. L.: *Anal.Chem.* **45**, 1272 (1973).

68. Warner T. B. and Bressan D. J.: *Anal.Chim.Acta* **63**, 165 (1972).

69. Slanina J., Bakker F., Lautenbag C., Lingerak W. A. and Sier T.: *Microchim.Acta* **1978**, 519.

70. Collis D. E. and Digens A. A.: *Water Treat.Exam.* **18**, 192 (1969).

71. Gavach G. and Bertrand C.: *Anal.Chim.Acta* **55**, 385 (1971).

72. Birch B. J. and Clarke D. R.: *Anal.Chim.Acta* **61**, 159 (1972).

73. Ishibashi N. and Kohava H.: *Bunseki Kagaku* **21**, 100 (1972).

74. Birch B. J. and Clarke D. E.: *Anal.Chim.Acta* **67**, 387 (1973).

75. Birch B. J. and Clarke D. E.: *Anal.Chim.Acta* **69**, 473 (1974).

76. Tanaka T., Hiiro K. and Kawahara A.: *Anal.Lett.* **7**, 173 (1974).

77. Fujinaga T., Okazaki S. and Hara H.: *Chem.Lett.* 1201 (1978).

78. Birch B. J. and Cockroft R. N.: Ion-Selective Electrodes Reviews, Vol. 3, p. 1 (1981).

79. Vytřas K.: Personal communication.

80. Smith V. and Newbery J. E.: *Colloid.Polymer.Sci.* **256**, 494 (1978).

81. Ciocan N. and Anghel D. F.: *Tenside* **13**, 189 (1976).

82. Ciocan N. and Anghel D. F.: *Anal.Lett.* **9**, 705 (1976).

83. Ciocan N. and Anghel D. F.: *Mikrochim.Acta* **1977**, 639.

84. Ciocan N. and Anghel D. F.: *Anal.Lett.* **10**, 423 (1977).

85. Ciocan N. and Anghel D. F.: *Z.Anal.Chem.* **290**, 237 (1978).

CHAPTER 7

SEMICONDUCTOR SENSORS

Josef Fexa

7.1   Introduction

A new group of sensors, denoted collectively as chemical transducers or CSSD (chemically sensitive semiconductor devices), are based on the concepts of the physics of solids and of the electrochemical theory of interfaces. Electric double-layer models have been applied to the insulator-test medium, dielectric-test medium, and semiconductor-test medium interfaces. This group of sensors can be classified on the basis of their underlying principles as follows:

- semiconductor gas detectors of the oxide type;
- ISFETs (ion-selective field-effect transistors of the MOS type);
- CSFETs (chemically sensitive field-effect transistors of the MOS type);
- sensors based on contact phenomena at metal-semiconductor and semiconductor-semiconductor interfaces (including, e.g., MIS detectors based on metal-insulator-semiconductor systems, Schottky diodes of the Pd-CdS type, etc.).

Since the discovery of transistors, the great sensitivity of semiconductor components to the presence of trace impurities has been studied. Following the discovery of the field-effect transistor, the main problem in its practical use has been the passivation of the SiO_2 surface. The charge distribution on the semiconductor surface is discussed in the key work [1].

The high sensitivity of semiconductor surfaces to external influences can be used to design new sensors for monitoring the concentrations of substances [2, 35]. However, the basic physicochemical processes characterizing changes in semiconductor surface states are still not sufficiently understood. Another problem is the question of the selectivity and reversibility of the processes occurring at the semiconductor-test medium interface.

Real semiconductors exhibit various surface states that are manifested by the formation of a surface charge. These states can be divided, depending on their origin, into intrinsic and nonintrinsic. Intrinsic surface states are related to unsaturated bonds of surface atoms and actually represent a perturbation in the periodical arrangement of the atoms in the crystal lattice. The atoms forming the last lattice plane have the same valences unsaturated because partners are lacking in the direction perpendicular to the surface. These defects appear in the band model as energy levels different from electronic levels in a regular lattice. These surface states are called Tamm's (1932) or Shockley (1939) states, after their discoverers.

Nonintrinsic surface states are caused by foreign atoms or molecules adsorbed on the semiconductor surface and by various surface defects, such as vacancies, dislocations, etc. Real semiconductor surfaces are mostly covered with a layer of a different composition, e.g., an oxide or nitride. The surface states of these layers and adsorbed gases are charge carriers that affect free current carriers in the semiconductor.

It is impossible to utilize changes in the surface states on the MOS structure for direct measurement of the partial pressures of adsorbed gases because of the insufficient stability of the layer of protective oxide. Freshly pretreated and stabilized layers have different surface states. Experimental results have led to classification of surface states into two groups [3]:

(a) Rapid surface states that are established within 10^{-8} s and depend mainly on the surface pretreatment,
(b) Slow surface states that are established within about 10^{-3} to 10^3 s and depend mainly on the medium in which the semiconductor is placed.

The total surface charge is the sum of the charges corresponding to rapid and slow surface states. The relationships between the components of this sum are determined by the physicochemical states of the surface, i.e., the chemical composition, structure, and the technology of the preparation of the protective layer, as well as by the composition and the partial pressures of the components of the gaseous atmosphere surrounding the test sample. The dependence of the sign of the surface charge on the composition of the gaseous atmosphere and of the charge magnitude on the partial pressures of the active components of the gas indicate that slow surface states depend on adsorption. This assumption has also been verified by the absence of slow surface states on atomically pure sample surfaces and on samples stored in a vacuum of 10^{-7} to 10^{-8} Pa.

For measurement of the concentration of substances, slow surface states of semiconductors are primarily useful. According to the contemporary concepts of the electronic theory of adsorption, most of the adsorbed particles (atoms or molecules) are bound by the forces of weak chemisorp-

tion on the semiconductor surface [4]. The nature of the adsorbed layer and the coverage of the semiconductor depend on the composition of the gaseous atmosphere and the partial pressures of the components, as well as on the character of the surface. Electrons from the adsorbed particles interact with the electrons and holes in the crystal lattice, the bond is stabilized, and the adsorbed particles acquire a charge that can be considered as that of a slow surface state.

The total number of surface states that contribute to the formation of the surface charge at thermodynamic equilibrium also depends on the location of the Fermi level in the bulk of the semiconductor and on the character of its surface. The mechanism of chemisorption on the semiconductor surface is discussed in Ref. 5.

7.2   Oxide-Type Semiconductor Sensors

These sensors comprise a large group of suitably doped oxides of the transition metals, on the surface of which a donor or acceptor reaction occurs with consequent curvature of the energy bands at the surface [5 – 10].

Molecules chemisorbed on the semiconductor surface may form a surface charge and thus alter the semiconductor properties. Changes in the surface charge cause changes in the conductance that are readily measured. The weak form of chemisorption, where the particle is bound to the crystal lattice of the semiconductor without participation of a free electron or hole in the crystal lattice, is rarely used, as this contribution to the change in the surface state is relatively small and difficult to measure.

The strong form of chemisorption [4], in which the chemisorbed particle accepts a free electron (n-bond) or a free hole (p-bond) from the crystal lattice, is much more important. Therefore, this type of semiconductor sensor is based on the existence of intrinsic and nonintrinsic surface states and the formation of a planar charge on the semiconductor surface. If the electroneutrality of the semiconductor is not to be disturbed, the opposite spatial charge must be located in the region adjacent to the surface. In this spatial charge region the two charges give rise to a macroscopic electric field with a high potential that causes curvature of the energy bands at the crystal surface. If the planar charge is negative, the bands are curved upwards; if it is positive, they are curved downwards. Depending on the character of the band curvature, the concentrations of the major charge carriers increase or decrease compared with their concentrations inside the crystal.

From the point of view of the character of the sorption of gaseous components on the semiconductor surface, these sensors can be classified into two types:

(a) The donor type; this is an n-type semiconductor on the surface of which a donor reaction occurs, the semiconductance increasing during the reaction. These include, e.g., ZnO, SnO_2, Fe_2O_3, TiO_2, V_2O_7, MnO_2, WO_3, ThO_2, CdO, etc.

(b) The acceptor type; this is a p-type semiconductor used for the detection of reducible gases. CoO, Cu_2O, NiO, Cr_2O_3, etc., belong in this group.

Some basic types of semiconductor sensors are surveyed in Table 7.1, together with the doping admixtures [6]. The number of papers and patents in this field has been rapidly growing since 1962 when Seiyama and Kato [7] published the first paper. Several dozen patents have been exploited by various manufacturers, the best known being Figaro Eng. Inc., Osaka (Taguchi Gas Sensor — TGS) [8]. It is estimated that the production in Japan alone exceeded 2 million detectors in 1979.

Some basic types of semiconductor sensors   *Table 7.1*

Sensor	Doping admixtures	Number of patents from 1972 to 1977
SnO_2	Sb, Al, Pd, Fe, Ni, Zn, Si, Ti, Zr, Th	61
ZnO	Fe, Cr, Li, Al, B, Mn, Gd, Ti, Co, In, Mg, Y	58
Fe_2O_3	Ni, Zn, Mg, Cr, Cu, Mn, Co, Sr, Ba, Li, Na	47
Cr_2O_3	Sn, Mn, Mg, Ti, Fe, Zn, Zr, Hf, Nb, Ta	17
$BaTiO_3$	Ti, Rh, Pd, Sn, Pt	9
CoO	Mg, Al, Sn, Li, Cu, Zn, Sb	9
MnO_2	C, Ti, Co, Cu, Ag	4

Semiconductor gas sensors from various manufacturers differ in their arrangement, in the materials employed, in the production method, and in their electric properties. The sintered oxides SnO_2, ZnO, and Fe_2O_3 are used in most sensors, with sensitivity to various gases and electric properties affected by the doping substances, treatment temperatures, etc. A thin layer of the semiconductor is usually deposited on an inorganic support (insulator) that can be shaped as a plate, tube, or cylinder or used as grains that are pressed together with the semiconductor.

The sensors are further provided with a heating element permitting heating of the semiconductor layer to a temperature between 30 and 300°C and usually consisting of a platinum or platinum-iridium spiral that simultaneously functions as the electrode for the conductance measurement. The other electrode is usually an analogous spiral. Hence, the heating and measuring circuits are not separated in most sensors. One of the arrangements of TGS sensors is shown in Fig. 7.1; pearl-type sensors are usually employed that can be prepared by a technology similar to that for the preparation of thermistors. The heating spiral is located inside the pearl of the sintered oxide mixture. A planar sensor is depicted in Fig. 7.2. The substrate is, e.g., a silicon or ceramic plate. The thin layer of semiconductor can be vacuum plated, deposited in a plasma or high-frequency discharge, electronically or chemically deposited, epitaxially grown, screen-printed, etc. [10]. A detector for methane indicating the concentration of the gas in the air is given in Fig. 7.3. Characteristic calibration curves for various gases, using the Figaro 711 and 308 sensors, are given in Figs. 7.4. and 7.5. The transient characteristic for a step change in the concentration of CO of 200 ppm is depicted in Fig. 7.6 for the Figaro 711 sensor.

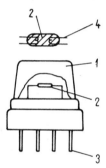

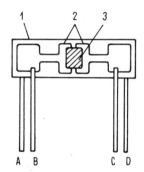

Fig. 7.1 TGS sensor. 1, protective stainless-steel gauze; 2, semiconductive element; 3, nickel rods; 4, heating spiral.

Fig. 7.2 Planar sensor. 1, substrate; 2, electrodes; 3, sensitive layer; A, B, C, D, leads.

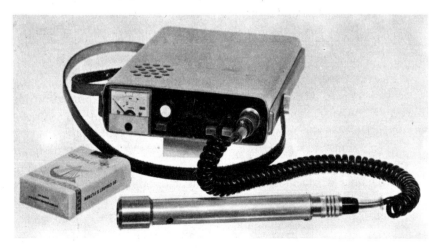

Fig. 7.3 Methane detector.

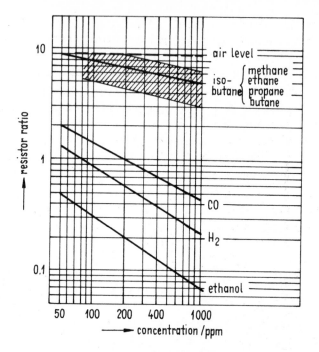

Fig. 7.4   Calibration curves for the Figaro 308 sensor [8]. R_{200} is the sensor resistance at a concentration of 200 ppm CO. The resistor ratio, R/R_{200}, is defined by the ratio of the sensor resistance in the test medium to that with a standrd, which is 200 ppm CO.

Many substances are used as binders of mixed oxides, e.g., organic substances, kaolines, and clays. The effect of the operational parameters, such as the heating and measuring voltages, and of various gases on the response of TGS sensors is discussed in detail in Ref. 9. A certain degree of selectivity for various gases can be obtained by judiciously choosing the doping and the heating voltage.

Sensors of this type are useful detectors for coal and natural gas leaks, for organic vapors, and for selective monitoring of carbon monoxide down to a concentration of tens of ppm. Some countries introduced these sensors by law into alarms in the chemical and coal power station industries.

Using semiconductor sensors, fire-alarm systems have been designed for the protection of gas plants, gas containers, stores of flammable substances, lacquer workshops, drying rooms, etc. Some companies market systems of various sizes, both for fire protection of residences and for large industrial or commercial enterprises where several hundred detectors are used, controlled centrally by a microcomputer [11]. Alcohol can also be determined in the breath of automobile drivers.

Common drawbacks of sensors based on sintered semiconducting oxides are polycrystallinity of the oxide aggregates, poor selectivity, considerable drift, and changes in the sensitivity with time. Some new thin-layer technologies promise a substantial suppression of these disadvantages [10, 12]. Research is becoming centered on a better understanding of the interaction of gas with the semiconductor surface. The poor selectivity

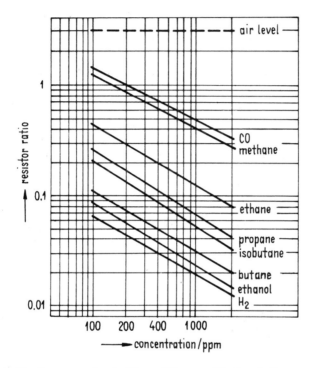

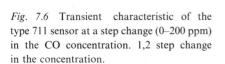

Fig. 7.5 Calibration curves for the Figaro 711 sensor [8]. R_{200} is the sensor resistance at a concentration of 200 ppm CO. The significance of the resistor ratio is explained in Fig. 7.4.

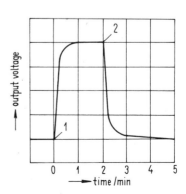

Fig. 7.6 Transient characteristic of the type 711 sensor at a step change (0–200 ppm) in the CO concentration. 1,2 step change in the concentration.

of these sensors will not be important as the various functions can be handled by a microcomputer, with a suitable doping and selection of the temperature regime, thus yielding the contents of the individual components in the gaseous mixture.

7.3 Ion-Selective Transistors of the MOS-ISFET Type

The function of field-effect transistors (FETs) depends on the effect of the voltage applied to the gate on the conductive channel between the source and drain. The gate is separated from the conductive channel by an insulator layer (SiO_2, Si_3N_4). Depending on the magnitude of the applied voltage, the concentration of the charge carriers and the semiconductor surface energy vary. Depleted, enriched, accumulated, and inverse layers are obtained, depending on the concentration of the major charge carriers at the surface. Transistors that are closed at zero gate voltage are called enriched and those that are open are called depleted. Such transistors can be employed to construct an ISFET or CSFET. The gate function is performed by a layer of ions or gas molecules adsorbed on the SiO_2 surface. The element geometry is shown in Fig. 7.7.

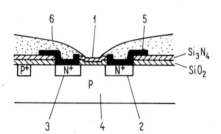

Fig. 7.7 ISFET geometry. 1, semipermeable membrane; 2, source; 3, drain; 4, substrate; 5, aluminum contact; 6, insulating case.

The first ISFETs were described by Bergveld [13] and no reference electrode was used. On immersion in water, the silicon dioxide is hydrated and the sensor response to the pH is roughly Nernstian. The structural type used was not useful for analytical applications.

In the ISFET there is an interface between the SiO_2 layer and the electrolyte solution. When the solution composition changes, the number of ions adsorbed at the interface also changes and thus the conductance of the channel on the opposite side of the gate is modulated. On adsorption of an ion at the interface, the ionic charge induces an opposite charge in the channel. The magnitude of the induced charge depends on the permittivity and thickness of the insulator, on the Debye length in the electrolyte, and on some other parameters. However, the effect of the adsorbed ions can be quantitatively described and calculated. In the resultant sensor, the changes at the electrolyte–solid phase (SiO_2) interface are converted into changes in the charge carriers in the semiconductor and thus the state of the interface can be indirectly monitored. These sensors are at present very interesting commercially.

The thickness of the active oxide layer is about 200 nm. The sensor is soaked in water for several days prior to use, similar to the glass electrode, to permit hydration of the SiO_2. Moss, Janata, and Johnson [15] applied a perm-selective membrane to the oxide layer that is permeable for a single kind of ion; in this way, e.g., potassium ions in serum can be selectively determined.

Although the mechanism of surface processes has not yet been sufficiently clarified [14], sensors have been applied to determinations of various ions, such as H^+, K^+, Na^+, Ca^{2+}, etc. Some drawbacks of ion-selective electrodes (high impedance, fragility) have been removed in this way.

During further development [16], a silver chloride reference electrode has been introduced and the structure of the element has been modified so that only the SiO_2 surface is in contact with the solution.

The construction of a pH-electrode is given in Fig. 7.8. Only the part of the chip with the gate conductivity channel is in contact with the solution and it is protected by an insulating layer of silicon dioxide or nitride. An Ag/AgCl reference electrode is also immersed in the solution. The other parts and leads are encapsulated and the solution penetrates to the electrodes through windows in the electrode holder. The calibration curves

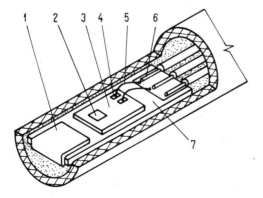

Fig. 7.8 Sensor design. 1, Ag/AgCl reference electrode; 2, hole for the test solution; 3, ISFET chip; 4, aluminum contact; 5, gold leads; 6, leads; 7, ceramic substrate.

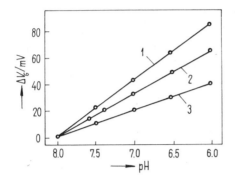

Fig. 7.9 Calibration curves for various surfaces. 1, Si_3N_4; 2, SiO_2 + Si_3N_4; 3, SiO_2.

given in Fig. 7.9 correspond to various insulators covering the conductive channel.

If the conductive channel is further covered with an ion-selective membrane, sensors selective for various ions in solution are obtained. The calibration dependences for various ions and biomedical applications can be found in a monograph [14]. The sensor dimensions are of the order of fractions of a millimeter, which is especially advantageous for medical use.

7.4   Chemically Sensitive Transistors of the MOS-CSFET Type

The principle of these sensors resembles that of the sensors of the previous group. However, the measurement is not carried out in an electrolyte, but in a gas in which the partial pressure of a certain component can be determined. For example, polar gaseous molecules can be adsorbed on the SiO_2 surface [17]. On an increase in the partial pressure of polar substances, maxima have been found on the conductance curves. The semiconductor basis was silicon doped with boron ($N_A = 10^{15}$ cm^{-3}) with an orientation of (111). The thickness of the oxide layer of the FET system was 50 to 100 nm. The transistor was exposed to various substances at increasing partial pressure in an evacuated cuvette. The conductance of the channel increases on sorption of polar molecules; a typical curve is shown in Fig. 7.10. A study of the heat of adsorption indicated that $CHCl_3$ molecules are adsorbed with their dipoles lying at right angles to the surface. Similar dependences have been obtained for alcohols (methanol, ethanol, and isopropanol). The position of the maximum was shifted to the right with increasing size of the alcohol molecules.

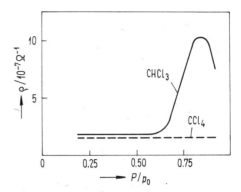

Fig. 7.10 Dependence of the conductance on the partial pressure for polar and nonpolar molecules.

The interfacial plane between the oxide layer and the gas contains surface states with various electric charges determined by the adsorption sites. Adsorption of alcohols on semiconducting oxides generally decreases the *p*-type conductance and increases the *n*-type conductance. At 20°C, alcohols are bound by weak physical adsorption combined with hydrogen bonding to the surface oxygens and strong chemisorption with formation of alkoxides [20].

When a layer of an organic substance is deposited on the thin layer of the oxide with the transistor conducting channel, FET systems can then also be used to monitor various gases and vapors. For example, β-carotene changes its electric conductance on adsorption of various gases and vapors on its surface [21] and forms weak donor–acceptor complexes with some gases, leading to reversible conductance variations. For example, on adsorption of oxygen the conductance δ increases by three orders of magnitude compared with the value in a nitrogen atmosphere. The conductance changes are generally large and amount to as much as four orders of magnitude.

Although not all the changes are completely reversible, these properties have been utilized for the construction of new sensors based on FETs [18]. During adsorption of, e.g., CO, a charge transfer complex is formed in β-carotene, causing a change in the conductance and polarization of the β-carotene molecule. The change in the charge magnitude induces the conductance in the conductive channel between the source and drain of a unipolar transistor. The drain current varies depending on the coverage of the organic surface with the adsorbed gas. It is necessary that the organic substance layer be uniform, applied, e.g., by sublimation. The length of the conductive channel must be many times greater than its width. The sensor sensitivity increases with increasing thickness of the organic layer (below 200 nm [18]). The active layer can also consist of cellulose acetate, *N,N,N',N'*-tetramethyl-*p*-phenyldiamine, tetracyanoethylene, and various organic semiconductors [22]. A certain selectivity can be attained by using perm-selective membranes [18].

A separate group of sensors is obtained when the FET gate is coated with a metal (Pd, Ag, Pt). Hydrogen detectors have been developed to a great degree of perfection [19, 23], using palladium gates. Palladium electrodes are readily permeable for hydrogen that penetrates into the SiO_2 layer and affects the escape work at the SiO_2–metal interface. The change in the threshold voltage is proportional to the hydrogen partial pressure. The response time constant depends on the temperature and equals a few seconds at 150°C. Therefore, the detector must be heated to a temperature above 100°C. The structure of such a transistor is given in Fig. 7.11. The oxide thickness between the gate and the semiconductor is

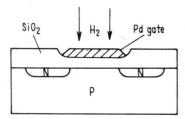

Fig. 7.11 Geometry of a CSFET with a palladium gate.

about 10 nm. In addition to the time constant, the threshold voltage is also affected by the temperature (0.8 mV/°C) and thus the system temperature must be stabilized in some simple way. The maximum sensitivity to hydrogen is 7 mV per ppm; this voltage can readily be handled by common operational amplifiers. It follows from the theoretical treatment that the double-layer threshold voltage depends on the hydrogen partial pressure. In addition to the dependence on the hydrogen partial pressure, the resultant function also depends on chemical reactions occurring at the Pd–SiO_2 interface in the presence of oxygen.

$$H_2(g) \underset{d_1}{\overset{c_1}{\rightleftarrows}} 2 H(ad) \tag{7.1}$$

$$O_2(g) + 2 H(ad) \overset{c_2}{\longrightarrow} 2 OH(ad) \tag{7.2}$$

$$OH(ad) + H(ad) \underset{d_3}{\overset{c_3}{\rightleftarrows}} H_2O \tag{7.3}$$

It is assumed that the rate of diffusion of hydrogen through the thin layer of Pd is comparable with the time constant of the chemical reactions at the interface.

A simple detector signaling a preset concentration of hydrogen is shown in Fig 7.12. A modified transistor with a palladium gate is powered by a constant current (100 μA). The transistor voltage then approaches

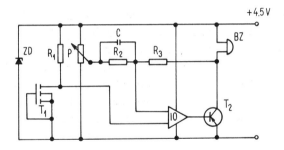

Fig. 7.12   Circuit for signaling of a required concentration of hydrogen.

the threshold voltage. The output signal is fed to an operational amplifier in a differential circuit. The level that is to be signaled is adjusted by potentiometer P and the sensitivity by the gain value, i.e., the resistance ratio, R_2/R_3. A circuit for temperature stabilization can be incorporated, e.g., by using an operational amplifier and a power transistor with a temperature feedback.

The hydrogen detector is surprisingly stable and selective, with simple construction. Instruments have been designed within a relatively short time for monitoring of hydrogen in the air, argon, nitrogen, ammonia, etc. An equally important application is the detection of coal gas leaks for pedestrian and mobile inspection, or even for private consumers, as the instrument is inexpensive.

To heat the MOS transistor chip, the required input is less than 1 W, the preparation time is about 20 s, and the time constant is 2 to 10 s, depending on the temperature. The weight of the instrument minus the sources is a few tens of grams.

7.5   Sensors Based on Interface Contact Phenomena

The effect of the surrounding medium on the contact potential has been known for a long time, but analytical applications are quite recent. A drawback involved in measuring the contact potential according to Fig. 7.13 is the necessity of using an electrometer.

The tested gas enters the chamber of measuring capacitor 1 and is directed between electrodes 2 and 3 as indicated by the arrow. The surface of active electrode 3 is coated with a thin layer of palladium and is heated to 120°C by electric spiral 4. Measuring electrode 2 is connected to electrometer 5 through a high-resistance lead-in insulator. The electrometer operates as a voltmeter and the values are displayed on meter 6.

Several methods have been developed for measuring the contact potential. The Zisman method is suitable for monitoring the effect of the medium on the value of the contact potential; this method is based on periodic changes in the distance between the active and reference electrodes that result in changes in the capacitance of the capacitor thus formed. The periodical changes generate an alternating current in the circuit and alternating voltage across the resistor in the circuit.

Good results have also been obtained [24] in measuring of the hydrogen partial pressure using a fixed electrode system. The surface of an active Pd electrode is placed against an inactive stainless-steel electrode. The results of measurements are illustrated in Fig. 7.14.

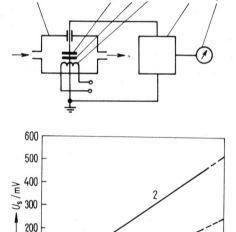

Fig. 7.13 Detector based on changes in the contact potential. 1, measuring capacitor; 2, 3, electrodes; 4, heating spiral; 5, electrometer amplifier; 6, voltmeter.

Fig. 7.14 Contact potential dependence on the concentration of hydrogen in the air at normal barometric pressure and a palladium electrode temperature of 120 °C. 1, Pd–stainless steel electrode pair; 2, Pd–Zn electrode pair.

Measurement of the hydrogen concentration using the Pd–CdS rectifying interface in a diode was first described by Steele and MacIver [25]. The system is actually a hydrogen-sensitive Schottky diode. It is assumed [23] that the ratio of currents (I) flowing through the diode is proportional to the hydrogen partial pressure according to the equation

$$\frac{I_1}{I_2} = k_1 \exp (k_2[H_2(g)]) \qquad (7.4)$$

An advantage of these types of detector is their high sensitivity (ppm) in the region of low concentrations and a lower sensitivity at higher hydrogen concentrations, as follows from the exponential dependence. Results analogous to those from the Pd–CdS diode have also been obtained using a Pd–Si diode [2].

Application of organic semiconductors to the measurement of gas and vapor concentrations is very promising. The detectors consist of layers of two organic semiconductors with different types of conductivity. The two layers are in contact and form an electrically nonlinear rectifying interface. The electrophysical properties of such a structure are decisively affected by processes occurring in the boundary layer between the two semiconductors. For example, the volt-ampere characteristic of such a diode changes considerably when the concentration of adsorbed gases changes by only a few dozen ppm [26]. Typical organic semiconductor systems are, e.g., phenazine (n-type) with fluorescein or chloroazine-p-phenylenediamine (p-type), or fluorescein with indigo.

7.6   Examples of Applications

TGS-type sensors [8] are based on tin dioxide doped with various admixtures. More than 10 million items have been manufactured since 1976, mostly for the detection of gases, vapors, and combustion products. The manufacturers recommend their use for the detection and determination of low concentrations of hydrocarbons, alcohols, ethers, ketones, esters, nitrated compounds, ammonia, carbon monoxide, hydrogen, and cyanogen. The detection limit is often lower than 0.1 ppm.

The calibration curves are stable for 2 or 3 weeks, but the sensors are poorly resistive to mechanical impurities. Temperature and humidity fluctuations cause considerable errors that, however, can be compensated using auxiliary circuits. These dependences are exactly defined for each type of sensor. The signal-handling circuitry depends on the character of the application.

Detectors and monitors from the SEMA company [27] have a thin sensitive layer placed on a silicon support that is resistance-heated to a working temperature of ca. 60°C. The sensor is protected against dust by a sintered-metal cover. The manufacturer recommends sensors for the detection of more than 200 gases and vapors.

Some sensors exhibit pronounced selectivity toward certain gases, such as CO, Cl_2, H_2S, and SO_2. For example, model 3900 can be used to detect methane in a concentration range of 0%–5%, with a lower range of 0 to 3 ppm. The instrument signal is independent of the content of silicones, chlorinated hydrocarbons, hydrocarbon vapors, hydrogen sulfide, sulfur dioxide, and carbon monoxide. The zero-line drift is $\pm 5\%$ of the range per year. The instrument has a digital display, the error in the determination of methane is $\pm 5\%$ of the range, and preset values can be signaled using the built-in auxiliary units.

Model 3908 has been designed to detect hydrogen sulfide in the chemical industry, steel mills, gas plants, gas pipelines, and oil refineries and in many other fields. The instrument is calibrated in three ranges, 0–2, 0–20, 0–200 or 0–20, 0–200 or 0–2000 ppm. Zero is adjusted automatically, which is important for production-line applications. The precision of the determination is $\pm 3\%$ of the range at temperatures from -5 to $+55°C$.

A fully automated SEMA instrument is the INDUSTRY MODULE 80, in which temperature maintenance, zeroing, and calibration curve construction are controlled by a microprocessor. The sensor operates under optimal conditions that are preprogrammed, including selection of the measuring range. The detection limit is from 1 to well below 0.01 ppm and the time constant lies between 1 and 3 s. The instrument is recommended especially for monitoring of CO, Cl_2, H_2S, SO_2, fluorinated hydrocarbons, tetraethyl lead, silicone vapors, and almost 250 other gases and vapors. The complete apparatus includes a calibration unit data printer and color display unit.

The model 2150 H_2S monitor (Drägewerk A. G., Lübeck, FRG) [28] consists of a sensor and a data-handling instrument. In the presence of H_2S the conductance of a semiconductor heated to a constant temperature changes. The sensor construction can be seen in Fig. 7.15. The

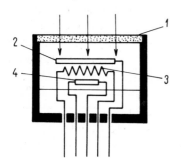

Fig. 7.15 Functional scheme of the H_2S detector. 1, sintered metal filter; 2, H_2S sensitive layer; 3, heater; 4, temperature sensor.

sensor space is separated from the test medium by a sintered-metal filter. The response is almost specific for H_2S and the presence of H_2, CO, SO_2, and CH_4 causes an error of less than 1 ppm. Mercaptans interfere to a certain extent; the instrument responds to mercaptans with a sensitivity about 12 times lower than that for hydrogen sulfide.

The time constant depends on the concentration and is from 15 to 30 s. Two measuring ranges are available, 1 to 3 and 0 to 100 ppm, with a logarithmic scale. The instrument is recommended for oil, natural gas, and paper, and for use in the chemical industry. Signaling outputs are also provided and the instrument is explosion-proof. The test medium temperature may fluctuate from -40 to $+85°C$.

NO_x detectors (NO, NO_2, and mixtures) can be prepared [29] by reactive high-frequency deposition of an SnO_x layer on an alumina substrate. Special ignition of the deposited layers improves the sensitivity to NO_x chemisorption. The dependence of the sensor resistance on the NO_x concentration in the air is logarithmic from 5 to 100 ppm. Interferences are small (50 ppm CO, H_2, C_3H_6, CH_4).

CO detectors can be prepared according to descriptions in three patents [30–32] by reactive deposition of an SnO_2 layer from an Sn target with a small amount of Sb or Pt. The SnO_2 layer can also be doped with platinum black [33, 34]. In addition to the basic material SnO_2, these sensors also contain $PdCl_2$ (0.5 to 10 wt. %) or chloroauric acid and platinum black formed in the layer during sintering. They can operate at normal ambient temperature. At higher temperatures the sensitivity to CO decreases, but the sensitivity to the water vapor partial pressure drops sharply. For example, the resistance decreases 150 times in the presence of 1000 ppm CO at a temperature of $25°C$, whereas the same concentration of H_2, hydrocarbons, alcohols, ketones, and aromatic hydrocarbons causes only a 1.6-fold decrease.

7.7   Conclusion

Semiconductor sensors are at present the subject of considerable research and commercial interest. These sensors are contained in detection and signaling instruments for low concentrations of CO, CH_4, coal gas, propane, many hydrocarbons, mercaptans, H_2S, and atmospheric pollutants. Especially promising is the use of these sensors in maintaining industrial safety limits, in fire protection, and in environmental protection. The technology of these sensors is based on techniques developed for the manufacture of semiconductors and thus these sensors can be manufactured cheaply on a large scale [35].

The simplicity of these instruments is analogous to that of analyzers based on thermal conductivity and heat of reaction, but their high sensitivity places them in the rank of sophisticated instruments, such as infrared analyzers, ionization detectors, etc. Contemporary research is mainly directed toward attaining long-term stability of the electric parameters and improved selectivity.

The study of reactions occurring on variously doped oxides has indicated that the conductance dependence on the temperature exhibits pronounced maxima for some substances. Therefore, possibilities of microcomputer treatment of these dependences are being studied for a large number of suitably doped oxidizable sites. A network of such sites can, e.g., be arranged in an addressable matrix on a single silicon chip, and the contents of components in a gaseous mixture can then be calculated.

REFERENCES

1. Many A., Goldstein Y. and Grover N. B.: Semiconductor Surface, Amsterdam 1965.
2. Zemel J. N.: Res.Develop. **28,** 38 (1977).
3. Marciniak W.: MIS Type Semiconductor Components, SNTL, Prague 1979 (in Czech).
4. Volkenshtein F. F.: Fiziko-Khimiya Poverkhnosti Poluprovodnikov, Nauka, Moscow 1973.
5. Fexa J. and Valenta S.: Scientific Papers, Institute of Chemical Technology, Materials Physics and Measuring Techniques, P4, 91 (1980).
6. Takata M. and Yanagida H.: Yogyo Kyokai Shi **87,** 13 (1979).
7. Seiyama T. and Kato F.: Anal.Chem. **34,** 1502 (1962).
8. General Catalogue Figaro Gas Sensors, Figaro Engineering Inc., Japan, Osaka 1976.
9. Watson J. and Tanner D.: Rad.Elec.Eng. **44,** 2 (1974).
10. Seiyama T., Fueki K., Shiokawy J. and Suzuki S. (Eds.): Chemical Sensors, Elsevier, Amsterdam 1983.
11. Cerberus Elektronik (Gasmelde system), Männedorf, Switzerland, manufacturer's literature.
12. Nagasaka T., Nitta M., Kanefusa S., Taketa Y. and Haradome M.: Nippon Daigaku Seisan Kogaku-bu Hokuku **10,** 145 (1977).
13. Bergveld P.: IEEE Trans.Biomed.Eng. BME-19, 342 (1972).
14. Cheung P. (Ed.): Theory, Design and Biomedical Applications of Solid State Chemical Sensors, CRC Press, Boca Raton, Florida 1979.
15. Moss S. D., Janata J. and Johnson C. C.: Anal.Chem. **47,** 2238 (1975).
16. Janata J. and Moss S. D.: Biomed.Eng. **11,** 241 (1976).
17. Thorstensen B., Fjeldly T. A. and Johannesen J. S.: 3rd Intern.Conf. Solid Surfaces, Vienna 1977.
18. FRG Patent 24 07 110 (1975).
19. Lundström I., Shivaraman M. S., Svenson C. and Lundkvist L.: Appl.Phys.Lett. **26,** 55 (1975).
20. Ponec V., Knor Z. and Černý S.: Adsorption on Solids, SNTL, Prague 1968 (in Czech).
21. Rosenberg B., Misra T. N. and Switzer R.: Nature **217,** 423 (1968).

22. Rexer E.: Organische Halbleiter, Academie Verlag, Berlin 1966.
23. Lundström I., Shivaraman M. S. and Svensson C.: *J.Appl.Phys.* **46**, 3876 (1975).
24. Fikec L.: *Elektrotech.Obzor* **3**, 162 (1980).
25. Steele M. C. and MacIver B. A.: *Appl.Phys.Lett.* **28**, 687 (1976).
26. FRG Patent 18 05 624 (1973).
27. SEMA Electronics Ltd., Shewalton Rd., Irvine, Great Britain, manufacturer's literature.
28. Dräger-General Monitors Model 2150, Drägerwerk AG Lübeck, FRG, manufacturer's literature.
29. Shih-Chia Chang: *IEEE Trans.Electron.Devices ED-26*, 1875 (1979).
30. FRG Patent 2831 394.
31. FRG Patent 2831 400.
32. FRG Patent 2832 828.
33. Brit.Patent 1464 415.
34. Patent Appl. FRG, 24 37 352 (1975).
35. Janata J. and Huber J. R. (Eds.): Solid State Chemical Sensors, Academic Press, New York 1985.

CHAPTER 8

ELECTROCHEMICAL DETECTORS AND MONITORS OF ATMOSPHERIC POLLUTION

Jiří Tenygl

8.1   Introduction

At present, more than 50 types of electrochemical monitors and detectors for various atmospheric and aquatic pollutants are manufactured. Their advantages include high sensitivity, simplicity and thus also reliability, simple operation, and low cost, permitting the use of many instruments forming a measuring network. Electrochemical detectors have low energy consumption, and light portable instruments or even personal monitors can be designed.

Pollutants are automatically monitored in the atmosphere and in water. Monitoring of the air pollution is somewhat simpler because the composition of the atmosphere is roughly constant, the concentrations of pollutants fluctuate within a certain known range, and the presence of various interferents and pollutants in the air can be predicted to a certain extent. Interferents can be suppressed using selective filters and by pre-treating the sample prior to the analysis.

There are varying requirements on the measuring precision and stability, the frequency of obtaining data, and the permitted delay. With personal monitors and portable monitors it is sufficient if the signal is stable for eight hours or even less. However, for continuous air pollution monitors, it is required that they measure with a precision of at least $\pm 2\%$ for one month without human interference. This is a rigorous requirement, but it has been fulfilled with a number of commercial monitors employing perfected sensors and devices for sample and reagent dosing, as well as through the introduction of automated calibration and zeroing methods.

When the sensor signal is used for regulation, a very short delay of a few seconds is required. However, when emissions and pollutant concentrations in the air are monitored at control stations, periodic measurements at five to ten minute intervals are usually sufficient and thus a longer data delay is unimportant. Most automatic monitors, however, operate continuously, even if the data delay is not important, because the sampling, sample treatment, and thus the whole apparatus are simpler in continuous measurements.

8.2   Problems of Measurement

The maximal permitted concentrations of pollutants in the atmosphere are very low and depend on the hygienic standard of a particular country and on the exposure time. For example, the Czechoslovak Standard permits an average concentration of sulfur dioxide of $0.15 \, mg/m^3$ over 24 h and $0.5 \, mg/m^3$ for an exposure of up to 30 min in 24 h. Therefore, to monitor the air pollution, it is necessary to continuously measure pollutant concentrations of 0.01 to 10 ppm. The concentration 0.01 ppm equals 1×10^{-6} vol. % or 4.46×10^{-10} mol liter^{-1}.* Such low concentrations are difficult to determine in the laboratory using the most sensitive methods and even more so continuously and automatically in the field.

Common electroanalytical methods readily determine concentrations from 10^{-5} mol liter^{-1} upwards. Therefore, the test pollutant must be preconcentrated from the air to obtain an easily measurable concentration in solution, e.g., simply by physical dissolution of the pollutant in an electrolyte according to Henry's law or by a chemical reaction, such as that with iodine at high air-to-electrolyte flow-rate ratios (10^3 to $10^4 : 1$). With periodical analytical methods, preconcentration is attained even by adsorption on a filter, followed by release of the adsorbed substance by heating. Methods of sample treatment prior to the analysis vary and are described in greater detail with the individual detectors.

The detection itself is based on potentiometry, voltammetry, coulometry, or, most often, on a combination of several methods. In the analysis of the atmosphere, oxygen is always present and interferes in measurements in the cathodic region. Oxygen can be removed, but this is disadvantageous in prolonged measurements. Therefore, the determination is carried out in the anodic region where oxygen does not interfere, which has an additional advantage in the prevention of the cathodic reduction

---

* Concentration conversions (under normal conditions, 0°C, 0.101325 MPa, i.e., 760 mm Hg):

100% gas $= 4.46 \times 10^{-2}$ mol liter^{-1}

1% gas $= 1 \times 10^4$ ppm $= 4.46 \times 10^{-4}$ mol liter^{-1}

1×10^{-4}% gas $= 1$ ppm $= 4.46 \times 10^{-8}$ mol liter^{-1}

1 ppm $SO_2 = 2.86 \, mg \, SO_2/m^3 = 2.60 \, mg \, SO_2/m^3$ (25°C)

1 mg $SO_2/m^3 = 0.35$ ppm SO_2

and deposition of metallic impurities on the electrode, leading to electrode passivation. Chemicals of common purity can then be used without danger of electrode poisoning, and prolonged voltammetric or coulometric measurements can even be carried out with solid electrodes without electrode surface regeneration, which is also made possible by the fact that in the determination of typical gaseous pollutants, such as SO_2, CO, NO_x, and O_3, the products of the anodic oxidation are soluble.

It is advantageous in the measurement when the test pollutant is directly oxidized or reduced on the electrode surface; then it is sufficient to bring the electrolyte with the dissolved pollutant to the measuring electrode in a defined way. Such a direct measurement can only be carried out with relatively higher concentrations of the order of tens of ppm where the analytical signal is higher than or at least comparable with the noise. Therefore, it is practically impossible to attain a substantial increase in the sensitivity by merely increasing the electrode surface area, because the signal-to-noise ratio remains virtually constant. For this reason, even electroactive pollutants are first preconcentrated by a preceding chemical reaction, usually with iodine or bromine. In this way, the selectivity of the measurement deteriorates somewhat, because it can no longer be affected by selection of the electrode potential. This disadvantage is circumvented by selectively filtering or pretreating the sample.

8.3   Principal Parts of Monitors

Monitors are instruments that measure the concentration of a given substance for a long time, completely automatically, from the sampling to the handling of the results. A substantial part of the monitor is the sensor (sometimes incorrectly termed the analyzer) that measures the concentration of the substance.

Monitors usually consist of four functional parts, which perform
(a) sampling and sample pretreatment,
(b) sample analysis,
(c) calibration and zeroing,
(d) data treatment.

This classification is schematic; one functional part often performs more operations, e.g., the calibrating and zeroing device also pretreats the sample and the data treating part also corrects for fluctuations in the temperature and the sample flow rate. This chapter will primarily deal with the sensors, i.e., the analytical part of the monitor.

A great deal of attention has been devoted to data treatment. Chart recordings are now only used for visual control and the measured values are immediately handled by a microcomputer, converted into appropriate units, and averaged. The results are stored in a memory or are telemetrically transmitted directly to a measuring center. The microcomputer also corrects the sensor data according to the results of the calibration and zeroing.

8.4   Sampling and Sample Treatment

Sampling often affects the precision and reliability of the whole measurement. In analyses of trace pollutants, there are many problems that do not occur at all or are of little importance in measurements of higher concentrations. Loss of the test substance must be prevented; losses occur, e.g., through sorption on the walls of the inlet tubing [1] and on filters, by condensation, or through a chemical reaction, which prevents the use of many plastics. The monitor inlet tube must thus be as short as possible and, together with the filters, is often heated above the dew point. For these reasons, the flow meter and pump are usually placed after the sensor, to prevent losses of the test substance. Interferents are removed on selective filters, with the active substance coated on glass wool or some other support with a large surface area.

Information on the construction and chemical composition of filters in commercial analyzers is scarce and the data vary greatly; therefore, only some literature data are given here. Inactive NO is oxidized to active NO_2 using a $MnO_2/KHSO_4$ mixture [2], CrO_3 crystals [3, 4], MnO_2 [5], and ozone [6–8]. Pure air is led around a uv lamp where it is enriched in ozone and is then mixed with the test air in which NO is oxidized quantitatively to NO_2. Oxidation is also used to remove NO_x, with the NO_2 formed being trapped selectively in tubes containing triethanolamine [6, 9–11], diethanolamine [5], or sodium acetate. Nitrogen dioxide, together with SO_2, O_3, CO_2, and water vapor, is trapped in filters connected in series and containing active charcoal and a mixture of sodium and calcium hydroxides with silica gel. A catalytic conversion of NO_2 to NO is described in Ref. 6. Sulfur dioxide is trapped in chemical filters, usually consisting of pumice coated with a mixture of Ag_2SO_4 and $KHSO_4$ [12] or with $NaHCO_3$ [13]. Sulfur dioxide is also removed, together with water vapor, on asbestos wool coated with NaOH. Hydrogen sulfide is removed on silver wool heated to 150°C [14]. The efficiency of selective filters is discussed in Ref. 15. Ozone that interferes in the determination of SO_2 is removed on $FeSO_4$ crystals. Carbon monoxide and hydrocarbons are catalytically oxidized on hopcallite (a mixture of copper and manganese oxides on a support).

In the preparation of pure air for zeroing or for the preparation of calibration mixtures, all interferents are removed. The filters are usually connected in the following order: air pump, dust filter, ozone generator (NO $\rightarrow$ NO$_2$ oxidation), active charcoal (sorption of SO$_2$, NO$_2$, and O$_3$), soda lime (CO$_2$), silica gel (H$_2$O), flow-rate regulator and flow meter. If necessary, filters with hopcallite and silver wool can be placed before the silica gel filter. Preparation of pure air by catalytic oxidation is described in Ref. 16.

A pure absorption solution or an analyzer base electrolyte is obtained by continuous regeneration of the spent solution that circulates through a purification device. Active charcoal or nylon fiber (removal of iodine) filters are used, or electrolysis is carried out for purification [17]. The evaporated water is replenished from a storage tank. In the Philips automatic SO$_2$ monitor, loss of water through bleeding with the outlet gas is prevented by condensation of water vapor in a cooler operating on the Peltier principle.

Flow meters and pumps for the air and the electrolyte are also part of the sampling device. The solution is circulated using a mammoth pump or a simple peristaltic pump.

Problems connected with the dosing and measuring of the reagents and of the air and with the preparation of calibration and zeroing mixtures are often underestimated. However, it is well known from practice that the long-term reliability of many monitors is less dependent on electro-chemical problems than on the auxiliary devices.

8.5   Calibration and Zero Adjustment

Most commercial monitors are provided with devices for calibration and zeroing, operated manually or automatically. In the manual operation, the operator adjusts the gas flow rate and the other parameters to the prescribed values and lets in the pure gas. A calibration mixture is then fed into the analyzer and the signal is corrected on the basis of the cali-bration results. The operator simultaneously checks the auxiliary devices.

In the automatic mode, the above procedure is carried out automa-tically, at preset intervals. Zeroing and the calibration with a single con-centration of the test substance are usually carried out daily. The gas flow rate, temperature, electrode potential, and other parameters are checked as well. The sensor signal is corrected by the microprocessor, depending on the values obtained. The methods of calibration and monitoring are surveyed in several monographs [18–21] and the preparation of the calibration mixtures in a sub-ppm range in Ref. 22.

8.6   Preparation of Calibration Mixtures

Gases that are chemically stable, do not react with the cylinder walls, and are not adsorbed can be stored at high concentrations (tens of ppm) in pressure cylinders (mainly CO and CO_2). Calibration mixtures containing SO_2, NO, and NO_2 at concentrations lower than 1 ppm cannot be stored for more than three months [23]. The pressure cylinders and valves are made of stainless steel or are coated with Teflon. Nitrogen oxides are stored in aluminum cylinders.

However, calibration mixtures mostly cannot be stored and are prepared *in situ* by dilution. To dose the test substance, permeation tubes [24–28] are usually employed, utilizing diffusion of the vapors of the liquefied gas through a plastic foil (a copolymer of fluorinated ethylene and propylene or Teflon) into the pure air stream (Fig. 8.1). The rate of the vapor penetration depends on the area and thickness of the tube walls, the molecular mass of the gas, and the temperature. In this way, mixtures containing 0.1 to 10 ppm of the test substance can be prepared with a precision of $\pm2\%$, provided that the permeation tube is thermostatted within $\pm0.1°C$. For the preparation of more dilute mixtures (down

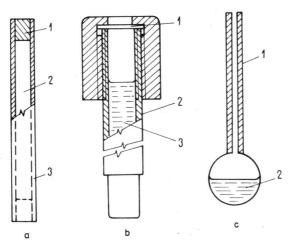

Fig. 8.1 Permeation apparatus. (a) Permeation tube: 1, stopper; 2, calibrating liquid; 3, plastic tube. (b) Permeation disk: 1, permeation disk; 2, metallic tube; 3, calibrating liquid. (c) Diffusion device: 1, glass capillary; 2, calibrating liquid.

to 0.005 ppm), permeation disks are used with a smaller diffusion area.

Permeation devices are suitable for dosing of gases with critical temperatures above $25°C$, i.e., SO_2, NO_2, H_2S, and NH_3. However, permeation tubes are also manufactured for propane, n-butane, chlorine, and methyl mercaptan.

Diffusion devices operate analogously to permeation apparatuses; however, the calibration substance diffuses through a glass capillary separating a storage tank containing liquified calibration substance from the pipe through which pure air flows. In this way, mixtures with concentrations of 0.1 to 100 ppm can be prepared even for substances with a low vapor pressure at normal temperatures [29–33]. A constant vapor pressure of chemical reagents is maintained by buffering of the solution [34].

Many substances can be prepared by electrolysis [35, 36]. The amount of substance produced is given by the Faraday laws and can be varied widely by varying the current. Hydrogen and oxygen are most often produced electrochemically, but other substances, such as O_3, CO_2, Cl_2, I_2, NO, NO_2, etc., can also be generated.

To produce ozone in the air at a low concentration, a uv lamp is used with a movable screen permitting variation of the ozone concentration over a wide range. The ozone concentration is very reproducible, so that the uv lamp is used as a secondary standard for the calibration of ozone, nitrogen oxide, and nitrogen dioxide analyzers. The reaction

$$NO + O_3 = NO_2 + O_2 \qquad (8.1)$$

is used, which is rapid and quantitative. It is called the "gas-phase titration" [37]. If a certain amount of ozone is added to an NO-containing mixture, then the amount of ozone consumed is equivalent to the amount of consumed NO or of NO_2 formed [7, 8, 38]. This reaction is employed in many commercial monitors.

8.7   Absorption Prior to Analysis

The classical arrangement, where the gas passes through a reagent solution, is used mainly with coulometric analyzers [39–41], in which the kinetic energy of the gas is also used to yield solution convection at the electrode or its circulation. The reagent is usually a solution of a halogen (I_2, Br_2) or a halide (I^-, Br^-). The contact area between the gas and the solution is increased by using absorbers operating on the principle of a mammoth pump that simultaneously circulates the solution [42, 43], or by letting the solution flow along the electrode surface [44]. In the Schulze monitor [45] the reagent is sprayed by an air stream into a spherical reaction space and flows along the walls toward the indicator electrode.

High efficiency can be attained with electrodes covered by a film of electrolyte and with porous electrodes. The test gas diffuses rapidly through the electrolyte film toward the electrode surface where it is subject to an electrochemical reaction. In this way the sensitivity is substantially increased even without preconcentration of the substance in the reagent (see below).

8.8   Principles of Measurement

In continuous measurements, potentiometric, voltammetric, coulometric, and related methods are employed. Continuous titrators with coulometric generation of the reagent and with potentiometric end-point detection are also common. It is difficult to classify commerical monitors, because they often combine several methods and the sensor construction is rather different from the classical textbook types.

8.8.1   Potentiometry

This method is used for the measurement of redox potentials with platinum electrodes, pH with glass electrodes, and various substances using ion-selective electrodes (ISEs). According to the Nernst equation, the electrode potential is a logarithmic function of the test substance activity, a_i,

$$E = E_o + \frac{RT}{nF} \ln a_i = E_o + \frac{59}{n} \log a_i \qquad (8.2)$$

where E_o is the standard potential, R is the gas constant, T is the absolute temperature, n is the number of electrons exchanged, and F is the Faraday constant (valid for 25°C and potential in mV). In dilute solutions it can be assumed that activity a_i approximately equals concentration c_i ($a_i = c_i \gamma_i$, where γ_i is the activity coefficient of substance i).

An advantage of potentiometry is its high sensitivity (the detection limit is usually 10^{-5} to 10^{-6} mol liter^{-1}) and the simplicity of the apparatus. A certain drawback is the logarithmic dependence of the signal on the concentration, which also limits the attainable measuring precision for higher concentrations. A gradual decrease in the signal is also often

observed, caused by leaching of the ion-selective membrane, and must
be corrected for by periodical calibration.

The measurement is carried out after absorption of the test substance
in a solution or a reagent. Membrane-covered electrodes have been developed for direct monitoring of gases and are characterized by a high
measuring selectivity; these electrodes are described below with voltammetric electrodes based on an analogous principle.

8.8.2   Voltammetry

The current is measured and is proportional to concentration C,

$$i = nFSD\frac{C}{\delta} = K\frac{C}{\delta}$$

(8.3)

where n is the number of electrons exchanged, F is the Faraday constant,
S is the electrode surface area, and D is the diffusion coefficient; these
quantities are included in the constant K. The diffusion layer thickness δ
at the electrode surface depends on the solution convection and must
be maintained constant, e.g., by movement of the electrode (rotation,
vibration), stirring, solution flow, etc. However, in analyzers of the atmosphere, convection is mostly maintained by bubbling the test gas through
the solution [46].

The current depends on the temperature ($2\%–4\%$ K^{-1}) and on the
electrode activity, which is usually stable in the anodic region. Solid
indicator electrodes (pyrolytic graphite, Ag, Au, Pt) of various shapes
and catalytically active electrodes (carbon impregnated with a Pt catalyst)
are used. The detection limit is about 10^{-5} to 10^{-6} mol liter^{-1}. An advantage is a wide linear dependence of the signal on the concentration (10^{-6}
to 10^{-1} mol liter^{-1}). The effect of temperature is compensated for electrically. Voltammetric sensors with bubbled electrolyte are used mainly
for measurements in a concentration range from 500 ppm to 10%. For
monitoring low concentrations, electrodes covered with an electrolyte
film, porous electrodes, and membrane electrodes are used and they
find application mainly in portable monitors and personal detectors.

8.8.3   Electrodes with an Electrolyte Film

These electrodes are based on the work of Hersch [35, 47–50], who utilized
an interesting effect analytically. A silver foil electrode partially immersed
in an electrolyte and in contact with gas containing ppm oxygen concentra-

tion yields a greater current than the completely immersed electrode, even if its surface area is smaller. Close to the liquid meniscus, the electrode is covered by a thin film of electrolyte through which oxygen rapidly diffuses to the electrode, where it is reduced. This principle is utilized by various sensors for oxygen and other gases. At present, the simple arrangement depicted in Fig. 8.2 is used, in which a silver wire or gauze is wound on a porous tube soaked with an electrolyte. The electrolyte film is maintained by capillary forces and its thickness must not vary, e.g., as a result of evaporation. Therefore, the gas is dampened by passage through the electrolyte before it enters the analyzer. However, this is impossible with reactive emissions. Hygroscopic substances are also used or the film is renewed by electrolyte flow and recirculation (in Beckman and Mast Corp. analyzers, USA).

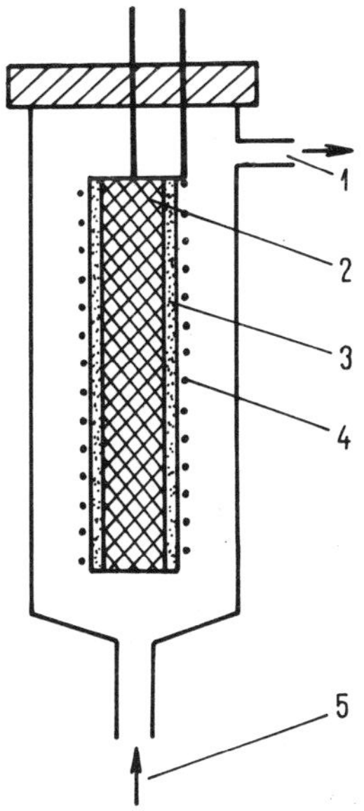

Fig. 8.2 Analyzer with an electrolyte film. 1, gas outlet; 2, reference electrode; 3, porous tube; 4, indicator electrode (a silver wire spiral wound over the porous tube); 5, gas inlet.

Electrodes with an electrolyte film have a high sensitivity (down to 0.1 ppm O_2 can be determined directly, without preconcentration by absorption) and are simple, and their response is rapid; they are mainly used in chemical production lines.

8.8.4   Porous Electrodes

Rapid gas diffusion through the electrolyte film on the pore walls results in the high sensitivity of porous electrodes. There are two measuring methods:

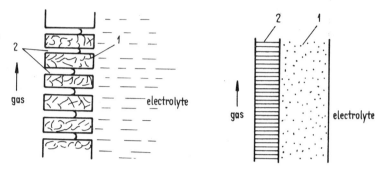

Fig. 8.3 Scheme of a porous electrode. (a) Porous electrode: 1, liquid pores; 2, gas pores. (b) Porous electrode with porous hydrophobic membrane: 1, membrane; 2, porous electrode.

(a) The test gas passes through the electrode pores and bubbles through the solution (similar to traps with glass frits). Good contact between the gas and the electrode permits quantitative reaction at low concentrations (up to ca. 100 ppm) and coulometric measurement is then possible [15, 51, 52]. The test gas must be pumped at an elevated pressure and constant flow rate.

(b) The test gas passes along one wall of the porous electrode (Fig. 8.3) and the other wall is in contact with the electrolyte (similar to galvanic cells with air depolarization). The test gas diffuses into the pores where the electrode reaction takes place. This method was used in some detectors as early as in the 1930s.

Porous electrodes have been perfected recently in connection with fuel cell research [53–55] and electrodes with reproducible properties suitable for electroanalysis have been made [56]. The main problem in the application of porous electrodes in electroanalysis is the necessity of maintaining a defined boundary between the gas and the electrolyte inside the pores, preventing concentration polarization by an inert gas inside the pores and eliminating ohmic polarization of the electrode. Thus, the electrode must only contain pores with two defined dimensions, small pores (ca. 3 µm) filled with the electrolyte and providing conductive connection and large pores (50 µm) into which the test gas diffuses [53, 56]. Opening of the gas pores is attained by an overpressure of the test gas or, without the danger of loss of the test gas in the pump, by creating a vacuum on the liquid side of the porous electrode [57]. Another method of preventing the flooding of the gas pores is hydrophobization of the surface (using Teflon or a naturally hydrophobic material, such as carbon black).

Porous electrodes are manufactured commercially and there is little information available concerning their construction and parameters.

It is difficult to compare them objectively. This is especially true of catalytically active electrodes impregnated with a "catalyst" and used for the determination of CO, hydrocarbons, and H_2S and for acceleration of the oxidation of SO_2 and NO_2. Platinum and palladium are mostly used as catalysts. The high catalytic activity of the two metals, however, also causes an increase in the residual current and catalysis of the reduction or oxidation of other substances that may interfere in the measurement. The manufacturer's literature is not sufficiently reliable in this respect.

Some types of electrode are covered with a porous Teflon membrane coated with a catalyst. To a certain extent, the Teflon membrane assumes the role of the gas pores; it prevents loss of the electrolyte but enables rapid diffusion of the gas. The sensitivity is thus improved and this system is used for the determination of medium concentrations of 1–100 ppm, especially in portable analyzers. An advantage is that the test gas can be dosed by free diffusion and need not be pumped, which is utilized in personal detectors of toxic gases. These sensors are sometimes called gas-diffusion electrodes.

The measurement is either potentiometric (e.g., in sensors for the determination of down to 10 ppm H_2 in the air [58, 59]) or, more often, voltammetric (amperometric), which is more suitable because of the wider measuring range.

8.8.5   Electrodes Covered with Membranes

Electrodes covered with membranes (ECM) can be used for measurements in gases and in liquids and are used for the determination of gases (O_2, CO, CO_2, SO_2, H_2S, NO_2, HCN, Cl_2), volatile substances (NH_3, amines), and, indirectly, of substances that can be converted into gaseous or volatile products (NH_4^+, amines, NO_3^-, CN^-, ClO^-).

ECMs are based on semipermeable plastic membranes placed over the indicator electrode and enclosing a thin film of the electrolyte. The test medium — gas or liquid — flows outside the membrane and the test substance diffuses through the membrane, dissolves in the thin electrolyte film, and is sensed by the indicator electrode. High-molecular-weight substances, impurities, colloids, and ions cannot penetrate through the membrane, but water vapor partially penetrates. The application of semipermeable membranes simply eliminated electrode passivation and substantially improved the measuring selectivity. ECMs are either potentiometric or voltammetric.

A typical representative of potentiometric ECMs is the sensor for continuous measurement of CO_2 containing a glass pH-electrode [60].

Carbon dioxide from the sample diffuses through the membrane and dissolves in the liquid film,

$$CO_2 + H_2O \rightleftharpoons HCO_3^- + H^+ \qquad (8.4)$$

The change in the pH is indicated by the glass electrode and is a logarithmic function of the CO_2 concentration. The reaction is reversible; on a decrease in the concentration, the CO_2 diffuses back from the film into the sample and a new equilibrium state is attained. The response rate is determined by the diffusion rate, which depends on the membrane and electrolyte film thickness, the temperature, and the rates of desorption and absorption. However, the diffusion rate has little influence on the sensor signal, which is independent of changes in the membrane permeability caused by aging or by nonsymmetrical mechanical strain. On the other hand, slow diffusion leads to slow data output, and the constructors of commercial sensors try to prevent this effect. Gases causing changes in the pH interfere depending on the acid–base equilibrium constants (SO_2, NO_2, HF, etc.). Sensors containing pH-electrodes are commercially available for CO_2, NH_3, SO_2, NO_2, Cl_2, and CH_3COOH. Sensors containing ISEs can be obtained for NO_2, H_2S, HF, Cl_2, Br_2, and F_2, where NO_3^-, S^{2-}, F^-, Cl^-, Br^-, and F^- are the monitored substances, respectively. Many substances can be determined after conversion into gases; e.g., nitrate can be determined as NH_3 after reduction to NH_4^+ followed by alkalization of the solution. (For details on the use of ISEs, see Chapter 6.)

Voltammetric ECMs are mainly used for the determination of oxygen in gases and liquids (Clark sensors [61, 62]). The indicator electrode is a silver disk on which oxygen is reduced, producing an electric current proportional to the oxygen concentration over a wide range, according to the equation

$$i = nFSC \, \frac{P_m}{b} \qquad (8.5)$$

where the symbols have the usual significance, P_m is the membrane permeability, and b is the membrane thickness. The signal stability primarily depends on P_m/b. The diffusional flux through the membrane must be constant and the various electrodes differ mainly in the method used to attain this requirement. The b value and the membrane tension must not change; the latter depends on the age of the membrane, as the membrane elasticity gradually decreases. For this reason, a thin separator is often placed between the membrane and the electrode and the pressure is regulated mechanically, or an elegant method involves coating of the membrane with a layer of porous metal [131] or pressing of the membrane onto a porous silver disk [115].

Voltammetric ECMs are also employed for determination of SO_2, NO_x, and Cl_2 ; when catalytic electrodes are used, H_2S, CO, some hydrocarbons, alcohol vapors, and H_2 can be determined.

Thin plastic membranes are used (polyethylene, Teflon, mylar, silicone rubber, etc.), differing in the permeability constant for various gases. With higher permeability a higher signal can be obtained and thus lower concentrations can be determined. However, the selectivity of the determination cannot be attained merely by suitable choice of the membrane, because the permeability for various gases never differs by more than one order of magnitude [63–65], and filters, solution pH adjustment, and variation of the indicator electrode potential must also be employed. The signal dependence on the temperature (2%–5% K^{-1}) is compensated. ECMs have no unified nomenclature and are called Clark electrodes, diffusion electrodes, gas sensors or probes, air-breathing electrodes, or membrane electrodes.

8.8.6 Coulometry

Emissions are coulometrically determined primarily in flow-through systems. The gas is fed at a constant velocity into the electrolysis vessel, in which the experimental conditions are adjusted to attain complete electrolysis of the test substance. The electrolysis current is then controlled by the feeding velocity and is given by the Faraday law,

$$i = nFN = knFvC \qquad (8.6)$$

where N is the number of moles of the test substance fed into the electrolysis vessel per time unit, v is the sample flow rate, and C is the test substance concentration. It is not easy to attain complete, or almost complete (at least 98%), electrolysis of the test substance in a flow-through system. For example, in the analysis of oxygen in the air this would mean that virtually pure nitrogen must leave the electrolysis vessel. Electrodes with large surface areas must be used, intense convection and the greatest possible contact of the sample with the electrode surface provided, and reliable and stable feeding of the test medium — solution or gas — into the electrolysis vessel ensured. The current is proportional to the velocity of the sample feed, which thus must be stable for long times. However, an advantage of flow-through coulometry is that the electrolysis current is independent of the temperature and of the electrode activity over a certain range.

Coulometric analyzers are used mainly for determinations of very low concentrations (of the order of ppm) when higher sample feed velocities

can be used with acceptable delays for complete electrolysis. Coulometric analyzers also employ preliminary chemical reactions of the test
substance with iodine, bromine, or another reagent, thus achieving quantitative trapping of traces of the test substance even in small absorbers.

The electric circuitry of coulometric and voltammetric sensors is
virtually identical and the two methods differ only in the experimental
conditions. Voltammetry employs electrodes with a small surface area
and a high sample flow rate, so that the test substance is not appreciably
depleted by electrolysis in the vessel. Coulometry, on the other hand, uses
electrodes with a large surface area and a small flow-rate of the test
substance and the experimental conditions are arranged so that the test
substance is completely electrolyzed.

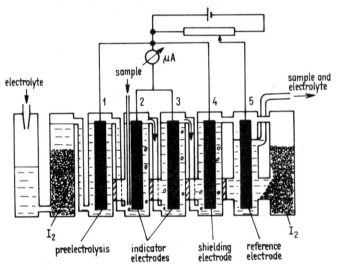

Fig. 8.4   Scheme of the Novák coulometric SO_2 analyzer [17, 46].

A typical representative of flow-through coulometry with gradual
electrolysis of the sample is Novák's coulometric SO_2 analyzer [17, 46]
(Fig. 8.4). The air is fed at a constant velocity into space 2 in which it
passes through an iodine solution and SO_2 is trapped as a result of the
reaction

$$SO_2 + I_2 + H_2O \ = \ SO_3 + 2HI \qquad (8.7)$$

The amount of iodide ions formed is proportional to the SO_2 concentration,
and these ions are determined by anodic oxidation on carbon electrodes
2 and 3, according to the reaction $2I^- = I_2 + 2e$. The current passing

through electrodes 2 and 3 is measured and is proportional to the SO_2 concentration. At electrode 1 the solution is preelectrolyzed and thus purified of iodide ions. Screening electrode 4 prevents penetration of I^- ions from reference cathode 5. The solution is pumped (0.1 ml/min) in the direction of the gas flow through the compartments of the individual electrodes that are conductively connected by diaphragms in the bottoms of the compartments. The sensitivity depends on the air flow rate and down to 0.04 mg SO_2/m^3 in air can be continuously measured.

In many commercial monitors only partial electrolysis of the test substance occurs (20%-60%) as a result of shortcomings in the construction. These sensors operate on the borderline between coulometry and voltammetry and their signal is a complicated function of the sample flow rate, temperature, and the electrode activity, which has a detrimental effect on the precision and stability of the measurement.

8.8.7   Galvanic Analysis

This name is used for methods that operate without an external current source. When the reference electrode is suitably selected, so that its potential corresponds to the reduction of the test substance on the indication electrode, then the current passing between the two electrodes connected through a microammeter (similar to a galvanic cell) is proportional to the test substance concentration at the indicator electrode. This system is mainly employed with electrodes with an electrolyte film and with membrane sensors, but also in voltammetric and coulometric analyzers. In galvanic analyzers, the indicator electrode is polarized at a constant potential and is thus maintained in the working state, and the slow signal stabilization occurring with sensors polarized from an external source on accidental interruption of the current is eliminated. An electronic potentiostat can thus be partially substituted. This is the greatest advantage of galvanic analysis, which has, therefore, found its most important application in analyzers for prolonged measurements.

The reference electrode must be capable of providing current for a long time without marked polarization; therefore, these electrodes are made of foils or sintered metals (Ag, Cd, Pb, Zn) with a large surface area and are usually regenerated by charging from an external current source against an auxiliary electrode.

8.8.8   Automated Electrochemical Titrators

Only the principle of the coulometric titrator originally designed at the beginning of World War II for continuous monitoring of yperite in the air [67], which is still being marketed under the trade name Titrilog [68–70], will be given here. The titrator has two basic circuits, an indicating and a generating circuit. The former consists of a reference electrode and a Pt indicator electrode by means of which the halogen concentration is determined potentiometrically. The latter circuit contains a large-area generating Pt anode, an auxiliary cathode placed in a separate compartment, and an amplifier controlled by the indicating circuit signal.

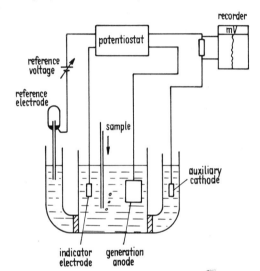

Fig. 8.5   Scheme of the Titrilog automatic coulometric titrator [67].

The test air is passed at a constant velocity (500 ml/min) through a solution of $1 \times 10^{-3} M$ Br_2 in 10^{-1} M KBr (Fig. 8.5). Sulfur dioxide or another oxidizable substance reacts with bromine

$$SO_2 + Br_2 + H_2O = SO_3 + 2HBr \tag{8.8}$$

The voltage signal coresponds to changes in the indicator electrode potential as a result of depletion of bromine and is amplified and controls the generating circuit connected in the feedback. Bromine is produced by electrolysis at the generating anode

$$2Br^- = Br_2 + 2e \tag{8.9}$$

The amount of bromine produced just compensates the loss of bromine through the reaction with SO_2. Therefore, a constant bromine concentration is maintained in the vessel and the generating current is proportional to the SO_2 concentration in the sample. It is advantageous that the difficult measurement of the reagent volume used in classical titrators is replaced by simple electrolysis and only the generating current is recorded.

The actual design of the automated titrator is much more complex. Loss of bromine through reduction on the auxiliary electrode must be prevented, losses of bromine vapors bleeding with the exit air must be compensated for, impurities must be removed from the circulating solution, and the two circuits that are galvanically connected by the electrolyte must be electrically separated. These problems have been solved; for example, Philips 1 (PW 9700 type) and Philips 2 (PW 9755 type) coulometric monitors [41] have been approved by the Environmental Protection Agency, USA [71] for use in a reference method for SO_2 monitoring, because of their precision and reliability.

An advantage of automated titrators is their universal applicability over a wide range of test substance concentrations and their reliability. However, even auxiliary devices are rather sophisticated and thus the instruments are relatively expensive. To improve the selectivity, the sample is passed through selective filters.

8.9   Determination of Various Pollutants

This section deals with detectors and monitors for various pollutants. The use of ISE sensors for these purposes is described in greater detail in Chapter 6.

8.9.1   Sulfur Dioxide

Electrochemical determination of the substance is easy, based, e.g., on the cathodic reduction [72] with $E_{1/2} = -0.42$ V (SCE; all the subsequent potential values are referred to the SCE). However, atmospheric oxygen interferes in the reduction and must be removed. Therefore, the cathodic reduction is rarely used analytically and is mostly used only in periodic determinations.

Methods based on the anodic oxidation of SO_2 to SO_3 in an acidic medium ($E_{1/2} = +0.58$ V [73], in which O_2 and CO_2 do not interfere, have found the widest application. The interference from other gases depends on the kind of electrode. Neither NO nor CO interferes when catalytically inactive carbon and bright platinum electrodes are used. However, on catalytically active electrodes (carbon with a platinum catalyst), CO, NO, NO_2, and H_2S can be simultaneously oxidized, and the interference is suppressed by a suitable selection of the electrode potential, the kind of catalyst, and the electrolyte composition and use of selective filters, so that interferents do not affect the precision of the determination when present at concentrations common in the atmosphere. Monitoring of SO_2 is carried out [74–77] at carbon [74] (higher concentrations of SO_2 in waste gases), at a PbO_2-covered electrode [75], and at other electrodes [76, 77]. A continuous coulometric analyzer for trace concentrations of SO_2 was described and evaluated in detail by Lindqvist [15]. The determination is carried out differentially at two carbon fiber electrodes with a Pt/Ni_2B catalyst, through which the test air passes.

Commercial instruments [24] often permit the determination of several substances with one sensor type, which is called, not particularly lucidly, the "multiparameter capability". The selectivity of the determination is attained by using filters, by selection of the electrode potential and of the reagent composition, and by other means. The substances determinable by various commercial sensors are given in parentheses below.

Continuous voltammetric monitors and portable instruments of various types (usually with electrodes covered with membranes) are manufactured by the Ecology Board [78] (SO_2, NO_2, CO, H_2S, O_2), Environmental Products [79] (SO_2, NO_2, NO_x, CO, HCOH), IBC [80] with the Faristor exchangeable membrane sensor (SO_2, NO, NO_2, NO_x, CO, H_2S), Interscan [81] (SO_2), Joy [82] (SO_2), and Theta Sensors [83] (SO_2, NO_x, O_2). Personal detectors and portable monitors (mostly with membrane and diffusion sensors) are manufactured by Blakell [84] (SO_2, H_2S, CO, O_2, NO, NO_2, Cl_2, NH_3, HCN), Bionics [85] (SO_2, H_2S, Cl_2, HCN, HCl, HF, Br_2, NH_3, O_2, F_2, H_2, CH_3COOH, PH_3, $C_2H_5NH_2$), Broadley-James Corp. [86] (SO_2, NO_2, CO_2, NH_3, O_2, biochemical oxygen demand (BOD)), Detection Instruments [87] (SO_2, H_2S, NO_2, Cl_2, NH_3), and Riken Keiki [88] (SO_2, HCN, Cl_2, NH_3).

Coulometric analyzers with an iodine or bromine solution are based on reaction given in Section 8.8.8. Hydrogen sulfide and mercaptans interfere by reaction with bromine and cause a positive error signal, whereas nitrogen dioxide produces bromine by oxidation of bromide and thus yields a negative error. As described above, the loss of bromine is compensated by its coulometric generation, which is the basis of the Titrilog instrument [89]. The titrator has been modified for monitoring SO_2 and H_2S [90] and SO_2 and O_3 [45] and its response has been accelerated by using a bromine solution instead of the original iodine [91]. The instrument has also been combined with gas chromatography [92] and applied to SO_2 monitoring in cellulose works [93].

An advantage of Titrilog-type instruments is their reliability and minimal maintenance requirements. It is common to measure one month or more without any servicing. Titrilog-type monitors of air pollution are manufactured by Barton [40] (SO_2, H_2S, mercaptans, organic sulfides), Beckman [68] (SO_2), Consolidated Electrodynamic Corp. [89] (SO_2), International Ecology Systems [69] (SO_2, Cl_2, O_3, HCl), Philips [41] (PW 9700 type, SO_2, H_2S, mercaptans; multicomponent type, SO_2, NO, NO_2, O_3, CO; PW 9755 type, SO_2), Process Analyzers [70] (SO_2), and in the USSR (ECHG [94], determination of sulfur-containing gases in the production of H_2SO_4).

A two-electrode coulometric monitor of SO_2 in the atmosphere [39] is manufactured by Elektronika [95]. The chemical reaction of SO_2 with iodine is used, similar to the Titrilog system. However, the SO_2 concentration is found from the concentration of the iodide produced that is coulometrically reoxidized to I_2 at the indicator anode. The SO_2 concentration is proportional to the anodic current of the oxidation of I^-.

Good results are obtained using electrodes covered with a semipermeable membrane, measuring the change in the pH of the electrolyte film [96–101]. Interferences are suppressed by suitable choice of the membrane and the electrolyte composition. These sensors are especially suitable for monitoring higher concentrations, e.g., in waste gases. Mascini and Cremisini [102] critically discussed membrane sensors and the factors controlling their selectivity.

Potentiometric atmospheric monitors are manufactured by Chemtrix [103] (SO_2, SO_3, H_2SO_4), Lazar [104] (SO_2, Cl_2, NO_x, H_2S, CO_2, NH_3), and Orion [105] (SO_2, SO_3, Cl_2, NO_2, HCN, CO_2, HF, H_2S, NH_3, amines). Methods for the determination of sulfur-containing compounds [106], the effect of interferences in the determination of SO_2 [107], and criteria for the choice of monitoring techniques [108, 143] have been reviewed.

So far, little information is available on sensors with solid electrolytes with ionic conductivity, e.g., LaF_3, that have been proposed for the determination of O_2, CO_2, SO_2, NO_2, and NO.

8.9.2   Hydrogen Sulfide and Sulfur-Containing Compounds

Hydrogen sulfide and mercaptans are not found among common atmospheric emissions and are monitored by instruments developed for the determination of SO_2. The ready oxidation of these substances on catalytic electrodes or their reaction with iodine or bromine is utilized. Hydrogen sulfide can further be very sensitively and selectively determined using an ISE.

A sensor with a diffusion catalytic electrode from Energetic Science [109] has been described in detail by Sedlak and Blurton [110]. Many companies manufacture personal detectors the size of a pocket calculator that emit an acoustic signal when a preset gas concentration is exceeded. Some of them integrate the measured concentration over time and determine the overall exposure. A number of them are discussed in the section on SO_2; these detectors are also manufactured by Compur [111] (H_2S, HCN, NO_2, phosgene) with an organic electrolyte and Energetic Science [109] with a reference electrode depolarized by air, contributing to the long lifetime and signal stability of this sensor.

8.9.3   Nitrogen Oxides

Nitrogen oxide, NO, nitrogen dioxide, NO_2, or, most often, the sum of the two, NO_x, is determined.

Nitrogen oxide is electroinactive on common electrodes, but can be determined by anodic oxidation at catalytic electrodes. Nitrogen dioxide can be determined cathodically after dissolution in a suitable electrolyte in the form of HNO_2, but atmospheric oxygen interferes. Therefore, the determination is carried out by anodic oxidation on noncatalytic or, most often, catalytically active electrodes. In the determination of NO_x, NO is oxidized prior to the analysis. The reaction with iodide,

$$NO_2 + 2HI \quad = \quad NO + H_2O + I_2 \tag{8.10}$$

is also used. Iodide is present in excess and thus the iodine forms I_3^- and cannot be carried away by the exit air. In the method of Hersch [35, 112], the I_3^- is reduced coulometrically to iodide on a Pt cathode over which an iodide solution flows. Ozone can be determined in the same way (see below). This system is employed in the Beckman 910 monitor of traces of NO_2 [68]. The Beckman 908 monitor [68] is designed for the determination of NO. The sample first passes through selective filters on which interferents (NO_2, O_3, SO_2, etc.) are removed. The NO passed is then quantitatively oxidized by ozone using gas-phase titration and the NO_2 obtained is determined by the Beckman 910 instrument. Beckman instruments have been described and characterized [113]. A new construction on the Hersch principle has been described by Allen [114].

The voltammetric reduction of the iodine obtained [44] is utilized in the Mast monitors [116] in which an iodide solution is recirculated over the electrode system by a pump. The principle of the Philips Multicomponent coulometric instrument [41] has not been described in detail.

Sensors covered with membranes measure both amperometrically (NO_2 oxidation) and potentiometrically (a change in the pH), as described above. They are manufactured by Dynasciences [117], EnviroMetrics [118], and Theta Sensors [83].

The Geomet monitor [119] operates periodically. The test air is passed through a solution and the NO_x trapped is then measured with a nitrate ISE, followed by automatic calibration and cell rinsing.

Chemoprojekt [121] manufactures a voltammetric continuous analyzer for monitoring 0.05% to 5% NO_x in gases during HNO_3 production [120]. NO oxidation is carried out before the analyzer using electrolytically generated O_2.

A sensor selective for NO_2 covered by a porous Teflon membrane has been described by Barna and Jasinski [122, 123]. The electrode is made of a chalcogenide glass oxidized at an elevated temperature, thus providing selectivity toward NO_2. The measurement can be performed potentiometrically or amperometrically. A personal detector of NO_x [124] is described in greater detail in Section 6.3.2. The cathodic reduction of NO_2 on a membrane-covered electrode and with cobalt phthalocyanine as a catalyst was utilized in the sensor for monitoring NO_2 in automobile exhaust gases [66].

8.9.4   Carbon Monoxide

Carbon monoxide is electroinactive on catalytically inactive electrodes (carbon, Au, bright passivated platinum) and cannot be determined directly. For this reason, it does not interfere in the determination of other substances. However, carbon monoxide can be directly anodically oxidized on catalytically active electrodes. Ovenden [125] has described an amperometric determination of CO on a vibrating palladium anode in a neutral buffered electrolyte; gradual loss of electrode catalytic activity is a problem that seems to be overcome when using catalytic porous electrodes that are sometimes covered with a porous or semipermeable membrane. Diffusion electrodes with a Pt catalyst [126–128] are very stable and are used in Ecolyzer portable monitors [109]. Monitors from EnviroMetrics [118] with Faristor exchangeable sensors, Dynascience [117], and Blakell personal detectors [84], as well as those from Detection Instruments [87] and Interscan

[81], are apparently based on an analogous principle. Their function has not been described for the Neutronics [129] and MiniCo [130] detectors.

Membrane amperometric sensors with thin layers of metals or catalysts deposited directly on plastic membranes [115, 131] exhibit a rapid response, and the effect of fluctuations in the test gas pressure is suppressed, which is advantageous for deep mine detectors. The test gas can be passed through the sensor without danger of membrane separation from the electrode leading to a change in the diffusion layer thickness or to the danger of electrolyte loss from the electrode system.

In an indirect determination, carbon monoxide reacts chemically with I_2O_5 in a heated (150°C) tube,

$$5CO + I_2O_5 = 5CO_2 + I_2 \qquad (8.11)$$

The iodine vapors are fed into an electrochemical analyzer of iodine where they are determined. Continuous analysis of iodine vapors is not difficult and can be carried out polarographically [132], potentiometrically, voltammetrically, coulometrically, or by coulometric titration. The use of the Hersch cell [133] is advantageous.

In voltammetric and coulometric sensors, I_2 is reduced to I^- on noncatalytic electrodes (bright Pt, C) without interference from O_2 and other substances. The preparation of stable I_2O_5 that does not spontaneously decompose and liberate I_2 vapors is a problem. Other substances are also oxidized by I_2O_5 (SO_2, NO_2, C_2H_2, C_2H_4, alcohols), but can be readily removed by filtering.

8.9.5   Photochemical Oxidants

These consist of a group of atmospheric pollutants that are produced by photochemical reactions in the atmosphere between hydrocarbons and nitrogen oxides, namely ozone, peroxyacetylnitrate (PAN) ($CH_3CO_2ONO_2$), and hydrogen peroxide. They are usually defined as substances that oxidize a neutral KI solution buffered with phosphate. This solution absorbs ozone completely and NO_x, PAN, and H_2O_2 partly. Reducing substances (SO_2) cause a negative error and must be removed. Therefore, oxidants can be determined iodometrically.

In electrochemical analyzers, iodine is determined by cathodic reduction to iodide. In Brewer's method [44] a 2% KI solution is used, buffered at pH 7. The solution flows down over a Pt wire cathode wound on a cylindrical support, reacts with the oxidants, and the iodine obtained is immediately reduced cathodically. The reduction current is proportional to the oxidant concentration. This system has been used in commercial monitors from Intertech [134] and Mast [116].

In Schultze's method [45], a 2.5% NaI solution is sprayed by the gas stream into the spherical reaction space. The solution flows down the walls back to the jet. This method has been applied in the Freeman monitor of oxidants and SO_2 [135]. Two independent detection systems are used: in one the oxidants are determined after filtering off the SO_2 and in the other the SO_2 is determined after removal of the oxidants.

In the Hersch ozone analyzer [42], a buffered KI solution is circulated around the sensor by the gas stream on the principle of a mammoth pump; the iodine produced is determined. This system was employed in the Beckman monitors manufactured earlier [43]. The coulometric reduction of I_2 is probably used in the Philips monitors [41].

The selectivity of the measurement is attained by filtering the sample prior to the analysis. Substances producing iodine (NO_2) cause a positive interference, whereas reductants (SO_2, H_2S) cause a negative interference. The extent of interference that depends on the absorption efficiency has been determined experimentally for various analyzers.

8.9.6   Chlorine

Chlorine can be determined directly by cathodic reduction at positive potentials ($E° = +1.11$ V) without interference from oxygen. Noncatalytic electrodes are used (bright Pt, carbon), so that other substances (H_2S, CO, H_2) are not reduced or oxidized and thus do not interfere. The determination of chlorine in water is one of the first analytical applications of electrochemistry and a galvanic analyzer with a Pt cathode and Cu anode was proposed some seventy years ago [136]. Later, many methods were developed, based on potentiometry, polarography, voltammetry, coulometry, and automatic titration. Chlorine is absorbed in a film of an electrolyte flowing down over a Pt electrode [137], or is trapped by the passage of the test gas through the solution or by free diffusion in porous and membrane electrodes. The reaction with an iodide solution is used as well, similar to ozone analyzers, which usually can be applied to the determination of chlorine, after a simple modification.

A continuous monitor of chlorine leaks is manufactured under the trade name Detachlor [138]. Some instruments used for monitoring of SO_2, O_3, and NO_x can also be used for the determination of chlorine; the same is true of personal detectors. A special detector for chlorine in the air at a ppm concentration is manufactured by Draeger [139]. Chlorine diffuses through a ceramic cylinder and is reduced on a Pt electrode with a calcium bromide electrolyte.

Membrane sensors, in which the change in the pH or in the redox potential is measured, are especially suitable for higher concentrations. A sensor with a membrane and a detector with an electrolyte film have been described by Kane and Young [140].

8.9.7   Cyanogen

This substance is not directly electrochemically oxidized or reduced. Its continuous monitoring is possible with membrane sensors, containing a cyanide ISE or a glass electrode. The methods and the interferences from other acidic gases, especially CO_2, have been surveyed by Mascini and Cremisini [102]. Commercial HCN sensors and detectors are discussed in Sections 8.9.1 and 8.9.2. Amperometric HCN detectors are based on the dissolution of a heavy metal electrode with the formation of stable cyanide complexes. Silver electrodes are chiefly used, where the atmospheric oxygen does not interfere. The necessity of oxygen removal is the main disadvantage of the very sensitive polarographic determination.

A single amperometric detector with an electrolyte film has been described by Fligier et al. [141]. A detector of poisonous gases used in warfare has been patented [142].

8.9.8   Hydrocarbons

These substances are electroinactive and cannot be determined directly. Electrochemical detectors have not been used for their determination.

8.9.9   Other Gases

In addition to the above typical emissions, other gases can also be measured electrochemically. Only brief information is given here because of the lack of space.

Electrochemical determination can be used for hydrogen, acidic gases (HCl), by the reaction with IO_3^- and subsequent determination of iodine, organic substances, using the reaction with I_2O_5 followed by determination of iodine, phosgene, ClO_2, and O_2. Gases that dissolve with a change in the solution pH (HCl, CO_2, NH_3, amines), etc., can be determined in sensors with a pH electrode.

Electrochemical detectors are often used in chromatography (see Chapter 9).

REFERENCES

1. Arbuzova A. S., Kutenova A. I., Prozorova R. G., Ovsyannikova L. R., and Finkelstein R. Y.: *Gig.Sanit.* **7**, 86 (1974).
2. Hartkamp H.: *Schr.Reihe Landesant.Imm.-u.Bodennutzungschutz Landes N.Rhein/Westfahlen* **18**, 55 (1970).
3. Dimitriades B.: *Health.Lab.Sci.* **13**, 63 (1976).
4. Levaggi D., Kothny E. L., Belsky T., de Vera E., and Mueller P. K.: *Environ.Sci.Technol.* **8**, 348 (1974).
5. Wagner B. and Pohl K.: *Z.Chem.* **15**, 111 (1975).
6. Munemori M. and Maeda Y.: Proc.Int.Clean Air Congr., 4th, 1977, p. 370.
7. Muldoon D. G. and Majahad A. M.: *NBS Spec.Publ.* **464**, 21 (1977).
8. Paur R. J.: *ibid.* **464**, 15 (1977).
9. Dimitriades B.: *Health Lab.Sci.* **12**, 371 (1975).
10. Bourbon P., Alary J., Esclassan J. and Lepert J. C.: *Atmos.Environ.* **11**, 485 (1977).
11. Willey M. A., McCammon C. S., Jr., and Doemeny L. J.: *J.Am.Ind.Hyg.Assoc.* **38**, 358 (1977).
12. Buck M. and Gies H.: *Staub-Reinhalt Luft* **26**, 379 (1966).
13. Adams D. F.: *Tappi* **52**, 53 (1969).
14. Hersch P. and Deuringer R.: 149th Meeting of Am.Chem.Soc., Detroit, April 1965.
15. Lindqvist F.: *J.Air Pollution Control Assoc.* **28**, 138 (1978).
16. Barcocchi A. T. and Knobel R.: *Am.Lab.* **12**, 81 (1980).
17. Novák J. V. A.: Coulometric Analyzer of SO_2 Traces, manuf. by Elektronika, Prague, Czechoslovakia.
18. Calibration in Air Monitoring, ASTM Spec.Techn.Publ. 598, Philadelphia, USA 1976.
19. Katz M. (ed.): Methods of Air Sampling and Analysis, 2nd ed., Amer.Public Health Assoc., Washington, USA 1977.
20. Axelrod H. D. and Lodge J. P., Jr.: Sampling and Calibration of Gaseous Pollutants, in Stern A. C. (ed.), Air Pollution, 3rd ed., Academic Press, New York 1976.
21. Stern A. C. (ed.): Air Pollution, Vol. 3, Measuring, Monitoring and Surveillance of Air Pollution, 3rd ed., Academic Press, New York 1976.
22. Popov V. A. and Pechennikova E. V.: *Zavod.Lab.* **40**, 1 (1974).
23. Fox D. L. and Jeffries H. E.: *Anal.Chem.* **53**, 1R (1981).
24. Instrumentation for Environmental Monitoring, Univ.of California, Lawrence Berkeley Lab. (1972); suppl. (1974).
25. O'Keefe A. E.: *Anal.Chem.* **49**, 1278 (1977).
26. O'Keefe A. E. and Ortman G. C.: *Anal.Chem.* **38**, 760 (1966).
27. Scaringelli E. P., O'Keefe A. E., Rosenberg E. and Bell J. P.: *Anal.Chem.* **42**, 871 (1970).
28. Hughes E. E., Rook H. L., Deardorff E. R., Margeson J. H. and Fuerst R. G.: *Anal. Chem.* **49**, 1823 (1977).
29. Tsang W. and Walker J. A.: *Anal.Chem.* **37**, 13 (1965).
30. Altshuller A. P. and Cohen I. R.: *Anal.Chem.* **32**, 802 (1960).
31. McKelvey J. M. and Hoelscher H. E.: *Anal.Chem.* **29**, 123 (1957).
32. Analytical Instruments Development, Inc., West Chester, Penn., USA.
33. Nelson G. O.: Controlled Test Atmospheres, Principles and Techniques, Ann Arbor Science Publ., Ann Arbor, USA 1971.
34. Hashimoto Y. and Tanaka S.: *Environ.Sci.Technol.* **14**, 413 (1980).
35. Hersch P.: Galvanic Analysis, in Delahay P. and Tobias C. W. (eds.), Advances Anal. Chem. and Instrum., Vol. 3, J. Wiley, New York 1969.

36. Hersch P. and Deuringer R.: *J.Air Pollut.Control Assoc.*, **13**, 538 (1963).
37. Hodgeson J. A., Baumgardner R. E., Martin B. E. and Rehme K. A.: *Anal.Chem.* **43**, 1123 (1971).
38. Stedman D. H. and Harvey R. B.: NBS Spec.Publ. 464, 393 (1977).
39. Novák J. V. A.: *Coll.Czech.Chem.Commun.* **30**, 2703 (1965).
40. Barton Inc. ITT, Process Instruments and Controls, Monterey Park, Calif., USA.
41. Philips Ltd., Eindhoven, The Netherlands.
42. Hersch P. and Deuringer R.: *Anal.Chem.* **35**, 897 (1963).
43. Beckman Instruments, Fullerton, Calif., USA.
44. Brewer A. W. and Milford J. R.: *Proc.Roy.Soc.* A **256**, 470 (1960).
45. Schulze F.: *Anal.Chem.* **38**, 748 (1966).
46. Novák J. V. A.: *Coll.Czech.Chem.Commun.* **21**, 662 (1956).
47. Hersch P.: Brit.Patent 707, 323 (1954).
48. Hersch P.: Brit.Patent 750, 254 (1956).
49. Hersch P.: *Nature* **180**, 1407 (1957); *Chem.Age* (London) **67**, 565 (1952).
50. Hersch P.: Brit.Patent 929,885 (1963).
51. Flook W. M. and Keidel F. A.: US Patent 2,898,282 (1959) (du Pont de Nemours Co.).
52. Keidel F. A.: *Ind.Eng.Chem.* **52**, 490 (1960).
53. Newman J. and Tiedemann W.: Flow-Through Porous Electrodes, Advances in Electrochem. and Electrochem.Engineering, Gerischer H. and Tobias C. W. (eds.), Vol. 11, p. 353, J. Wiley, New York 1978.
54. Berger C.: Handbook of Fuel Cell Technology, Prentice Hall, Hampstead 1968.
55. Breiter M. W.: Electrochemical Processes in Fuel Cells, Springer Verlag, Berlin 1969.
56. Jansta J. and Dousek F.: *Coll.Czech.Chem.Commun.* **36**, 1212 (1971).
57. Tenygl J. and Fleet B.: *Coll.Czech.Chem.Commun.* **38**, 1714 (1973).
58. Bianchi G., Faita G., and Mussini T.: *J.Sci.Instr.* **42**, 693 (1965).
59. Brill K.: Les Journées Int.d'Etude des Piles à Combustible, Compt.Rend., Brussels 1969.
60. Stow R. W., Baer R. T. and Randall B. F.: *Arch.Phys.Med.Rehabil.* **38**, 646 (1957).
61. Clark L. C., Jr.: *Trans.Am.Soc.Artif.Internal Organs* **2**, 41 (1956).
62. Sawyer D. T., George R. S. and Rhodes R. C.: *Anal.Chem.* **31**, 2 (1959).
63. Stern S. A.: Gas Permeation Processes, in Lacey R. E. and Lorb S. (eds.), Industrial Processing with Membranes, Wiley-Interscience, New York 1972.
64. Chand R. and Marcote R. V.: AICHE Meeting, C 68 (1971).
65. Lucero D. P.: *Anal.Chem.* **40**, 707 (1968).
66. Dietz H., Haecker W. and Jahnke H.: Electrochemical Sensors for Analysis of Gases, in Advances in Electrochem. and Electrochem.Engineering, Gerischer H. and Tobias C. W. (eds.), Vol. 10, p. 1, J. Wiley, New York 1977.
67. Shaffer P. A., Jr., Briglio A., Jr. and Brockman J. A., Jr.: *Anal.Chem.* **20**, 1008, (1948).
68. Beckman Instruments Inc., Fullerton, Calif., USA.
69. International Ecology Systems, City of Industry, Calif., USA.
70. Process Analyzers, Inc., Princeton, N. J., USA.
71. Environmental Protection Agency, Washington, D.C., Federal Register 41, 36 245 (Aug. 27, 1976); 41, 34 105 (Aug. 12, 1976).
72. Kolthoff I. M. and Miller C. S.: *J.Am.Chem.Soc.* **63**, 2818 (1941).
73. Rozental K. I. and Veselovskii V. I.: *Zh.Fiz.Khim.* **27**, 1163 (1953).
74. Novák J. V. A.: *Coll.Czech.Chem.Commun.* **21**, 662 (1956).
75. Belanger G.: *Anal.Chem.* **46**, 1576 (1974).
76. Konoplev Y. I., Shkatov E. F., and Shcherbakova O. L.: *Mekh.Avtomat.Proizvod.* **3**, 27 (1975).

77. Gauthier M. and Chamberland A.: *J.Electrochem.Soc.* **124**, 1579 (1977).
78. Ecology Board, Inc., Chatsworth, Calif., USA.
79. Environmental Products, Whittaker Corp., Chatsworth, Calif., USA.
80. International Biophysics Corp., Irvine, Calif., USA.
81. Interscan Corp., Chatsworth, Calif., USA.
82. Joy Manufacturing Co., Los Angeles, Calif., USA.
83. Theta Sensors, Inc., Altadena, Calif., USA.
84. Blakell Ltd., Blandford Forum, Dorset, England.
85. Bionics Instruments Co., Ltd., Higasiyamato-City, Tokyo, Japan.
86. Broadley-James Corp., Santa Ana, Calif., USA.
87. Detection Instruments Ltd., Wokingham, Berkshire, England.
88. Riken Keiki Fine Instruments Co., Ltd., Tokyo, Japan.
89. Consolidated Electrodynamic Corp., Pasadena, Calif., USA.
90. Washburn H. W. and Austin R. R.: Air Pollution, Proc.U.S.Techn.Conf. on Air Pollution, McGraw-Hill, New York 1952.
91. Adams D. F., Jensen G. A., Steadman J. P., Koppe R. K., and Robertson T. J.: *Anal. Chem.* **38**, 1094 (1966).
92. Klads P. J.: *Anal.Chem.* **33**, 1851 (1961).
93. Adams D. F. and Koppe R. K.: *J.Air Pollut.Control Assoc.* **17**, 181 (1967).
94. Váňa J.: Gas and Liquid Analyzers, Elsevier, Amsterdam 1982.
95. Elektronika, Prague, Czechoslovakia.
96. Tenygl J.: in West T. S. (ed.), Ion-Selective Electrode Analysis, International Review of Science, Butterworth, *Phys.Chem.,* Ser. 1, Vol. 12, 132 (1973).
97. Tenygl J.: *ibid.,* Ser. 2, Vol. 13, 1 (1976).
98. Young M., Driscoll J. N., and Mahoney K.: *Anal.Chem.* **45**, 2283 (1973).
99. Ross J. W., Riseman J. H., and Krueger J. A.: *Pure Appl.Chem.* **36**, 473 (1973).
100. Krueger J. A.: *Anal.Chem.* **46**, 1338 (1974).
101. Nash T.: *Air Water Pollut.* **8**, 121 (1964).
102. Mascini M. and Cremisini C.: *Chim.Ind.* (Milan) **62**, 222 (1980).
103. Chemtrix, Inc., Hillsboro, Oreg., USA.
104. Lazar Research Lab., Los Angeles, Calif., USA.
105. Orion Res., Inc., Cambridge, Mass., USA.
106. Chiagneau M. and Santarromana M.: *Rev.Ins.Fr.Pet.* **29**, 697 (1974).
107. Marshale G. B.: *Clean Air* **5**, 15 (1975).
108. Neuscheler R. C.: *Am.Soc.Testing Mater.Spec.Publ.* **555**, 9 (1974).
109. Energetic Science, Inc., Elnesford, N.Y., USA.
110. Sedlak J. M. and Blurton K. F.: *Talanta* **23**, 445 (1976).
111. Compur-Electronic GmbH, Munich, FRG.
112. Hersch P. and Deuringer R.: *Anal.Chem.* **35**, 897 (1963).
113. Harman J. H.: *Advan.Instrum.* **26**, Pt. 1, 554 (1971).
114. Allen J. D.: *Analyst* **99**, 765 (1974).
115. Tenygl J.: Czechoslovak Patent 164,600 (June 26, 1973).
116. Mast Development Co., Davenport, Iowa, USA.
117. Dynasciences Comp., Chatsworth, Calif., USA.
118. EnviroMetrics, Inc., Marina Del Rey, Calif., USA.
119. Geomet Inc., Pomona, Calif., USA.
120. Tenygl J.: Czechoslovak Patent 113,881 (Sept. 5, 1962).
121. Chemoprojekt, Prague, Czechoslovakia.
122. Barna C. and Jasinski R.: *Anal.Chem.* **46**, 1834 (1974).

123. Jasinski R., Barna C., and Trachtenberg I.: *J.Electrochem.Soc.* **121,** 1575 (1974).

124. Ferber B. I., Sharf F. A., and Freedman R. W.: *J.Am.Ind.Hyg.Assoc.* **37,** 1 (1976).

125. Ovenden P. J.: *J.Electroanal.Chem.* **2,** 80 (1961).

126. Bay H. W., Blurton K. F., Lieb H. C., and Oswin H. G.: *Am.Lab.* **4,** 57 (1972).

127. Blurton K. F. and Bay H. W.: *Am.Lab.* **6,** 50 (1974).

128. Bay H. W., Blurton K. F., Sedlak J. M., and Valentine A. M.: *Anal.Chem.* **46,** 1837 (1974).

129. Neutronics Ltd., Stansed, Essex, England.

130. Mine Safety Appliances Comp., Ltd., East Shawhead, Coatbridge, England.

131. Bergman I.: *Ann.Occupational Hyg.* **18,** 53 (1975).

132. Novák J. V. A.: *Chem.Listy* **49,** 277 (1955).

133. Hersch P. and Sambucetti C. J.: Pittsburgh Conf.Anal.Chem.Spectrosc., March 1963.

134. Intertech. Corp., Princeton, N.J., USA.

135. Freeman Laboratories, Inc., Rosemount, Ill., USA.

136. Rideal E. K. and Evans U. R.: *Analyst* **38,** 353 (1913).

137. Haller J. F.: US Patent 2,651,612 (1953).

138. Fischer and Porter Comp., Warminster, PA 18 974, USA.

139. Draeger Safety, Chesham, Bucks., England.

140. Kane P. O. and Young J. M.: *J.Electroanal.Chem.* **75,** 255 (1971).

141. Fligier J., Czichon P., and Gregorowicz Z.: *Anal.Chim.Acta* **118,** 145 (1980).

142. Vertes M. A. and Oswin H. G.: US Patent 3,470,071 (1969).

143. Hollowell C. D., Gee G. Y., and McLaughlin R. D.: *Anal.Chem.* **45,** 63 (1973).

CHAPTER 9

ELECTROCHEMICAL DETECTORS FOR LIQUID CHROMATOGRAPHY AND OTHER ANALYTICAL FLOW-THROUGH SYSTEMS

Antonín Trojánek

9.1 Introduction

In many modern analytical methods the final operation, the detection, is performed in a flowing liquid in which the components to be detected form zones. This is true primarily of liquid chromatography and of continuous-flow analyses, both direct and titration analysis. The requirements placed on flow detectors differ somewhat depending on the method used and the type of apparatus.

9.2 Flow-Through Analytical Systems

All liquid chromatographic techniques are based on the separation in time of the components of mixtures on the basis of their different physico-chemical properties and the subsequent detection of the separated components in the mobile phase stream. High-performance liquid chromatography (HPLC) employs columns with internal diameters of a few millimeters and sample volumes of the order of tens of microliters and thus the elution volumes are small. For good separation of the components, detectors with small internal volumes must thus be used; this requirement is even more stringent in work with microcapillary columns, which have internal diameters of 50 to 200 µm.

Liquid chromatography employs both universal and specific detectors. The latter, which detect substances selectively depending on their properties, involve mainly variable wavelength photometric, fluorescence, and electrochemical detectors. Their properties are compared in several reviews [1, 2].

In continuous flow-through analyzers, the samples are aspirated or injected into a stream of liquid which transports them into the detector. During the transport the substances are often modified in various ways

[3–5]. In view of the precise timing of the reaction and measurement conditions, it is important that these systems should not place stringent requirements on the detector response rate.

Continuous titrations in flowing liquids combine the simplicity of the reading of the test substance concentration, characteristic of direct analytical methods, with the precision of titration methods. All these techniques are based on the attainment of an equivalent material ratio during continuous mixing of the reagent and sample solutions and differ in the selection of the variable parameter and thus also in the experimental arrangement [6–9]. Electrochemical (especially potentiometric) detectors are widely used in these systems. Again, the requirements on the detector response rate and, except for certain cases [9], even on the response reproducibility are not especially stringent.

9.3 Electrochemical Detection Methods

Of electroanalytical methods (potentiometry, voltammetry, coulometry, conductometry, high-frequency impedance methods), the first two are by far the most often used for detection in flow-through systems. Therefore, potentiometric and voltammetric flow-through detectors are discussed in this chapter; information on other types of detectors can be found in several reviews [2, 10]. The properties of potentiometric detectors are mainly determined by the properties of the electrodes used; the most important electrodes, ISEs, are discussed in Chapter 6. Therefore, the selectivity, sensitivity, and dynamic characteristics are influenced little by the detector design with a given kind of electrode. On the other hand, the basic properties of voltammetric detectors can be varied by varying the experimental and design properties.

9.3.1 Potentiometric Detection

9.3.1.1 Ion-Selective Electrodes

Ion-selective electrodes, ISEs [11, 12], make it possible to construct detectors with very small internal volumes, but still their direct use in analytical flow-through systems is rather limited. The generally high selectivity of ISEs permits determinations of some components without prior separations, but excludes use of these electrodes in universal detectors. Liquid-membrane ISEs (e.g., those for Ca^{2+}, K^+, and NO_3^-) are less selective than solid-membrane electrodes and their response rate is slower. Therefore, ISE detectors are rarely used in liquid chromatography, but mainly in ion-exchange and gel chromatography [2].

Although slow electrode response is a principal drawback from the point of view of the detector design, this effect can be compensated to a considerable extent by suitable handling of the signal [13]. As already mentioned, potentials can be reproducibly measured in continuous flow-through analyzers even before establishment of a stationary state, and more than a hundred samples per hour can be handled in such systems.

Direct measurements with ISEs are mostly performed on natural samples, where the activity of a particular ion is usually required rather than the overall content of an element or a functional group, e.g., in flow-through systems for the determination of calcium [14] and potassium [15]. The difficulties stemming from the presence of high-molecular-weight substances that are adsorbed on the sensor surface and alter the liquid junction potential are largely removed in the vessel depicted in Fig. 9.1a. The electrolyte is brought to the electrode orifice, thus stabilizing the liquid-junction potential and preventing contact of the samples with the liquid junction [16]. Detectors with tangential flow of the test solution over a common ISE membrane have surprisingly small effective internal volumes (5 to 10 µl) and have yielded good results in the determination of nitrate in soil extracts and waste waters, of nitrate and potassium in fertilizers, etc. Two components can be determined simultaneously when using a detector with two electrodes arranged in a cascade (Fig. 9.1b) [17].

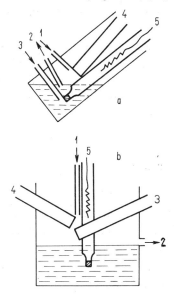

Fig. 9.1 Potentiometric detectors. (a) detector for work with natural samples [16]; (b) detector with two different ISEs [17]. 1, test solution; 2, waste; 3, auxiliary electrolyte; 4, ISE; 5, reference electrode.

In flow titrations, ISEs are among the most common sensors. For example, halide ISEs have yielded good results in titrations of halides by electrolytically generated silver ions [18], the determination of calcium by EDTA titration can be carried out using the flow-injection analysis (FIA) technique and a Ca^{2+}-ISE, etc. An interesting application is the use of a flow-through detector with an ISE that responds to both the titrand and the titrant [7].

In addition to common ISEs, special electrodes are used, e.g., with a small surface area [19], with a relatively rapid response [20], or constructed as flow-through devices [21].

The ISE selectivity can be modified considerably by placing additional membranes over the electrode membrane, forming gas probes and enzymatic potentiometric sensors.

Gas probes respond to variations in the composition of the gaseous phase on the basis of a reaction of the test gas with a thin layer of an electrolyte that is fixed on the surface of the ISE membrane by another membrane permeable to the test gas components. In the air-gap electrodes a porous material on the electrode surface is soaked with the electrolyte and is separated from the test medium by air; the sensing membrane is thus protected against direct contact with the test medium, which is a great advantage in work with natural samples.

Enzyme sensors have a thin layer of an immobilized enzyme over the ISE membrane. The test substrate diffuses through this layer and the reaction products are detected by the ISE. These sensors are practically limited to those employing enzymatic reactions accompanied by variations in the concentration of carbon dioxide and ammonia and suffer from many drawbacks, such as memory effects and very sluggish response. Therefore, combination of an ISE with a selective enzyme system based on the detection of the products of the enzymatic reaction taking place in the solution away from the sensor surface is mostly preferred in flow systems [22, 23]. The enzyme layer on the ISE surface can sometimes be replaced by a layer of live bacteria with a suitable assimilation system [24–26].

9.3.1.2   Redox Electrodes

Simple measurement of redox potentials of samples is rare in flowing systems, because of the poor selectivity and sensitivity. Redox potentials are more often measured in continuous titrations and in flow-through analyzers. Indirect measurements are common in FIA and are based on changes in the potential of a redox system present in the carrier solution [27].

When combined with enzyme systems, redox measurements attain selectivity [28]; there are also enzyme redox electrodes, analogs of enzyme sensors with ISEs [29].

9.3.1.3   Electrodes of the Second Kind

Electrodes of the second kind are more selective than inert redox electrodes. However, only a few ions can be detected, virtually only halide ions. Halide concentrations are obtained either from the change in the electrode potential on replacing a standard solution by the test solution or from the equilibrium voltage of a cell consisting of one electrode placed in the sample stream and another placed in a stream of reference solution. Porous silver electrodes are mostly used, as they have large surface areas and thus also decreased polarizability. In view of the slow potential stabilization (of the order of minutes), the use of electrodes of the second kind in flowing systems is not very significant.

9.3.2   Voltammetric Detection

The most important properties of voltammetric detectors that can be affected by the choice of the experimental and design parameters are the selectivity, sensitivity, linear dynamic range, response stability, and reproducibility and dynamic properties.

9.3.2.1   Selectivity

Voltammetric detectors are substance-specific and one of their greatest advantages is the possibility of varying the detection selectivity by controlling the working electrode potential.

The measurement of the current at a constant potential (amperometry) is still the most common technique. To improve the detection selectivity, various voltammetric techniques that yield current peaks are advantageous, e.g., differential pulse, ac, and square-wave voltammetry. It is often possible to select the dc component of the potentials in these techniques so that the detection has a maximum sensitivity for a given component, with minimal interference from the other components. In contrast to dc amperometry, where the detection selectivity always decreases with increasing potential, in ac techniques it is unimportant whether the component of interest undergoes an electrode reaction at a potential higher or lower than that of the interfering components. The use of ac techniques with a constant dc component suffers from certain drawbacks, namely, the necessity to repeat the measurement at various dc potentials to detect all the test substances and the dependence of the signal on the precision with which the dc potential is adjusted, which may be affected in practice by fluctuations in the reference electrode and the liquid-junction potentials.

Therefore, the optimal application of pulse and ac techniques involves periodical and continuous variation of the working electrode potential over a preset interval. This substantially improves the selectivity of, e.g., liquid chromatography, because continuous separation of the test components in time is accompanied by continuous separation in terms of the potential [30, 31].

9.3.2.2   Sensitivity

The sensitivity is usually defined as the slope of the dependence of the detector response on the concentration of the test substance. With voltammetric detectors this is a function of the working potential, flow rate, and electrode surface area. The importance of the working potential is obvious and need not be discussed here. As the faradaic current increases with increasing flow rate and electrode surface area, it seems reasonable to use high measuring sensitivities: however, the problem is more complicated than may appear at first glance.

The possibilities of increasing the flow rate in the whole system are limited in practice. In liquid chromatographs, the mobile phase flow rate is chiefly determined by the optimal conditions for the column separation, and in continuous analyzers an increase in the flow rate leads to increased consumption of the reagents and to higher pressure in the liquid pathways. It is thus better to attain a high velocity of the liquid only in the close vicinity of the electrode; these types of detectors are discussed below.

In considering the resultant effect of an increase in the working electrode surface area it must be borne in mind that an increase in the signal alone is practically insignificant and can be more readily obtained by electrical amplification. Practically attainable sensitivity is given by the signal-to-noise ratio (noise is any output quantity that carries no information on the input function, i.e., the concentration). As the amplitude of the noise in voltammetric detectors is usually proportional to the electrode surface area, whereas the amplitude of the faradaic signal is proportional only to a fractional power of this quantity, an increase in the electrode surface area mostly causes a decrease in the signal-to-noise ratio and thus a decrease in the attainable sensitivity. The advantages of using a small surface area electrode have been demonstrated in voltammetric and polarographic detectors. In the latter, an optimal signal-to-noise ratio is obtained with small mercury droplets from horizontal capillaries [32].

When using pulse and ac voltammetry techniques, the signal-to-noise ratio has only rarely been improved compared with dc voltammetry [33, 34]. This is apparently because of unsuitable electrode geometries that lead to large time constants of the detectors (see below).

9.3.2.3   Linear Dynamic Range

It is usually stated in the literature [1, 2] that the concentration linear dynamic range of voltammetric detectors is five orders of magnitude. This is a relatively large value compared with other detectors (uv, ca. 5×10^3; refractometric, ca. 10^3; fluorescence, ca. 10^3–10^6), but can only be attained with proper detector design. Simplified designs which ingore the basic condition for proper function of the potentiostatic circuit are rather common in practice. The reference electrode (and sometimes also the counter electrode) is often placed in the waste vessel connected with the working electrode by an electrolyte stream. The three-electrode circuit then loses its significance and only reference electrodes with high internal resistances can be used. The calibration curves are often nonlinear because of potential instability at higher concentrations, i.e., at higher currents.

9.3.2.4   Response Stability and Reproducibility

These parameters are generally the weakest point of voltammetric detectors, as they are directly connected with the sensor activity. The electrode is in direct contact with the test medium and is often passivated, with a gradual loss of activity. Many procedures have been developed for preventing loss of electrode activity, but none is quite reliable and universally applicable. Of course, detectors using a dropping mercury electrode are free of these drawbacks.

9.3.2.5   Dynamic Behavior

The dynamic behavior of detectors, reflected in their response rate, is a complex property of the whole detection system, not only of the detector. As mentioned above, in liquid chromatography and in continuous analyses the test substances form more or less separated zones that move together with the flowing liquid. Therefore, the detector output signals usually appear as peaks. The dynamic response of the detection system limits the frequency with which the zones can enter the detector. Peak broadening (as a result of slow response) thus limits the velocity with which the analytical system can generate independent concentration data. The overall peak broadening is given by the contributions from various parts of the system appearing in series [35].

The contribution of the electronics (signal handling and recording) can usually be determined by feeding an artificial signal to the input, and it can often be predicted from the known time constants.

In voltammetric detectors, the rate of the electrochemical reaction is usually sufficiently fast and the overall peak broadening is primarily given by zone dispersion in connecting tubes and inside the detector. It has been demonstrated that the sample zone dispersion during passage through a narrow tube increases with the fourth power of the tube radius and linearly with its length. Therefore, the connecting tubing must have a small internal diameter (0.5 or even 0.25 mm) and a minimal length. The zone broadening inside the detector depends not only on the detector internal geometric volume (which is often considered as the only decisive parameter), but also on the liquid velocity profile, the shapes of the detector interior and of the sensor, etc. The overall effect is difficult to predict, because the velocity profile and the effective internal volume are often unknown. However, in certain simple cases it is possible to make some estimates and compare them with real systems [36].

9.3.2.6   Detectors

There are a great variety of detectors for voltammetric flow measurements and their construction is largely determined by the working electrode material.

9.3.2.6.1   Mercury Electrodes

In spite of some drawbacks of the use of mercury as an electrode material, these electrodes are widely used, because the cathodic polarization range is wide and it is possible to utilize the enormous volume of experimental material collected since the invention of polarography and obtain an almost ideal electrochemical sensor — the dropping mercury electrode (DME).

The DME is a sensor with a continuously renewed surface, so that its contamination and passivation are considerably suppressed. On the other hand, the periodical dropping of the mercury gives rise to current oscillations that contribute to the noise and complicate the handling of polarograms and of the signal in general. Various kinds of damping are employed to suppress the oscillations. As discussed above, the use of any effective filter distorts the recorded curves and decreases the response rate, because this is equivalent to the introduction of a further time constant into the detection circuit. Therefore, the oscillations are best removed by sampling the current at the end of the drop time, with this time controlled by a mechanical hammer or by the frequency of auxiliary potential pulses applied to the electrode. The latter technique utilizes the potential dependence of the mercury surface tension [37]. On applying a potential pulse to the drop at the end of its natural lifetime, the surface tension sharply decreases and the drop is disconnected. Another method of current sampling uses a detector of peak values that is periodically zeroed and does not require synchronization of the instant of current sampling with the disconnection of the drop [38].

The first works with polarographic flow-through detectors were carried out in the 1950s in connection with a technique called chromato-polarography [39] and the design of automatic analyzers. During subsequent development, these instruments became progressively more different from detectors designed for quiescent solutions. The first polarographic detectors were vessels with a volume of a few milliliters in which electrode systems of normal dimensions could readily be placed. However, progressive miniaturization caused by increased demands on the detector quality led to problems with the placement and geometry of the electrodes. These problems were satisfactorily solved in one of the first detectors [40] in which a conical DME capillary and a reference electrode were placed in series, at right angles to the test liquid flow (Fig. 9.2a). The conical shape of the capillary permitted the design of a detector with an internal geometric volume of 2 to 6 µl [41], schematically depicted in Fig. 9.2b. The separation of the test solution into two branches at the working electrode permits electrolytic connection with the reference and auxiliary tubular electrodes. An important point in the detector design was the introduction of a DME with a very short drop time. A decrease in the drop time attained by placing the capillary horizontally, by increasing the mercury flow rate, or by modifying the detector internal space leads to a decrease in the current signal dependence on the liquid flow rate and effectively suppresses the amplitude of the current oscillations. The detector schematically depicted in Fig. 9.2c [42] has outstanding properties in this respect. A movable piston is placed opposite a horizontal capillary at a distance from the capillary orifice that can be continuously varied. The drop time is exceptionally short (about 50 ms) and can be controlled to a certain extent by the choice of this distance. The current is

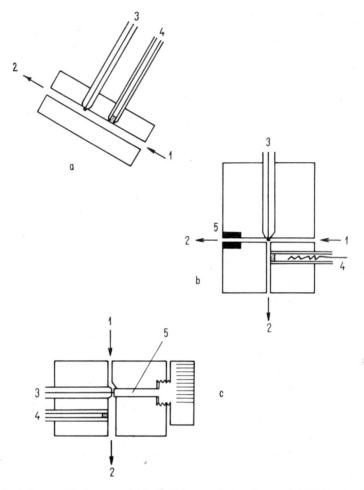

Fig. 9.2 Polarographic detectors. (a) Ref. 40. 1, test solution; 2, waste; 3, DME; 4, reference electrode; (b) Ref. 41. 5, auxiliary electrode; (c) Ref. 42. 5, moveble piston.

sampled at a relatively high frequency and thus the concentration profile is followed exactly and the oscillations are suppressed, even with a recorder with a common time constant. A detector of this type is also available commercially [43].

The trend towards the development of detectors with short drop times is also illustrated in Refs. 36 and 44. The DME capillary is horizontal and a short drop time is attained by using a conical capillary and sometimes by placing the reference electrode liquid junction opposite the capillary. In this way, a small internal volume of the detector is attained while the test liquid flow is disturbed minimally.

The PAR-310 polarographic detector from Princeton Applied Research, USA, has a very simple design. The three-electrode system is placed in a vessel containing a suitable base electrolyte and virtually does not differ from systems used in a quiescent solution. The test solution is fed by a tube to the surface of a DME and its components are detected without perceptible interference from the components contained in the base electrolyte, as a result of shielding of the working electrode by the flowing liquid; therefore, dissolved oxygen need not be removed from the base electrolyte in the vessel.

The polarographic detectors described so far employ a DME with a glass capillary and are actually flow-through versions of the classical vessel. A different design principle is employed in a detector based on feeding the test solution through a porous jet into a space filled with mercury [45]. The detector does not employ a glass capillary and exhibits rapid response.

Detectors with static mercury electrodes (SMEs) are similar to those with solid electrodes and are used only exceptionally, e.g., when it is necessary to work at high cathodic potentials, when the electrode material reacts with the test substances (e.g., in the detection of some sulfur-containing compounds), etc. New possibilities of application of SMEs in flow detectors appeared when mercury electrodes became commercially available, permitting the design of detectors in which the sensor surface can be regenerated periodically or at selected intervals.

9.3.2.6.2   Solid Electrodes

Porous Electrodes. In detectors with porous electrodes the test solution is percolated through the electrode [46]. As the electrode surface area is large, large currents are generally obtained, which are, however, subject to high background currents. For this reason and because of the impossibility of mechanically cleaning the electrode surface, these detectors are not often used in practice. A modern material for electrode preparation is reticulated vitreous carbon, RVC. A simple detector with an RVC electrode is, e.g., a column packed with this material whose bottom is immersed in a vessel in which the other electrodes are placed [47]. More complex detectors have more homogeneous potential distribution and permit both voltammetric and coulometric measurements (Fig. 9.3) [48].

Tubular Electrodes. These electrodes are tubes of an electrically conductive material whose inner walls are the electrode working surface. In spite of their simplicity and excellent hydrodynamic properties, these electrodes are not often used because of difficulties connected with cleaning and polishing of the electrode surface in detectors with a small internal volume. The detector manufactured by Chromatix (USA) is interesting in this connection. It contains tubular electrodes made of plastic-matrix carbon that are prepared for use; if necessary, the electrodes are simply replaced.

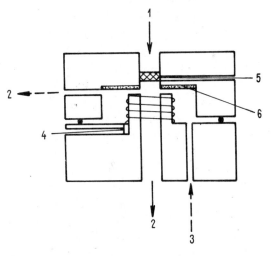

Fig. 9.3   Detector with a porous electrode [48]. 1, test solution; 2, waste; 3, reference electrode electrolyte; 4, lead to the reference electrode; 5, lead to the working electrode; 6, ion-exchanger membrane.

Figure 9.4 shows a detector with a flowing liquid junction between the electrodes that was used, e.g., with a modulated flow rate of the test liquid [49]. The ac signal obtained was handled by a band filter, thus improving the detection sensitivity as a result of suppression of nonconvective current components.

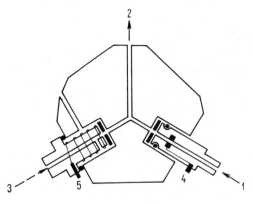

Fig. 9.4   Detector with a tubular electrode [49]. 1, test solution; 2, waste; 3, reference electrode electrolyte; 4, lead to the working electrode; 5, lead to the reference electrode.

A detector with a turbulent tubular electrode inside which a stirring device is placed [50] has the basic properties of detectors with rotating working electrodes that are mentioned below.

Wall-Jet and Thin-Layer Systems. In liquid chromatography, where a minimal internal volume of the detector is a principal requirement, these two hydrodynamic systems have exhibited good performance.

In the wall-jet system (Fig. 9.5) the test solution is directed at right angles to the electrode surface by a jet. The intense mass transport permits measurement with high sensitivity, while the effect of surfactants is decreased, because the liquid stream partially removes them mechanically from the electrode surface. The detector performance is very dependent on the dimensions of the jet and the electrode and on their relative positions [51]. The distance of the jet from the electrode determines the internal volume of the detector and affects the signal-to-noise ratio [41].

In thin-layer detectors, the liquid layer thickness is generally determined by a plastic spacer between two insulator blocks, with the working electrode placed in one of the blocks. Figure 9.6 depicts the working

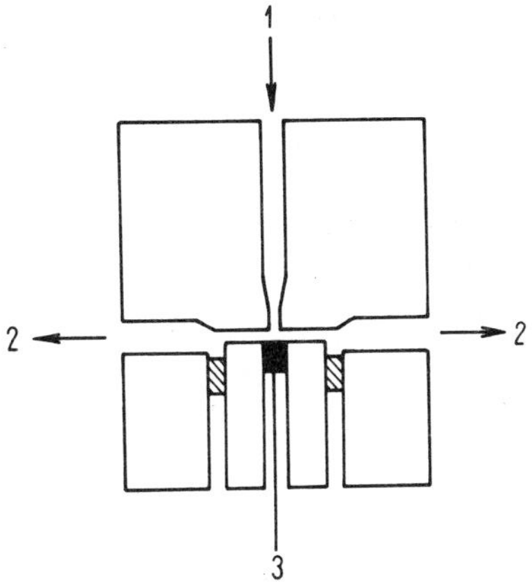

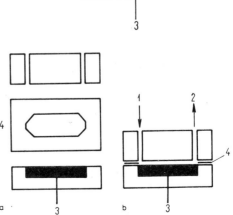

Fig. 9.5 Wall-jet detector [41]. 1, test solution; 2, waste and connection to the reference and auxiliary electrodes; 3, lead to the working electrode.

Fig. 9.6 Working space of a thin-layer detector. (a) taken apart; (b) assembled. 1, test solution; 2, waste and connection to the reference and auxiliary electrodes; 3, lead to the working electrode; 4, Teflon spacer.

space, with a reference and an auxiliary electrode placed outside this space. In work with low currents (nanoamperes) and with small requirements on the rate of stabilization of the applied potential, no difficulties have been reported as a result of this disadvantageous electrode arrangement. Because of their small internal volume, these detectors can contain one or several working electrodes. Detectors with two working electrodes in series can, e.g., be used for selective detection of the products generated on the preceding electrode. Detectors with two working electrodes in parallel maintained at different potentials enable separate detection of various components of the test mixture, etc. [52–54]. Modern detectors of the wall-jet and thin-layer types [55–57] are characterized by good electrode symmetry and also permit the use of ac and pulse techniques.

Rotating Electrodes. Introduction of moving mechanical parts complicates the design and maintenance of detectors, but may bring some advantages in certain applications. Detectors with a rotating (usually disk) electrode enable work with high sensitivity even at low flow rates, which is important for flow-through analyzers where a decrease in the flow rate saves reagents; moreover, the current signal is independent of the test liquid flow rate at common flow and rotation rates [58, 59].

In a detector with an RVC rotating disk electrode (Fig. 9.7), a high measuring sensitivity is attained by combining efficient mass transport with a large electrode surface area. The high background current is suppressed by using the "pulse rotation technique" in which the electrode rotation rate is periodically varied between two values and the difference in the corresponding current values is recorded [60]. Similar to the modulation of the liquid flow rate, this technique suppresses nonconvective components of the current that usually form a substantial part of the background signal.

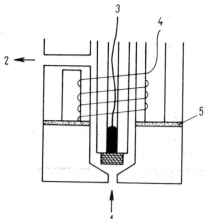

Fig. 9.7 Detector with a rotating porous disk electrode [60]. 1, test solution; 2, waste; 3, lead to the working electrode; 4, lead to the reference electrode; 5, ion-exchanger membrane.

9.4   Conclusions and Applications

Flow-through potentiometric and voltammetric detectors are substance-specific. Their main practical advantages are their high sensitivity and the ability to selectively detect various substances on the basis of their electrochemical behavior. It follows from the examples given above that potentiometric detectors (except for enzyme sensors) are applicable only to the detection of inorganic ions and are thus mainly used in continuous-flow analyzers and titrators. Voltammetric detectors are much more universal. When the parameters discussed above, including the base electrolyte composition and the working electrode material, are judiciously selected, many inorganic and organic substances can be detected.

Among inorganic substances, the determination of heavy metals (Cd, Cu, Pb, Fe, Ni, etc.) on mercury electrodes is especially important [61, 62]. However, the main use of voltammetric detectors is in liquid chromatography, where organic substances are primarily detected owing to the presence of electroactive functional groups or bonds. Reducible substances involve, e.g., quinones, olefins, nitro, nitroso, and azo compounds, unsaturated and aromatic ketones, aldehydes, oximes, imines, hydrazones, halides, organometals, some heterocyclic compounds, sulfur compounds, etc. [63]. Aromatic amines, phenols, aminophenols, enolates, organometallic, azo and hydrazo compounds, disulfides, etc., are usually rapidly oxidized anodically [64]. A separate group includes compounds detectable on the basis of their adsorption on the electrode surface, involving various surfactants [65].

The great volume of published results indicates that most practical applications of voltammetric detection in flowing systems are in clinical and biochemical analyses.

Among applications in environmental protection, the determination of pesticides [66–68], various nitro compounds [69–71], benzidine [72], and aromatic amines [73] in environmental samples can be mentioned. A selective detection of phenolic compounds, including chlorinated derivatives, has been described [74–76], as well as the detection of phenolic growth stimulants in biological materials [77]. The determination of chloroaniline and its methyl derivative in an industrial atmosphere [78] is a good example of utilization of the high sensitivity of voltammetric detection; a detection limit two orders of magnitude lower than that obtained with a photometric detector has been attained.

REFERENCES

1. Váradi M., Balla J., and Pungor E.: *J.Pure Appl.Chem.* **51**, 1175 (1979).
2. Štulík K. and Pacáková V.: *J.Electroanal.Chem.* **129**, 1 (1981).
3. Snyder L., Levine J., Stoy R., and Conetta A.: *Anal.Chem.* **48**, 942A (1976).
4. Růžička J. and Hansen E. H.: Flow Injection Analysis, J. Wiley, New York 1981.
5. Trojánek A.: *Chem.Listy* **75**, 1020 (1981).
6. Nagy G. and Fehér Zs.: *Anal.Chim.Acta* **91**, 87 (1977).
7. Fleet B. and Ho A. Y. W.: *Anal.Chem.* **46**, 9 (1974).
8. Růžička J., Hansen E. H., and Mosbaek H.: *Anal.Chim.Acta* **92**, 235 (1977).
9. Blaedel W. J. and Laessig R. H.: *Anal.Chem.* **37**, 332, 1255, 1650 (1965).
10. Rucki R. J.: *Talanta* **27**, 147 (1980).
11. Veselý J., Weiss D., and Štulík K.: Analysis with Ion-Selective Electrodes, E. Horwood, Chichester 1978.
12. Koryta J.: *Anal.Chim.Acta* **61**, 329 (1972); **91**, 1 (1977); **111**, 1 (1979).
13. Skogberg D., Richardson R., and Blasczyk T.: *Anal.Chem.* **51**, 2054 (1979).
14. Maier H., Röckel A., Heidland A., Schneider D., Steffen Ch., Aziz O., Dennhardt R., Lindt H. O., and Schindler J. G.: *Res.Exp.Med.* **172**, 75 (1978).
15. Schindler J. G., Riemann W., and Schäl W.: *Biomed.Technik* **21**, 135 (1976).
16. Hansen E. H., Růžička J., and Ghose A. K.: *Anal.Chim.Acta* **100**, 151 (1978).
17. Růžička J., Hansen E. H., and Zagatto E. A.: *Anal.Chim.Acta* **88**, 1 (1977).
18. Nagy G., Fehér Zs., Tóth K., and Pungor E.: *Anal.Chim.Acta* **91**, 97 (1977).
19. Fehér Zs., Nagy G., Tóth K., and Pungor E.: *Anal.Chim.Acta* **98**, 193 (1978).
20. Van der Linden W. E. and Oostervink R.: *Anal.Chim.Acta* **101**, 419 (1978).
21. Blaedel W. J. and Dinwiddie D. E.: *Anal.Chem.* **47**, 1070 (1975).
22. Růžička J., Hansen E. H., Ghose A. K., and Mottola H. A.: *Anal.Chem.* **51**, 199 (1979).
23. Mascini M. and Giardini R.: *Anal.Chim.Acta* **114**, 329 (1980).
24. Kobos R. K., Rice D. J., and Flournoy D. S.: *Anal.Chem.* **51**, 1122 (1979).
25. Hikuma M., Obana H., Yasuda T., Karube I., and Suzuki S.: *Anal.Chim.Acta* **116**, 61 (1980).
26. Matsumoto K., Seijo H., Watanabe W., Karube I., Satoh I., and Suzuki S.: *Anal.Chim. Acta* **105**, 429 (1979).
27. Karlberg B. and Thelander S.: *Analyst* **103**, 1154 (1978).
28. Gorton L. and Bhatti K. M.: *Anal.Chim.Acta* **105**, 43 (1979).
29. Shinbo T., Sugiura M., and Kamo N.: *Anal.Chem.* **51**, 100 (1979).
30. Samuelson R., O'Dea J., and Osteryoung J.: *Anal.Chem.* **52**, 2215 (1980).
31. Trojánek A. and De Jong H. G.: *Anal.Chim.Acta* **141**, 115 (1982).
32. Hanekamp H. B. and van Nieuwkerk H. J.: *Anal.Chim.Acta* **121**, 13, (1980).
33. Hanekamp H. B., Voogt W. H., Frei R. W., and Bos P.: *Anal.Chem.* **53**, 1362 (1981).
34. Swartzfager D. C.: *Anal.Chem.* **48**, 2189 (1976).
35. Poppe H.: *Anal.Chim.Acta* **114**, 59 (1980).
36. Hanekamp H. B., Bos P., Brinkman U. A. Th., and Frei R. W.: *Z.Anal.Chem.* **297**, 404 (1979).
37. Hanekamp H. B., Voogt W. H., and Bos P.: *Anal.Chim.Acta* **118**, 73 (1980).
38. Trojánek A. and Holub I.: *Anal.Chim.Acta* **110**, 161 (1979).
39. Kemula W.: *Rocz.Chem.* **26**, 281 (1952).
40. Blaedel W. J. and Strohl J. H.: *Anal.Chem.* **36**, 445 (1964).
41. Fleet B. and Little C. J.: *J.Chromatogr.Sci.* **12**, 747 (1974).
42. Michel L. and Zátka A.: *Anal.Chim.Acta* **105**, 109 (1979).

43. Solea-Tacussel, DELC Electrochemical Detection System for HPLC, 1981 (manufacturer's literature).

44. Kutner W., Debowski J., and Kemula W.: *J.Chromatogr.* **191,** 47 (1980).

45. Trojánek A. and Křesťan L.: *J.Liq.Chromatogr.* **6,** 1759 (1983).

46. Ostrovidov E. A.: *Zh.Anal.Khim.* **37,** 1703 (1982).

47. Strohl A. N. and Curran D. J.: *Anal.Chem.* **51,** 353 (1979).

48. Strohl A. N. and Curran D. J.: *Anal.Chem.* **51,** 1045 (1979).

49. Blaedel W. J. and Yim Z.: *Anal.Chem.* **52,** 564 (1980).

50. Blaedel W. J. and Schieffer G. W.: *Anal.Chem.* **46,** 1564 (1974):

51. Yamada J. and Matsuda H.: *J.Electroanal.Chem.* **44,** 189 (1973).

52. MacCrehan W. A. and Durst R. A.: *Anal.Chem.* **53,** 1700 (1981).

53. Roston D. A. and Kissinger P. T.: *Anal.Chem.* **54,** 429 (1982).

54. Roston D. A., Shoup R. E., and Kissinger P. T.: *Anal.Chem.* **54,** 1417A (1982).

55. Schieffer G. W.: *Anal.Chem.* **52,** 1994 (1980).

56. Štulík K. and Pacáková V.: *J.Chromatogr.* **208,** 269 (1981).

57. Štulík K., Pacáková V., and Stárková B.: *J.Chromatogr.* **213,** 41 (1981).

58. Brunt K., Bruins C. H. P., Doornbos D. A., and Oosterhuis B.: *Anal.Chim.Acta* **114,** 257 (1980).

59. Oosterhuis B., Brunt K., Westerink B. H. C., and Doornbos D. A.: *Anal.Chem.* **52,** 203 (1980).

60. Blaedel W. J. and Wang J.: *Anal.Chim.Acta* **116,** 315 (1980).

61. Joynes P. L. and Maggs R. A.: *J.Chromatogr.Sci.* **8,** 427 (1970).

62. Slezko N. I., Chashchina O. V., and Synkova A. G.: *Zavod.Lab.* **41,** 13 (1975).

63. Zuman P. and Perrin C. L.: Organic Polarography, J. Wiley, New York 1969.

64. Weinberg N. L. and Weinberg H. R.: *Chem.Rev.* **68,** 449 (1968).

65. Lankelma J. L. and Poppe H.: *J.Chromatogr.Sci.* **14,** 310 (1976).

66. Koen J. G., Huber J. F. K., Poppe H., and den Boef G.: *J.Chromatogr.Sci.* **8,** 192 (1970).

67. Anderson J. L. and Chesney D. L.: *Anal.Chem.* **52,** 2156 (1980).

68. Mayer W. J. and Greenberg M. S.: *J.Chromatogr.* **208,** 295 (1981).

69. Kemula W. and Sybilska D.: *Anal.Chim.Acta* **38,** 97 (1967).

70. Wasa T. and Musha S.: *Bull.Chem.Soc.Japan* **48,** 2176 (1975).

71. Bratin K., Kissinger P. T., Briner R. C., and Bruntlett C. S.: *Anal.Chim.Acta* **130,** 295 (1981).

72. Armentrout D. N. and Cutie S. S.: *J.Chromatogr.Sci.* **18,** 370 (1980).

73. Rice J. R. and Kissinger P. T.: *Environ.Sci.Technol.* **16,** 263 (1982).

74. Armentrout D. N., McLean J. D., and Long N. W.: *Anal.Chem.* **51,** 1039 (1979).

75. Weisshaar D. E. and Tallman D. E.: *Anal.Chem.* **53,** 1809 (1981).

76. Shoup R. E. and Mayer G. S.: *Anal.Chem.* **54,** 1164 (1982).

CHAPTER 10

MEASUREMENT OF OXYGEN IN BIOLOGICAL SYSTEMS

Lubomír Šerák

10.1   Introduction

Oxygen is one of the most important components of the atmosphere because it is a necessary component of the earth's biosphere in its present form, whose aerobic part includes an immense wealth of organisms, from single-cell formations to man. None of these organisms can live without oxygen and thus they also need green plants, which play a very important and unique role in the development of life on earth, i.e., photosynthetic accumulation of solar energy. This complex process that can be considered as a rare example of a perfectly wasteless technology leads to the production of oxygen, in addition to the synthesis of complex organic compounds from carbon dioxide and water during production of the biomass. Animals that are exclusive consumers of oxygen are simultaneously producers of carbon dioxide and thus are also useful to plants, which they supply with a nonnegligible part of their basic building material. Plants also consume oxygen by respiration, often even during photosynthesis and always in the dark or when photosynthetically active radiation is deficient, and thus also change from a producer into a partial or full consumer of oxygen, with simultaneous production of carbon dioxide.

At present, human activity is, among other things, connected with the production of an enormous and ever growing amount of carbon dioxide and the consumption of an even greater amount of oxygen. If debatable side effects of this situation on the environment are disregarded here, it is certainly surprising that the composition of the atmosphere is still almost constant and more perceptible changes in the contents of oxygen and carbon dioxide appear only in localities with a high concentration of industrial and urban activity.

One of the tools for monitoring and quantifying the responses of various components of the environment, i.e., various organisms and their parts, such as organs and organelles, to external influences is the measurement of oxygen in biological systems.

10.2   Methods and Sensors

The information content of the results of measurements carried out on a sensitive system depends largely on the magnitude of the perturbation of the system during the measurement. For the measurement and monitoring of oxygen in biological systems this is one of the foremost criteria in the selection of a measuring method.

Polarography and voltammetric methods derived from polarography belong among important nondisturbing methods for the determination of oxygen and have an advantage in their perfectly utilizing the general properties of oxygen, such as primarily the solubility of oxygen in water and plant and animal body fluids, as well as in solutions that are compatible with these fluids. The oxygen concentrations in these solutions are optimum for polarography. The determination is simple, sufficiently precise and accurate, provided that the principal methodological requirements are met. It is also advantageous that polarographic and voltammetric instrumentation for the determination of oxygen is cheap and single-purpose instruments can even be assembled in the laboratory (see Chapter 13).

The polarographic determination of oxygen is based on its electrochemical reduction that proceeds in two steps at a dropping mercury electrode. The first step is the almost reversible reduction to hydrogen peroxide with the exchange of two electrons in a steep and readily measurable polarographic wave, followed immediately by the irreversible, two-electron reduction of the hydrogen peroxide to water in a wave with the same height as the previous wave. Therefore, four electrons are consumed per molecule of oxygen during the reduction and the determination is thus sufficiently sensitive. Oxygen may be reduced directly to water in a single step at other metal or carbon electrodes, depending on the experimental conditions, again with exchange of four electrons. However, hydrogen peroxide is often formed as a side product of the electrode reaction, although it should appear only as an intermediate that is constantly removed by the subsequent electrode reactions, and may interfere with the function of some electrochemical sensors, especially in prolonged measurements.

10.2.1   The Dropping Mercury Electrode

The dropping mercury electrode (DME) [1] that yields a current signal proportional to the concentration of the active substance in solution is still the most perfect electrochemical sensor and permits determinations of reducible or oxidizable substances. It is characterized by the following advantages:

(i) As the electrode surface is constantly renewed, the electrode properties are not affected by the electrolysis products and thus the current–potential curves are perfectly reproducible and independent of time. The electrode construction is simple and thus the function is reliable, which cannot be simply achieved with other electrodes.

(ii) The evolution of hydrogen on the perfectly smooth mercury surface occurs at substantially more negative potentials than on other metals even in acidic solutions, i.e., its overvoltage is high. A potential of about -1.2 V (SCE) can be attained in acidic solutions without perceptible evolution of hydrogen.

(iii) The amount of substance electrochemically converted on the small surface area of the DME is so small that the concentration in the solution is virtually unchanged, provided that an extremely small solution volume is not used (less than 0.1 ml). Usually it is possible to repeat the measurement many times in a few milliliters of solution with the same results.

The electrode suffers from certain drawbacks so that it is sometimes replaced by less perfect electrodes whose surface is not renewed. The system of the DME must be completely at rest, because movements and vibrations cause disturbances on the curves. The electrode is rather large and this disadvantage is not removed even in modern designs that are technologically much more complicated. Mercury is toxic and may also poison sensitive biological materials. However, the latter drawback can sometimes be avoided by a judicious design of the apparatus.

The DME has undergone great development. The present version, a much perfected form of the glass capillary that is the most important part of the electrode, is actually based on the classical procedure for the DME preparation that J. Heyrovský adopted from his teacher, B. Kučera. Fine capillaries were then obtained after heating thick-walled capillaries in a glassblower's burner [2]. The capillaries obtained were narrowed to an external diameter of less than 1 mm for a length of a few centimeters, the internal diameter decreased slightly to the capillary tip, to 0.01 to 0.03 mm. This procedure required a certain skill and experience and excluded the preparation of two capillaries with exactly the same properties.

Therefore, blunt capillaries that were commercially available for thermometers were later introduced by Maas, Siebert, and Langer [3, 4]. These capillaries were cylindrical with an external diameter of 1.5 to 6.0 mm and an internal diameter of 0.05 to 0.1 mm. An advantage of these capillaries was their easy availability and replaceability and almost identical properties.

Smoler studied the properties of the DME in detail, proposed horizontal electrodes [5], dealt with instantaneous polarographic currents during drop growth, and examined the reliability of the electrode as a function of the capillary shape [6, 7]. He critically evaluated the findings of other authors [8-10] and found that a conical capillary shape is a necessary condition for reliable electrode function as in this way creeping of the solution into the capillary along the mercury column is suppressed. It is noteworthy that the original hand-made capillaries of J. Heyrovský [2] satisfied this requirement.

A perfectly functioning DME was prepared by Novotný [11], who extended Smoler's work. The capillary channel is expanded at a certain spot and the cavity formed is drawn out to obtain a spindlelike shape. The geometry of this cavity is decisive for the resultant properties of the capillary. By cutting off the drawn-out end at a site with a particular internal diameter, very variable drop times can be obtained that are determined by the capillary internal diameters above the widened part and at the cut. This capillary can also be used for the construction of a simple and reliable hanging drop electrode [12].

The Novotný capillary improved the DME reproducibility by several orders of magnitude compared with the blunt capillaries introduced in the late 1930s [3, 4], and it is now evident that commercial blunt capillaries eliminated the difficult manual preparation, but functionally represented a step backwards. It can be expected that modern spindle-shaped capillaries will contribute to the present renaissance of polarography, resulting from progress in semiconductor analog and digital measuring techniques and the pioneering work of Barker [13], on which modern pulse polarographic methods are based.

10.2.1.1   Determination of Oxygen Using the Dropping Mercury Electrode

The course of the electrochemical polarographic reduction of oxygen, i.e., the current–potential curve referred to a silver chloride reference electrode, is shown in curve 2 in Fig. 10.1. Curve 1 is the charging current in the same solution from which oxygen was expelled by the passage of a polarographically inert gas, e.g., nitrogen. This polarogram can be used for calibration, provided that it is recorded in a solution saturated with the air at known, constant temperature and pressure and that the oxygen concentration in the solution is found in a table of oxygen solubilities (Table 10.1). All the other measurements must, of course, be carried out under the same conditions, i.e., with the same electrolyte composition, temperature, pressure, mercury reservoir height, and recorder sensitivity and using the same DME.

Table 10.1
Dependence of the dissolved oxygen concentration (mg liter^{-1}) in distilled water on the temperature

Temp. (°C)	0.0	0.1	0.2	0.3	0.4	0.5	0.6	0.7	0.8	0.9
0	14.65	14.61	14.57	14.53	14.49	14.45	14.41	14.37	14.33	14.29
1	14.25	14.21	14.17	14.13	14.09	14.05	14.02	13.98	13.94	13.90
2	13.86	13.82	13.79	13.75	13.71	13.68	13.64	13.60	13.56	13.53
3	13.49	13.46	13.42	13.38	13.35	13.31	13.28	13.24	13.20	13.17
4	13.13	13.10	13.06	13.03	13.00	12.96	12.93	12.89	12.86	12.82
5	12.79	12.76	12.72	12.69	12.66	12.62	12.59	12.56	12.53	12.49
6	12.46	12.43	12.40	12.36	12.33	12.30	12.27	12.24	12.21	12.18
7	12.14	12.11	12.08	12.05	12.02	11.99	11.96	11.93	11.90	11.87
8	11.84	11.81	11.78	11.75	11.72	11.70	11.67	11.64	11.61	11.58
9	11.55	11.52	11.49	11.47	11.44	11.41	11.38	11.35	11.33	11.30
10	11.27	11.24	11.22	11.19	11.16	11.14	11.11	11.08	11.06	11.03
11	11.00	10.98	10.95	10.93	10.90	10.87	10.85	10.82	10.80	10.77
12	10.75	10.72	10.70	10.67	10.65	10.62	10.60	10.57	10.55	10.52
13	10.50	10.48	10.45	10.43	10.40	10.38	10.36	10.33	10.31	10.28
14	10.26	10.24	10.22	10.19	10.17	10.15	10.12	10.10	10.08	10.06
15	10.03	10.01	9.99	9.97	9.95	9.92	9.90	9.88	9.86	9.84
16	9.82	9.79	9.77	9.75	9.73	9.71	9.69	9.67	9.65	9.63
17	9.61	9.58	9.56	9.54	9.52	9.50	9.48	9.46	9.44	9.42
18	9.40	9.38	9.36	9.34	9.32	9.30	9.29	9.27	9.25	9.23
19	9.21	9.19	9.17	9.15	9.13	9.12	9.10	9.08	9.06	9.04
20	9.02	9.00	8.98	8.97	8.95	8.93	8.91	8.90	8.88	8.86
21	8.84	8.82	8.81	8.79	8.77	8.75	8.74	8.72	8.70	8.68
22	8.67	8.65	8.63	8.62	8.60	8.58	8.56	8.55	8.53	8.52
23	8.50	8.48	8.46	8.45	8.43	8.42	8.40	8.38	8.37	8.35
24	8.33	8.32	8.30	8.29	8.27	8.25	8.24	8.22	8.21	8.19
25	8.18	8.16	8.14	8.13	8.11	8.10	8.08	8.07	8.05	8.04
26	8.02	8.01	7.99	7.98	7.96	7.95	7.93	7.92	7.90	7.89
27	7.87	7.86	7.84	7.83	7.81	7.80	7.78	7.77	7.75	7.74
28	7.72	7.71	7.69	7.68	7.66	7.65	7.64	7.62	7.61	7.59
29	7.58	7.56	7.55	7.54	7.52	7.51	7.49	7.48	7.47	7.45
30	7.44	7.42	7.41	7.40	7.38	7.37	7.35	7.34	7.32	7.31

Practically, the current is usually measured amperometrically, i.e., at a constant voltage applied to the electrodes. With a silver chloride reference electrode, a voltage of 1.5 V is suitable and corresponds to the region of the limiting current of the second oxygen reduction wave, as can be seen in Fig. 10.1. Correction for the capacity current must be carried out by subtracting it from the overall current, thus obtaining a value proportional to the oxygen concentration in the solution according to the Ilkovič equation [14].

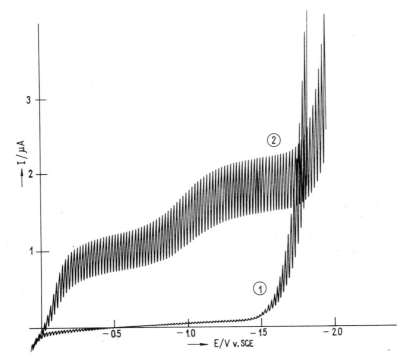

Fig. 10.1 Calibration polarogram. 1, dependence of the charging current on the voltage; 2, polarographic curve of oxygen in the physiological solution saturated with air.

The first paper on the polarographic determination of oxygen was published by Vítek [15] in 1935, followed in 1938 by a paper on biological applications by Petering and Daniels [16]. The latter authors measured the current alternately at voltages of 0.1 and 1.0 V and constructed a calibration curve from the differences in the two currents obtained for various oxygen concentrations. In this way, oxygen was determined during photosynthesis and respiration in algae and in the respiration of blood cells, yeast, and other systems. A disadvantage was direct contact of the mercury with the biological material, which was placed directly in the polarographic vessel with a DME and a mercury pool reference electrode.

A pioneering work on the polarographic determination of oxygen for biological purposes was carried out by Baumberger [17, 18], who measured oxygen concentrations in blood. The DME was also in contact with the test material, but the reference calomel electrode was separated from the sample by an agar bridge.

In many other papers (see, e.g., Refs. 19, 20) the studied material was in contact with the mercury, as the measurement was performed in a constant volume of a solution containing a cell suspension or a tissue section. In addition to the possible toxic effect of the mercury, the test material was also leached, the sample structure was damaged by the stirring of the solution, and the oxygen concentration in the solution gradually decreased, even beyond the limit required for sustaining of aerobic processes. Therefore, conditions were similar to those in the Warburg manometric method and were additionally made worse by contamination with mercury.

In an effort to make the experimental conditions more similar to physiological conditions, a flow-through continuous polarographic method for monitoring the oxygen consumption by tissue sections has been developed [21]. The tissue studied is separated by a cellophane membrane from a slowly flowing physiological solution containing a known concentration of dissolved oxygen and the DME is placed in the solution leaving the analyzer. The cellophane membrane behaves basically as a fixed boundary between the tissue and the oxygen source, i.e., the flowing solution. It is readily permeable for gases, ions, and low molecular mass substances, but is impermeable for compounds with a molecular mass above 5 000, so that leaching of the tissue is prevented. A thin tissue section (at most 0.2 mm thick) is enclosed in a cylindrical contact vessel at a complete rest and adheres on at least one side to the cellophane membrane. A physiological solution saturated with air flows in a thin layer under the membrane and may also contain other substances whose effect on the tissue respiration is of interest. Membranes of other materials can be used, e.g., of poly-

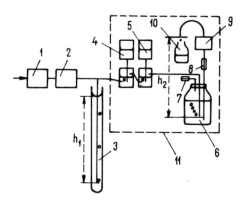

Fig. 10.2 Scheme of the arrangement of the apparatus for the measurement of the oxygen consumption by tissue sections. 1, air pump; 2, air filter; 3, manostat; 4, dampener; 5, air-cleaner; 6, reservoir with the aerated solution; 7, capillary valve; 8, capillary regulator of the solution flow rate through the analyzer; 9, analyzer of the oxygen consumption; 10, drop flow meter and the collector of the solution that passed through the analyzer. h_1, the height of the liquid column in the manostat; h_2, the height of the liquid column during the passage through the analyzer; the flow rate through the analyzer is proportional to the difference, h_1-h_2; 11, thermostated oven.

Fig. 10.3 Scheme of the analyzer of the oxygen consumption by tissue sections. 1, horizontal dropping mercury electrode; 2, insert; 3, thread; 4, seal; 5, siphon mercury closing; 6, solution inlet; 7, solution outlet and an Ag/AgCl reference electrode; 8, two-way stop-cock leading the solution either underneath cellophane membrane 9 or directly to dropping electrode 1; 10, O-ring for fixing the cellophane membrane; 11, thread for the O-ring; 12, contact chamber for placement of the test tissue section; 13, lid of the contact chamber; 14, a piece connecting the analyzer with the piece 15; 16, rod for fixing the analyzer in the holder.

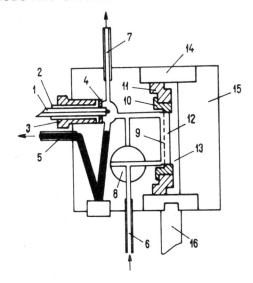

Fig. 10.4   Analyzer of the oxygen consumption by tissue sections.

ethylene, PTFE, or silicone rubber, which is important in specialized experiments in which only gases are to be transported between the tissue and the solution. The experimental arrangement, the analyzer scheme, and the actual analyzer are shown in Figs. 10.2, 10.3, and 10.4.

The time dependence of the oxygen concentration in the solution leaving the analyzer obtained by continuous or periodical measurement yields not only the respiration activity of the test material, but also other important information, namely, the pathway of establishment of the equilibrium oxygen concentration in the final solution, yielding important diagnostic data characteristic of the rate of the metabolic processes (Fig. 10.5).

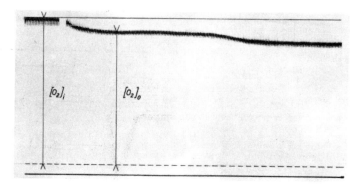

Fig. 10.5 Time dependence of the oxygen concentration in the solution leaving the analyzer ($[O_2]_o$) during periodically interrupted recording.

The respiration activity of a biological material can be expressed in terms of the amount of oxygen consumed by a unit amount of the test material per unit time; the dimension can be, e.g., mg O_2 cm^{-2} h^{-1}. To calculate this value at any instant during the experiment, the inlet and outlet oxygen concentrations, $[O_2]_i$ and $[O_2]_o$, must be known, as well as the solution flow rate, V, and the quantity K to which the respiration activity, RA, is related by the calculation, and the simple formula

$$RA = [O_2]_i - [O_2]_o \cdot \frac{V}{K} \qquad (10.1)$$

is used. Quantity K can be, e.g., the dry matter in the sample, the number of cells or other particles, the amount of some active substance in the sample, or the surface area of the tissue section that is in contact with the separating membrane.

This method has, e.g., been used for monitoring the respiration activity of healthy and pathologically changed rabbit aorta [22, 23] and some human and animal tissues in relation to inflammation and to external factors [24–28]. It can also be employed to monitor the respiration of cell cultures and suspensions of subcellular particles, after modification of the contact vessel of the analyzer to permit the introduction of a particle suspension and magnetic stirring [29].

On removal of the contact vessel and placement of the separation membrane on the analyzer surface, the analyzer can be used to monitor *in situ* the uptake and consumption of oxygen by human skin and the external effects on it [30–33]. The experience gained over several years of research in this field has provided a somewhat deeper insight into the function of the human skin under normal and pathologically altered conditions and has permitted a better understanding of the mechanism of the effect of some externally used pharmaceuticals and of some irritants to which human skin is often exposed. A significant finding is the existence of a barrier function of skin that prevents penetration of oxygen from the body through the skin, even during the respiration of pure oxygen [30]. However, Baumberger and Goodfriend [34] have shown that on immersion of a finger into an electrolyte heated to 45°C, the oxygen pressure in the solution becomes equal to the pressure in the arterial blood within 15 to 60 min and based a determination method on this fact. Local heating of skin was used [35, 36] for the construction of a transcutaneous sensor for the determination of oxygen pressure in arterial blood, employing the Clark sensor described below.

10.2.2   Sensors Other than the Dropping Mercury Electrode

Right from the beginning of biological applications of polarography, a sensor has been sought that would have the advantages of the DME, without its drawbacks. As no insoluble products are formed in the electrochemical reduction of oxygen that would alter the properties of the electrode, it seemed that any sufficiently electrochemically inert material would be suitable for the determination of oxygen. The results obtained in measurements in simple, electrochemically "ideal" solutions were also promising. However, the reproducibility of the measured values deteriorated on contact of the indicator electrode (e.g., platinum) with a biological material.

The first success in the search for a practically useful indicator electrode for the measurement of oxygen pressure in tissues was attained by Davies and Brink [37], who proposed a recessed platinum microelectrode. A thin platinum wire was sealed in a thin-walled capillary with the glass extending over the sealed metal, thus obtaining an approximately cylindrical recess that was filled with a physiological solution with an isotonic sodium chloride concentration with respect to the test tissue. The electrical circuit of this indicator and a suitable reference electrode was connected for 20 s to a source of a 0.5-V polarization voltage and the deflection of a mirror galvanometer in the circuit was measured at the end of this interval. Then the polarization circuit was opened and 10 min were allowed for equilibration of the oxygen concentration in the recess solution with that in the surroundings before the measurement was repeated. These times were used for a 0.6-mm-deep recess. With deeper recesses it was necessary to wait longer; with shallower recesses it was necessary to decrease the polarization time, which made the reading of the galvanometer deflection more difficult. Assuming that the length of the polarization pulse was selected so that the oxygen concentration gradient did not extend beyond the recess, the authors considered their method as an absolute one.

Development logically continued with efforts to simplify the measurement, such as protecting the indicator electrode by layers permeable for water and electrolytes, not only for oxygen, e.g., by dipping a platinum electrode in an ether solution of collodion prior to the measurement. A thin layer of cellulose nitrate remained on the electrode after evaporation of the solvent and had to be removed and formed again after each experiment [38].

10.2.2.1   The Clark Sensor

The result of these efforts and a basic solution to the problem is the idea proposed by Clark [39] of separating the electrode system with the electrolyte from the test medium by a membrane that is only permeable for gases and not for water and ions (Fig. 10.6). At present, polyethylene, polypropylene, PTFE, and silicone rubber membranes are mostly used. The Clark sensor, often incorrectly termed the Clark electrode, is a qualitative step forward in the solution of the problem of determining oxygen in biological materials that involve such complicated media as blood in measurements *in vivo*. The specific properties of blood inspired Clark. One of the main problems in the *in situ* determination of oxygen in blood is blood inhomogeneity, as blood is not a solution but — very imprecisely stated — a suspension of various particles including erythrocytes in which oxygen is bound to haemoglobin. A single erythrocyte deposited accidentally on an uncovered indicator microelectrode renders the measurement worthless, because the electrode ceases to yield a signal corresponding to the average analytical concentration of oxygen in the whole blood, but responds in an undefined way to the dissociation of oxyhaemoglobin contained in

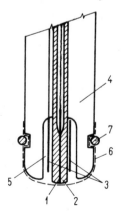

Fig. 10.6 Scheme of the Clark sensor. 1, indicator electrode; 2, insulation body of the indicator electrode; 3, reference electrode; 4, sensor body; 5, electrolyte space; 6, covering membrane; 7, O-ring for fixing the membrane.

the erythrocyte adhering to the electrode. This is accompanied by strong adsorption of high-molecular-weight substances, leading to rapid and irreproducible changes of the electrode activity toward oxygen.

These undesirable effects cannot occur with the Clark sensor, because the separating membrane offsets the effect of electrostatic attractive forces exerted by the electrode double layer on the surrounding charged particles and because the membrane repulses polar species, including water, because of its hydrophobicity.

The above great advantages of the Clark sensor are partially offset by the strong dependence of the current signal on the temperature, caused by the high temperature coefficient of the membrane permeability for oxygen; the latter increases exponentially with increasing temperature. This effect considerably exceeds the effect of temperature on the oxygen diffusion coefficient in water and the resultant average temperature coefficient of the Clark sensor is 6% to 8% per °C, depending on the kind of membrane and the sensor construction.

Since Clark's first publication a quarter of a century ago, the sensor has been modified by a great many authors for many purposes. Kreuzer, with his pupils and co-workers, who have contributed substantially to the development in this field, should at least be mentioned [40].

In comparison to uncovered electrodes, the diffusion transport of oxygen toward the indicator electrode surface in Clark sensors is more complicated, as the oxygen concentration gradient involves both the membrane and the thin layer of electrolyte between the surface of the indicator electrode and the internal surface of the membrane during constant polarization of the sensor (the oxygen concentration at the electrode surface is zero during the measurement).

When sensors with large indicator electrodes are used, the oxygen concentration gradient even extends beyond the membrane into the test solution. These sensors are sometimes favored because they yield large signals, but the results are only accurate if the concentration gradient outside the sensor is eliminated by making measurements in flowing or stirred solutions or by moving the sensor itself. The difference in the signals obtained in quiescent and moving media is a simple practical criterion for the usefulness of the sensor in media where the oxygen consumption by the electrode process should be as small as possible. Only sensors with small signals (e.g., of the order of 10^{-9} A in water saturated with air), which are the same in quiescent and moving media, are suitable for these purposes.

The adverse properties of the membrane, especially its effect on the temperature coefficient of the sensor signal and the effect of the external hydrodynamic conditions, can be efficiently eliminated, as shown recently on the basis of consideration of conditions in the sensor immediately after the beginning of the polarization. The concentration gradient due to the consumption of oxygen in the electrode process begins to extend from the electrode surface toward the membrane. After rapid decay of the charging current, the electrolytic current passing through the indicator electrode changes with time according to the equation

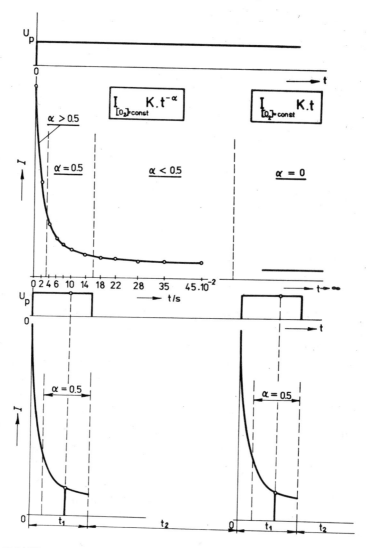

Fig. 10.7 The polarization pulses and the current signal during the polarization in the measurement described in the text.

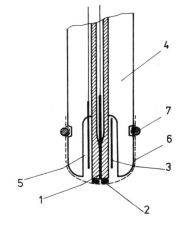

Fig. 10.8 Schematic representation of a modified Clark-type sensor for pulse measurements. 1, indicator electrode; 2, spacer providing defined diffusion space between the indicator electrode and the membrane; 3, reference electrode; 4, sensor body; 5, electrolyte space; 6, covering membrane; 7, O-ring fixing the membrane.

$$I = KCD^{1/2}t^{-1/2} \tag{10.2}$$

where I is the current (A), K is a proportionality constant, D is the oxygen diffusion coefficient ($m^2\ s^{-1}$), C is the dissolved oxygen concentration (mol liter^{-1}), and t is time (s).

Therefore, the instantaneous current is inversely proportional to the square root of time, as follows from the laws governing planar diffusion. This equation holds until the concentration gradient reaches the internal surface of the membrane. The Davies and Brink recessed electrode [37] described above obeys the same dependence, provided that the polarization interval is sufficiently short.

Detailed study has shown that a Clark sensor can be adapted for measurement with polarization by short voltage pulses (Figs. 10.7, 10.8). The conditions can be selected so that the concentration conditions inside the sensor are restored so quickly that the polarizing pulse can be repeated after only 10 s [41, 42, 48]. The sensor modification consists in the formation of a defined space between the indicator electrode and the covering membrane, e.g., using a perforated insert, as shown schematically in Fig. 10.8. This new method practically removes the drawbacks of the Clark sensor while retaining its advantages. The price paid for this improvement is not large, as the only new disadvantage involved in this method is a decrease in the response rate of the sensor by about a factor of 10, compared with the value for stationary measurements at a constant polarization of the electrodes. A programmable and reproducible pulse switch is required for this method.

The consumption of oxygen by biological materials can be measured using a contact sensor with a gauze electrode, covered by a membrane, on which oxygen consumed by the test material is generated. The indicator electrode of the oxygen sensor is placed immediately below the generating electrode and its signal controls the oxygen generation on the gauze electrode; the generating electrode current is a measure of the oxygen consumption [43].

A device for determining the BOD (biochemical oxygen demand) of waste waters is marketed by Leeds and Northrup. The sensor contains a system of alternate multiple cathodes and anodes covered by a membrane and these electrodes are separated by thin layers of an electrolyte. The oxygen reduced at the cathodes during the measurement is replenished by generation on the anodes and thus no oxygen is consumed from the sample. Therefore, no stirring is required and the manufacturer states that the signal is also independent of fouling of the membrane. The overall surface area of all the electrodes is large and thus the signal need not be amplified [44].

An important field of application is offered by bioelectrochemical sensors combining an oxygen sensor with auxiliary membranes on which various substances or microorganisms can be fixed. Suzuki and Karube [45] reviewed some bioelectrochemical sensors, e.g., a sensor for the determination of sucrose employing a mixture of invertase, mutarotase, and glucose oxidase enzymes immobilized in a collagen membrane. The freshness of meat can be checked using a sensor with immobilized mono-amino oxidase. Free cholesterol can be determined in serum using cholesterol oxidase.

For a rapid BOD determination, a sensor has been developed with immobilized microorganisms isolated from activated sludges. The decrease of the sensor signal to an equilibrium value that occurs within only 15 min after the beginning of the measurement is proportional to the results of the conventional five-day test on standard samples containing up to 350 ppm of glucose and glutamic acid. Good agreement was also obtained for distillery waste waters. Sensors for determinations of acetic acid and ethanol were constructed using the yeast *Trichosporon brassicae*.

Using the method developed by Karube and co-workers, the microbiological test for chemical mutagens and carcinogens can be accelerated and its sensitivity substantially increased. *Salmonella typhimuri* is used, which requires histidine in the nutrient medium for its growth. However, mutagens produce a form that can procreate even without histidine, and the conventional cultivation test is based on this fact; this lasts 2 days and down to 10 μg of a mutagen can be determined in 1 ml of solution. In the new method, the microorganism is placed in a suspension on a membrane filter that is fixed to an oxygen sensor membrane and its growth can be detected in a phosphate buffer saturated with air. After 10 h, a decrease in the sensor signal due to oxygen consumption by the growing microorganisms can be detected in the presence of as little as 0.001 µg mutagen in 1 ml solution [46].

10.3   Conclusion

In view of the necessity of cautious *in situ* measurements in tissue research and clinical practice, noninvasive contact methods are very important, as they do not produce undesirable traumatic reactions at the site investigated and do not affect the attitude of the patient to the examination, which often may be decisive. The same holds for measurements on laboratory animals. However, direct intervention into the organism cannot always be avoided, e.g., in catheter examinations where direct measurement of saturation of blood with oxygen may enable reliable diagnosis of an intra-cardial or arterial defect. Direct measurement in the circulatory system is also important in some surgical operations and in many other cases, for example, in monitoring of the saturation of arterial blood in premature babies, which is necessary for the vital control of the incubator operation, as not only a lack but also an excess of oxygen may cause irre-parable damage, here to the retina of the baby's eye. Direct measurement with a sensor permanently fixed in the umbilical artery is very reliable and the babies are disturbed very little. Of course, the transcutaneous sensor cited above [35] and now also commercially available is more gentle and easier to apply.

In research into the photosynthetic activity of plants and the effects of pollutants on it, the gentleness and sensitivity of polarographic and voltammetric techniques for the determination of oxygen are also advanta-geous. In measurements on aqueous plant segments, the possibility of measuring in a small volume of circulating liquid [47] is also advantageous.

In hydrobotanical and limnological research, the vertical stratifica-tion of the oxygen concentration in water reservoirs must also be known. Here only oxygen sensors can be used in which changes in the hydrostatic pressure do not interfere in the measurement.

Electrochemical methods of measurement of oxygen metabolism provide many possibilities in the solution of ecological questions and will definitely be gradually used more to obtain reliable data on oxygen circula-tion for application of mathematical models of various biological systems. The results of modern research also indicate the importance of metabolic measurements with oxygen indication in toxicology and in a number of other experimental fields whose development is reflected in the increasing number and importance of the published works.

REFERENCES

1. Heyrovský J.: *Chem.Listy* **16**, 256 (1922).
2. Heyrovský J.: Polarographie, Springer Verlag, Vienna 1941.
3. Maas J.: Thesis. University of Amsterdam, 1937.
4. Siebert H. and Langer I.: *Chem.Fabrik* **11**, 141 (1938).
5. Smoler I.: *Coll.Czech.Chem.Commun.* **19**, 238 (1954).
6. Smoler I.: Thesis. Polarographic Institute, Czechoslovak Academy of Sciences, 1958.
7. Smoler I.: *J.Electroanal.Chem.* **6**, 465 (1963).
8. Barker G. C.: *Anal.Chim.Acta* **18**, 118 (1958).
9. Cooke W. D., Kelley M. T., and Fisher D. J.: *Anal.Chem.* **33**, 1209 (1961).
10. Gouy M. G.: *Ann.Chim.Phys.* **7**, 145 (1903).
11. Novotný L.: Thesis. J. Heyrovský Institute of Physical Chemistry and Electrochemistry. Czechoslovak Academy of Sciences, Prague, 1979.
12. Novotný L.: Proc. J. Heyrovský Memorial Congress on Polarography, p. 129, Prague 1980.
13. Barker G. C. and Jenkins I. L.: *Analyst* **77**, 685 (1952).
14. Ilkovič D.: *J.Chim.Phys.* **35**, 129 (1938).
15. Vítek V.: *Coll.Czech.Chem.Commun.* **7**, 537 (1935).
16. Petering H. G. and Daniels F.: *J.Amer.Chem.Soc.* **60**, 2796 (1938).
17. Baumberger J. P.: *Amer.J.Physiol.* **129**, 308 (1940).
18. Baumberger J. P.: *Amer.J.Physiol.* **129**, 8 (1940).
19. Longmuir I. S.: *Biochem.J.* **65**, 378 (1957).
20. Stüben J.: *Deutsche Zahnärtzl.Z.* **9**, 1171 (1954).
21. Šerák L.: Advances in Polarography, I. S. Longmuir (ed.), Pergamon Press, London 1960, p. 1057.
22. Šerák L., Krčílek A., and Janoušek V.: Metabolismus Parietis Vasorum, Proc. IV. Int. Congress of Angiology, p. 483, SZN, Prague 1962.
23. Krčílek A., Janoušek V., and Šerák L.: *ibid*, p. 51.
24. Zajíček O.: *Odontol.Rev.* (Sweden) **16**, 1 (1965).
25. Zajíček O.: *Čs. Stomatologie* **69**, 241 (1969).
26. Zajíček O. and Kindlová M.: *J.Periodont.Res.* **7**, 242 (1972).
27. Zajíček O., Škach M., and Kindlová M.: *Čs. Stomatologie* **75**, 1 (1975)
28. Zajíček O.: Thesis. Charles University, Prague 1968.
29. Šerák L.: Proc.Biochem.Conference, p. 110, Bratislava 1964.
30. Hybášek P., Lejhanec G., and Šerák L.: *Acta Univ.Palackianae Olomucensis* **45**, 174 (1967).
31. Lejhanec G., Šerák L., and Hybášek P.: *Čs.Derm.* **36**, 205 (1961).
32. Lejhanec G., Šerák L., and Hybášek P.: *Čs.Derm.* **39**, 78 (1964).
33. Lejhanec G., Hybášek P., Šerák L., Smolan S., and Viktorinová M.: *Čs.Derm.* **49**, 3 (1974).
34. Baumberger J. P. and Goodfriend R. B.: *Fed.Proc.Fed.Am.Soc.Exp.Biol.* **10**, 10 (1951).
35. Eberhard P., Hammacher K., and Mindt W.: Proc.Medizin-Technik 1972, p. 26, Stuttgart 1972.
36. Huch A., Huch R., Meinzer K., and Lübbers D. W.: Proc.Medizin-Technik 1972, p. 26, Stuttgart 1972.

37. Davies P. W. and Brink F., Jr.: *Rev.Sci.Instrum.* **13**, 524 (1942).
38. Jacob H. E. and Horn G.: Abh.der DAW zu Berlin, Kl. für Chemie, Geologie und Biologie, Jrg. 1964, 1, 320, Academie Verlag, Berlin 1964.
39. Clark L. C., Jr.: *Trans.Amer.Soc.Art.Int.Organs* **2**, 41 (1956).
40. Kreuzer F., Kimmich H. P., and Březina M.: in J. Koryta (ed.), Medical and Biological Applications of Electrochemical Devices, J. Wiley, New York 1980, p. 173.
41. Šerák L. and Herout M.: will be published.
42. Šerák L. and Čáp J.: Czechoslovak Patent 231026/1984, British Patent 2,127.977/1986.
43. Šerák L.: Czechoslovak Pat. 215730/1982.
44. Phelan D. M., Taylor R. M., and Fricke S.: *Intern.Lab.* **12**, No. 7, 60 (1982).
45. Suzuki S. and Karube I.: in L. B. Wingard, E. Katschalski-Katzir and L. Goldstein (eds.), Applied Biochemistry and Bioengineering, Vol. 3, Academic Press, New York 1981.
46. Karube I., Nakahara T., Matsunaga T., and Suzuki S.: *Anal.Chem.* **54**, 1725 (1982).
47. Šerák L., Čáp J., and Pokorný J.: Proc.J.Heyrovský Memorial Congress on Polarography, Prague 1980, Vol. 2, p. 158.
48. Šerák L.: *J.Total Environ.* **37**, 107, 1984.

CHAPTER 11

THE USE OF ELECTROCHEMICAL MEASUREMENTS IN THE ECOLOGY OF SHALLOW WATER RESERVOIRS

Jaroslav Čáp, Jan Pokorný, and Lubomír Šerák

11.1   Description of Processes Occurring in Reservoirs

The quality of water has changed substantially during the last ten to fifteen years as a result of intensification of agriculture, modification of rivers, and an increasing number of pollution sources. Eutrophication generally occurs, i.e., the content of nutrients in water increases disproportionately, causing increased production of organic matter, especially plant biomass. The balance within the whole aquatic biocenosis, involving both plants with roots and freely floating plants, fiber and microscopic algae and other microorganisms, and zooplankton and other animals, is altered.

The rapid growth is the response of the vegetation to the increased supply of nutrients, coming mostly from fields fertilized with excessive or unsuitably timed amounts of mineral fertilizers and due to unsuitable melioration interferences and the removal of zones of protective vegetation around ponds. Depending on the amount and intensity of atmospheric precipitation, the concentrations of nitrogen- and phosphorus-containing substances in water increase; a deficiency of these substances is the main factor limiting growth and thus also the intake of other elements. Other important sources of pollution are toxic industrial wastes and pesticides washed out of fields. The determination of organic and inorganic substances in water is difficult and costly. Therefore, suitable organisms and methods are sought for the evaluation of eutrophy and toxicity. For example, the trophic potential − the ability of the test water sample to induce the growth of a known organism − is determined using biotests. More detailed monitoring of the conditions, i.e., also of the solution composition, is necessary for explanation of the observed phenomena, and thus rapid, precise, and cheap determination methods are required.

Growth of higher aquatic plants and the development of some algae populations in shallow water reservoirs are manifestations of eutrophication that are fought by fishermen and water economists, because the hygienic properties of water and the conditions for fish breeding deteriorate. Aquatic plants bind nutrients; if they are harvested and removed, the quality of the water improves. If they are not harvested, they decompose, especially at the end of their vegetation period, and thus consume oxygen. The nutrients are partly liberated into the water and are partly bound in the organic and mineral components of the sediments from which they are gradually released. The growth of aquatic plants is not limited by a lack of water and in eutrophic waters it is usually not limited by a lack of nutrients. With a sufficient supply of solar radiation, some marsh and aquatic plants exhibit a net yearly production of dry matter that is equal to or even higher than the production of agricultural plants. For comparison, the following values of the maximum production of dry matter per year can be given:

Reeds (above-ground biomass), 50 tons hectare^{-1} [6]

Elodea canadensis (total biomass), 6 tons hectare^{-1} [6]

Typha angustifolia (above-ground biomass) 30 tons hectare^{-1} [6]

Lemma cultures (total biomass) 7.5 tons hectare^{-1} [15]

Sugar beet (total biomass) 23 tons hectare^{-1} [9]

Wheat (above-ground biomass) 11 tons hectare^{-1} [9]

Maize (above-ground biomass) 14 tons hectare^{-1} [9]

To produce one ton of dry matter (polysaccharides), 1.6 tons of carbon dioxide are consumed, corresponding to 2.8×10^{10} J of bound solar energy. Additional energy required for the cultivation of agricultural plants must also be considered; aquatic cultures obviously do not require it because they grow spontaneously and are supplied with nutrients washed off the fields. Therefore, marsh and aquatic plants in fertile locations often yield more energy for man than agricultural plants. In the cultures of aquatic plants with submerged assimilation organs, free carbon dioxide is depleted from the water with the growth of the plant mass during the vegetation period and is later also taken from the hydrogen carbonate ions. Carbon dioxide is often not sufficiently replenished from the atmosphere, from the bottom where it is produced by breathing of microorganisms and plant roots, and from the respiration of higher plants and animals. This leads to higher pH values in the morning, during the most intense growth of the culture. For example, in cultures of *Elodea canadensis,* the morning pH minimum was about pH 7 in May and about pH 8 in August. During the growth, the overall photosynthetic release of oxygen predominates over the oxygen consumption for respiration. The afternoon maximum dissolved oxygen concentrations are often more than 200% saturation and the morning minimum does not decrease below 90% saturation (100% saturation corresponds to water in equilibrium with the air at a given temperature and pressure). After the period of the greatest development of the culture, the decay of the organic matter begins to predominate over its photosynthetic production; the morning concentrations of oxygen are often lower than 2 mg liter^{-1}. The decrease in the oxygen concentration in the summer and early autumn is further enhanced by the increased water temperature, which accelerates the decay process. A prolonged period of oxygen deficiency in water favors the development of anaerobic aquatic moulds, pathogenic microorganisms whose metabolism leads to decay processes. Under such conditions the organic matter is not degraded completely to carbon dioxide, water, nitrate (or nitrogen), and mineral compounds of phosphorus and other elements. Nitrification does not occur and increased amounts of ammonium ions and organic amines remain in the water. Under anaerobic conditions, the solubility of phosphates increases in the bottom sediment and ferrous and manganese (II) ions are released from the bottom, because they form more soluble compounds than the oxidized forms of these elements. The oxygen concentration in the root zone also largely determines the species composition of the plant biocenosis. However, there is still no method available for the measurement of oxygen concentrations below 0.1 mg liter^{-1} in a medium as heterogeneous as the bottom of a eutrophic pond or flooded soil with root plants. The direct determination of low oxygen concentrations is sometimes circumvented by measurement of the "redox" potential, but the results are often poorly reproducible.

Oxygen concentrations in water below 2 mg liter^{-1} are dangerous to fish life. At dissolved oxygen concentrations around 4 mg liter^{-1} fish do not die, but do not accept food. An increased concentration of free carbon dioxide in the water also has negative effects. During intense growth of plants, the afternoon pH maxima are dangerous and can attain a value of as high as pH 11 in cultures of some species. Under these extreme conditions, the ammonium ions in the water are converted into free ammonia that is especially toxic for fish. Fish eliminate nitrogen through their gills in the form of ammonia; if the concentration gradient of ammonia between the gills and the water is insufficient, the fish die as a result of ammonia autointoxication, or the gills are almost incurably damaged. The diurnal cycles of the concentrations of oxygen and carbon dioxide in cultures of submerged higher plants are mainly controlled by the metabolism of the plants themselves and of the attached algae (phytoplankton) and algae in the growth (periphytone), bacteria, and lower animals, by the exchange of oxygen and carbon dioxide between the water and the atmosphere, and between the water and the bottom. The photosynthetic activity depends on the kind of plant and on its physiological state and is mainly controlled by the input of photosynthetically active radiation with a wavelength of 380 to 720 nm. The photosynthetic activity depends, in turn, on the oxygen concentration, decreasing with an increase in this value, and on the concentration of carbon dioxide, where the dependence is reversed. The metabolic balance of oxygen is given by the combination of three processes, namely, photosynthesis in which oxygen is liberated as a side product of water decomposition, photorespiration (respiration in the light), and dark respiration. These processes depend differently on external factors, i.e., irradiation intensity, temperature, and the concentrations of oxygen and carbon dioxide. Although there is a certain adaptation region, the scattered data available suggest that these dependences are characteristic for individual plant species. Submerged plants generally exhibit the properties of shade-loving plants, i.e., their photosynthesis is saturated with light at a relatively low input of the photosynthetically active radiation (50 to 80 $W\,m^{-2}$). The light-compensation point, i.e., the radiation input at which respiration is in balance with photosynthesis and thus the oxygen and carbon dioxide concentrations around the plant are constant, is also low (ca. 5 $W\,m^{-2}$). The response of the individual parts of the plant is not constant and the extent of adaptation to the shade increases from the top to the base. Even then, the strongly shaded parts of plants usually have a passive photosynthetic balance with prevalent respiration during the greatest development of the culture.

With aquatic plants, the dependence of the photosynthesis on the pH, i.e., on the form of the carbon dioxide available, is especially important. Some species can produce oxygen up to a pH of 11 (e.g., *Potamogeton pussillus*); other species that utilize only free carbon dioxide are not photosynthetically active at a pH higher than 8.5 (e.g., aquatic moss). These properties to a certain extent determine the occurrence of plant species in various biotopes.

With increasing oxygen concentration, the intensity of radiation required for photosynthetic saturation and for attainment of the light-compensation point increases. The dependence of the photosynthetic activity on the oxygen concentration is primarily given by photorespiration that directly utilizes the photosynthesis products.

Most aquatic plants belong among C_3 plants [8] for which the carboxylation reaction of the Calvin cycle is inhibited by oxygen. Oxygen is bound to ribulosobisphosphate instead of carbon dioxide, and instead of two molecules of phosphoglyceric acid, one molecule of phosphoglyceric acid and one molecule of glycolic acid are formed, the latter being degraded in peroxisome through photorespiration with consumption of oxygen. With C_4 plants, the atmospheric carbon dioxide is bound in the chloroplasts of the internal leaf tissue (mesophyll) to phosphoenolpyruvic acid.

This reaction is not inhibited by oxygen. The four-carbon-atom acid formed is transported into the chloroplasts through veins, where the Calvin cycle is efficient in the absence of oxygen. C_4 plants do not exhibit photorespiration, have a higher photosynthesis temperature optimum, and are, under optimum conditions, photosynthetically more efficient than C_3 plants. These dependences of photosynthesis on the external conditions are determined in the laboratory using plant segments and permit estimation of the changes in respiration and photosynthetic activity during ontogenesis under known conditions, as well as comparison of the activities of individual parts of a plant or of various species. These data cannot be simply recalculated to obtain the photosynthetic production of the whole culture, because many variables are involved and the conditions in a natural culture cannot be exactly simulated in the laboratory. This is true, e.g., of the position of the leaves with respect to the light source, the ratio of direct and diffuse radiation, the spectral composition of the light, and the water flow. Higher aquatic plants have internal tissues filled with gases that are developed to various degrees and permit connection between the upper, photosynthetically active parts of plants that produce oxygen and parts with a negative photosynthetic balance, whose internal atmosphere contains more carbon dioxide and less oxygen. It would be ideal to measure the production of oxygen and the consumption of carbon dioxide directly in cultures and to know the contributions of all the components of the ecosystem to the changes in the concentrations of the two gases in water (the respiration of other organisms and the gas flux among the water, the atmosphere, and the bottom), thus determining the metabolic activity of the plants in their natural environment. For proper knowledge of the changes in the oxygen concentration in a culture, a single value at one depth is insufficient and the whole vertical profile from the surface to the bottom must be known, from which the average concentration or the changes in the overall amount of oxygen in the water column can then be calculated. The concentration profile yields information on the direction and, to a certain extent, on the intensity of oxygen exchange between the water, the bottom, and the atmosphere, and characterizes biological conditions in the water, indicating the relative contents of aerobic and anaerobic organisms. In submersion cultures, a constant vertical concentration profile is formed, with a characteristic uniform decrease in the concentration from the surface to the bottom, provided that the structure of the culture permits the penetration of the radiation even to the bottom, so that a few percent of the photosynthetically active radiation incident on the surface is transmitted as far as the bottom. A constant light and oxygen profile is typical for pure water and may extend to a depth of several meters. In polluted, turbid waters the plant biomass is concentrated at the surface and less than 1% of the photosynthetically active radiation penetrates to the bottom, which is often less than 1 meter under the surface. The oxygen concentration profile corresponds to this biomass distribution. In the water layer containing photosynthetically active parts of plants, the oxygen concentration can be as much as 200% of the saturation value during the afternoon, while it can be less than 50% at a depth of only 0.2 m. In a half-meter-thick profile of a submersion culture, the oxygen concentration gradient may attain values of from 18 to 20 mg liter^{-1} at the surface in the layer of photosynthetically active plants to unmeasurable values (i.e., below 0.1 mg liter^{-1}) at the bottom in the afternoon. These differences are eliminated at night. Floating cultures that cover the water surface with their leaves, e.g., water lilies or *Lemna* (minor), decrease the passage of radiation even more, and these plants liberate oxygen only into the atmosphere and not into the water. Under such cultures the oxygen concentration rarely attains 100% saturation. An important role in the balance of oxygen and carbon dioxide in water is played by the fluxes of these gases between the water and the atmosphere. Water absorbs infrared radiation and its relatively motionless upper layer is heated in dense cultures of aquatic plants and the temperature cannot be equilibrated with respect to the ambient air; therefore, the oxygen solubility decreases and the oxygen passes from this layer into the atmosphere. At night, when the oxygen concentration in the water decreases as a result of plant respiration, the upper layer of the water is cooled, the oxygen solubility increases, and oxygen passes from the atmosphere into the water. However, more detailed measurements are required for a quantitative evaluation of these processes.

There is little quantitative data on the consumption of oxygen and the production of carbon dioxide by living bottom sediments, obtained by measurements *in situ*. A small oxygen consumption at the bottom is manifested by the occurrence of oxygen in measurable concentrations at the surface of the bottom sediment or even a few centimeters inside the sediment. This situation, with a relatively high oxygen concentration over the whole water profile, is typical of oligotrophic pure waters. In eutrophic waters, the bottom, which is rich in reduced organic and inorganic substances, acts as an important consumer of oxygen and thus the concentration of oxygen at and in the bottom is not measurable, because of the metabolic activity of microorganisms and lower animals.

11.2   Measuring Methods

Contemporary knowledge of daily fluctuations in the concentrations of oxygen, carbon dioxide, hydrogen carbonate, carbonate, and ammonia over the water profile and especially of the possibility of predicting these values is insufficient. The changes in these concentrations are immediately related to photosynthesis and respiration, which are the principal processes in the primary production and decay of biomass and thus are also directly related to botanical problems, such as the conditions for deceleration and acceleration of plant growth, the ability of plants to accept and bind mineral substances, and the roles of various plant species in these processes. The correctness of the answers to these questions depends on experimental knowledge of these processes occurring under natural conditions and in laboratory experiments.

The rate of photosynthesis — the conversion of light energy into energy bound chemically in adenosine triphosphate and reduced nicotin-amidodinucleotide phosphate, used to reduce carbon dioxide – is expressed in terms of the amounts of oxygen liberated or carbon dioxide consumed per time unit. The amount of absorbed carbon dioxide can be measured using the ^{14}C radioactive isotope or with an infrared carbon dioxide analyzer. The use of the latter requires passage of the gas through a solution and is only possible when free carbon dioxide is present, i.e., at a pH below 8.5. Free carbon dioxide can also be measured potentiometrically.

The dissolved oxygen concentration was usually determined in hydrobiology by the Winkler titration method in samples that were taken periodically. Polarographic methods, either classical with the dropping mercury electrode or the derived voltammetric method with a Clark-type sensor, are very useful for the determination of oxygen; the oxygen concentration in the environment can be monitored continuously and thus also the metabolic activity of aquatic organisms can be followed as a function of the external conditions. As this measurement is simple and sufficiently reliable, it is used more often than the more difficult direct measurement of carbon dioxide concentration changes, which are usually estimated from the measured pH values that are considered as the product of an ideal equilibrium among free carbon dioxide, hydrogen carbonate, and carbonate. This assumption is subject to some limitations, especially in eutrophic waters, where the equilibrium between ammonium ions and free ammonia may exert a significant effect.

Various instruments have been employed for laboratory monitoring of photosynthetic activity. For example, a glass circulation apparatus has been employed in which the pH, the dissolved oxygen concentration, and possibly other components can be measured electrochemically under various external conditions. The plant is placed in the apparatus in a vertical tube and is immersed in a nutrient solution. The circulating and saturating devices are connected to the upper and lower ends of the tube by short horizontal arms. The sockets for the placement of pH, oxygen, and any other sensors are located in the outlet connecting arm [21].

Free carbon dioxide, free ammonia, and ammonium ions can be determined potentiometrically, especially with gas sensors (free carbon dioxide, free ammonia, and the sum of ammonia and ammonium ions after sample pretreatment) and ion-selective electrodes (ammonium ions).

A gas sensor for the determination of carbon dioxide was first used almost 30 years ago by Stow and Randall [19]. The sensor was later perfected by these authors [20] and especially by Severinghaus [17], whose name was later connected with this device. This is a glass pH electrode covered with a thin film of an aqueous solution of an alkali hydrogen carbonate that is separated from the sample by a membrane permeable for carbon dioxide and impermeable for ions and water. Carbon dioxide molecules penetrate from the sample into the sensor solution until the same concentration of carbon dioxide is attained on both sides of the membrane. This leads to a change in the ratio of the concentrations of free carbon dioxide and hydrogen carbonate in the sensor solution, with a consequent change in the pH that is monitored by the glass electrode. The sensor was originally manufactured by Radiometer in Denmark and is now supplied by most companies producing potentiometric sensors.

An analogous sensor can be used for the determination of ammonia [1, 16] and applied to measurements in surface waters [2, 3, 7, 10, 22]. Other potentiometric sensors for the determination of ammonia and ammonium ions have been described, e.g., by Scholer and Simon (a liquid-membrane ion-selective electrode containing a mixture of nonactin and monactin) [18], Guilbault and Nagy (a silicone-rubber matrix ion-selective electrode containing nonactin) [5], Morf and Simon (a PVC-matrix ion-selective electrode with nonactin) [13], and Meyerhoff (a gas sensor consisting of a PVC-matrix ion-selective electrode with nonactin immersed in an ammonium salt solution and separated from the sample by an ammonia-permeable membrane) [12]. Sensors with nonactin ion-selective electrodes have also been used to determine ammonium ions in surface waters [4, 11].

In the project of the Hydrobotanical Department of the Botanical Institute of the Czechoslovak Academy of Sciences, the pH, the dissolved oxygen, the free carbon dioxide, and the sum of the free ammonia and ammonium ions were monitored. The pH was measured using single-stem cells containing standard glass sensors (Chemoprojekt, Satalice, Czechoslovakia); in some applications these electrodes were too large (diameter, 12 mm; length, 120 mm) and were replaced by nonstandard, smaller types (diameter, 5 mm). Prolonged pH measurements exhibited a precision of better than 0.05 pH, provided that the temperature dependence of the signal was automatically corrected.

Clark-type oxygen sensors were used for the monitoring of dissolved oxygen, using a platinum indicator electrode (a surface area of ca. 0.01 mm^2) and an Ag/AgCl reference electrode (the SOPS type from Chemoprojekt, Satalice, Czechoslovakia). In water saturated with air at 20°C, the sensor signal is about 2×10^{-9} A and is almost independent of the sample motion around the sensor. The signal strongly depends on the temperature and thus the latter must be measured and an automatic correction must be introduced. In prolonged measurements, the signal is decreased by the periphyton that hinders the oxygen diffusion toward the working electrode. The sensor is suitable for measuring concentrations higher than 0.1 mg O_2/liter (i.e., ca. 1% of the equilibrium oxygen content).

Severinghaus-type gas sensors (Chemoprojekt, Satalice, Czechoslovakia) were employed for monitoring free carbon dioxide at concentrations higher than 1 mg CO_2/liter.

The sum of ammonia and ammonium ions was also determined by using gas sensors that differ from the CO_2 sensors only in the kinds of membrane and in the internal electrolyte (Chemoprojekt, Satalice, Czechoslovakia) and a 95-10 sensor from Orion Research, USA. Concentrations higher than 1 mg NH_3/liter were readily measured after adjusting the solution pH to a value above 12.

Electrochemical methods are attractive for ecological research, because they enable immediate measurement of changes in the component concentrations, can sometimes be used for continuous monitoring, and are suitable for field work, where the systematic error caused by the transport and storage of the samples is avoided. The low cost of the instrumentation and the possibility of using a single instrument with various types of potentiometric sensor are also advantageous.

11.3   Ecological Evaluation

Measurements with electrochemical sensors have contributed to a deeper understanding of primary production processes in the aquatic environment and thus they have enabled more precise predictions to be made of changes in oxygen, carbon dioxide, and ammonia concentrations due to plant photosynthetic activity. As shown above, the mechanism by which the growth of plants causes fluctuations in the dissolved gas concentrations in water is the sum of many processes that can be experimentally monitored with varying precision. This is a polyfunctional dependence that can be evaluated in terms of a mathematical model [14]. To a first approximation, the system variables involve the concentrations of oxygen, carbon dioxide, hydrogen carbonate, and total carbon dioxide and the pH. The controlling variables are the total radiation incident on the water surface and the water temperature. The time changes in the system variables are described by two differential and two algebraic equations. The input data of the model are the biomass, the initial concentration of oxygen, the total carbon dioxide and its forms, the initial pH, and the constants that are contained in the functions describing the photosynthesis and the individual mass fluxes. The computation is carried out as follows: During a short selectable time interval, the oxygen concentration changes by a value corresponding to the production or consumption of oxygen during photosynthesis or respiration, respectively, and by a value given by the fluxes of oxygen between the water, the atmosphere, and the bottom substrate. Within the same time interval the total carbon dioxide concentration changes by a fraction corresponding to the production or consumption of oxygen and a value corresponding to the flux of free carbon dioxide between the water and the atmosphere. At the end of this time interval, the changes in the oxygen and carbon dioxide concentrations are determined over a selected time interval of the integration step and a set of five equations is solved, involving three equations for dissociation equilibria and two equations expressing the total carbon dioxide concentration and the total alkalinity. The concentrations of all the components of the total carbon dioxide content and the oxygen concentration participate in the subsequent time interval in controlling the values of the photosynthesis and the fluxes of oxygen and carbon dioxide. The stratification of the oxygen and carbon dioxide concentrations is also included, by considering the diffusion exchange between arbitrarily defined layers of the water column, in addition to the fluxes among the atmosphere, the water, and the bottom. The model employs one-hour timing and each hour the water temperature and the intensity of the total radiation are supplied, relating to the beginning and the end of the one-hour interval.

REFERENCES

1. Bailey P. L. and Riley M.: *Analyst* **100**, 145 (1975).
2. Banwart W. L., Tabatabai M. A., and Bremmer J. M.: *Comm.Soil Sci.Plant Anal.* **3**, 449 (1972).
3. Barica J.: *J.Fish.Res.Board Can.* **30**, 1389 (1973).
4. Dewolfs R., Broddin G., Clysters H., and Deelstra H.: *Z.Anal.Chem.* **275**, 337 (1975).
5. Guilbault G. G. and Nagy G.: *Anal.Chem.* **45**, 417 (1973).
6. Hejný S., Květ J., and Dykyjová D.: *Folia Geobot.Phytotax.* **16**, 73 (1981).
7. Helfgott T. and Mazurek J. S.: *Prag.Water Technol.* **8**, 443 (1977).
8. Hough R. A. and Wetzel R. G.: *Aq.Bot.* **3**, 297 – 313 (1977).
9. Hruška L., Janíček J., and Bednářová E.: The Use of the Solar Energy in Some Agricultural Plants. Proceedings of the Symposia of the Agricultural Faculty in Brno, 1975 (in Czech).
10. Ip S. Y. and Pilkington N. H.: *J.Water Poll.Control Fed.* **50**, 2778 (1978).
11. Kahr G. and Kissling B.: *Chem.Rundschau* **29**, 1 (1976).
12. Meyerhoff M. E.: *Anal.Chem.* **52**, 1532 (1980).
13. Morf W. E. and Simon W.: Ion-Selective Electrodes in Analytical Chemistry, Vol. I (H. Freiser, ed.), Plenum, New York 1978, p. 211.
14. Ondok J. P. and Pokorný J.: Czechoslovak Ecology, in press.
15. Rejmánková L.: The Role of Lemna in Pond Ecosystems. Thesis, Botanical Institute, Czechoslovak Academy of Sciences, Průhonice 1979 (in Czech).
16. Ross J. W., Riseman J. H., and Krueger J. A.: *Pure Appl.Chem.* **36**, 473 (1973).
17. Severinghaus J. W. and Bradley A. F.: *J.Appl.Physiol.* **13**, 515 (1958).
18. Scholer R. P. and Simon W.: *Chimia* **24**, 372 (1970).
19. Stow R. W. and Randall B. F.: *Amer.J.Physiol.* **179**, 678 (1954).
20. Stow R. W., Baer R. F., and Randall B. F.: *Arch.Phys.Med.* **38**, 646 (1957).
21. Šerák L., Čáp J., and Pokorný J.: Proc. J. Heyrovský Memorial Congress on Polarography, Vol. II, Prague 1980, p. 158.
22. Thomas R. F. and Both R. L.: *Environ.Sci.Technol.* **7**, 523 (1973).

CHAPTER 12

ELECTROCHEMICAL ANALYSIS OF WATER TOXICITY

Pavel Hofmann

12.1   Introduction

A practical example will be given of the application of oxygen concentration measurements with a Clark-type amperometric sensor to the detection of toxic substances in water, using inhibition by toxic substances of aerobic degradation of organic substances by a culture of indicator organisms [1].

Detection of toxicity as the property of matter that exerts a harmful effect on living organisms is a daily task in all laboratories directly or indirectly connected with environmental problems. The information obtained is important for industrial control and research and for related control, managerial, and planning activities, from a local to a global scale.

The approach to and methods for detecting toxicity are determined by three basic factors, namely, the object on which toxic substances act (man, animal, plant, microorganism, etc.), the environment in which they occur (the atmosphere, water, useful materials, soil, etc.), and the way of acting or type of contamination (alimentary, respiratory, contact, etc.). It must not be methodically overlooked that there are differences in the character of the interaction of the toxic substances with the object (specificity of effects on certain life processes, rate of action, accumulation in the organism, the effect of the composition and the physical properties of the environment on the intensity of this action, an increase or a suppression of the effect in mixtures, etc.), in the sensitivities of indicator organisms of various types even within a single species, and in the adaptability of indicator organisms to the substance. For selection of the method, instrumentation, and experimental conditions, the question of similarity or identity of the substance behavior in the analytical system and in a natural or technological macrosystem is important.

This broad field will be confined here to methods for determination of toxicity of water (or of substances contained in water) and methods for determination of toxicity of substances, especially wastes, that can enter water. The best known and most widely used methods are biological procedures based on monitoring indicator organisms exposed to the sample (water sample, diluted sample, a solution of the test substance, or samples pretreated by filtration, neutralization, etc.) and comparing their response or state with those of individuals exposed to a nontoxic environment. As indicators, various kinds of fish, worms, plankton, algae, microorganisms, and aquatic and cultivated plants are used. Depending on the indicator type, one or more parameters are studied, e.g., the velocity of motion, position, and orientation in the medium, a change in the color, the rate of gill movement, closing of mussel shells, damage to tissues, plasmolysis and decomposition of cells, the growth velocity, death, and other parameters that are usually macro- or microscopically detectable. For most methods, an exact statistical evaluation has been developed and in some methods the monitoring is automated, e.g., with fish movement, respiration frequency, procreation of microorganisms, etc.

Another group of methods consists of biochemical-physiological tests based on comparison of the parameters of the normal manifestations of an organism or a culture of organisms with those for the behavior affected by the sample. As indicators, fish or especially monocultures or mixed cultures of microorganisms, algae, and yeasts are used, either as a developed culture or in the growth stage. The parameter that is monitored is either a substance that is consumed, e.g., oxygen during respiration or aerobic biochemical degradation of organic substances by microorganisms, or a substance that is produced, e.g., carbon dioxide in complete aerobic mineralization, methane, fatty acids in anaerobic degradation, oxygen during assimilation, etc. Biochemical processes are monitored using either classical manometric and volumetric methods that have been highly unified during their development or electrochemical methods whose useful development was possible only after the discovery of the Clark amperometric membrane oxygen sensor. This chapter describes the range of the use of electrochemical analyzers for determination of toxicity and the types of values and parameters obtained in the measurements.

12.2   Methods of Determination of Water Toxicity

The determination of toxicity described here is based on comparison of characteristic parameters of aerobic biochemical degradation of organic substances by a culture of indicator organisms in a medium containing the test substance with those obtained in an unaffected reference medium. This biochemical process, the analytical basis of the method, is monitored by measuring the consumption of oxygen as a summary parameter.

The method employs two different exposure techniques. In the first, the test substance is added simultaneously with the indicator organism inoculum (a culture of the indicator organisms) and the effect of the test substance on the initial adaptation stage, on the stage of full development of the culture accompanied by the degradation of the organic substrate, and on the final stage of the process after the decomposition of the organic nutrient is monitored. For this purpose the first two analyzer types described below are used. The third analyzer type is employed for the other technique based on the exposure of a fully developed indicator culture to the test substances, monitoring the effect of the substance on the intensity of the life functions of the organisms, manifested by a change in the rate of oxygen consumption or even by its cessation.

The characteristic parameters of the analytical process (in the first technique the length of the lag stage, i.e., the first stage after the inoculation, the rate of oxygen consumption during the full development stage and possibly the stage of degradation of the organic substrate, and in the second technique the rate of oxygen consumption) determined in the presence of the test substance are expressed as functions of the concentration of this substance, together with the values of the reference blank experiment for zero concentration. In the second technique, the reference values can also be obtained from the measurement in the system before exposure to the test substance. These dependences yield the concentration range within which the test substance does not affect the biochemical process or some of its stages, regions of various intensities of the effect, and the limiting concentration above which the indicator culture of organisms is destroyed or its development is stopped.

12.2.1   The Principal Analytical Process

The above principle and basic techniques for the determination of toxicity are common to many methods, and variants have been developed for various purposes and conditions. The basic biochemical analytical process is therefore modified in various ways, especially the composition and concentration of the organic substrate, the kind, state, and the development stage of the culture of indicator organisms, and the selection of the analysis conditions. Most of these factors affect the demand of the process for the measured substance, oxygen, which is decisive for instrumentation selection.

As substrates, original organic substances from water samples (surface, waste, etc.), artificial mixtures modeling the compositions of certain types of water (Weinberg sewage, etc.), or individual, readily biologically degradable substances (glucose, glutamic acid, etc.) are used. The first two kinds of substrate are significant for making the test similar to real conditions, the second and third type enable general comparison of the results and unification of the procedures.

The organic nutrient concentration varies widely with various tests, from milligrams to grams per liter, corresponding to the compositions from unpolluted surface waters to concentrated waste waters. The concentration is selected from the point of view of the model similarity in the composition but also depending on the type of indicator organism used.

The analytically nonreactive other substrate — mineral nutrients — is added in an amount ensuring uninhibited development of the culture during the test. Unified or well-known solutions are used, containing mainly phosphorus and nitrogen compounds (e.g., the Knop solution or diluting water for determination of BOD).

A decisive role in the determination of toxicity and the interpretation of the results obtained is played by the species or type of indicator organism. In the selection, the species or type sensitivity of the organisms toward various substances must be considered, as well as the model analogy of the organism properties with respect to the experiment, advantages and drawbacks of using monocultures and mixed cultures, and the prescriptions and recommendations of standardized methods. The following types of indicator organism cultures are usually employed:

- Cultures of organisms taken directly from the environment, minimally treated and used shortly after being obtained, modeling the biological state and properties of the given test system under the toxicity test conditions (e.g., microorganisms concentrated by centrifugation from surface or waste waters, activated sludge, etc.).
- Cultures taken from a given environment and pretreated in a defined way for use in a toxicity test (e.g., by cultivation for adaptation to the test conditions, to the substrate composition, temperature, etc.).
- Mixed cultures cultivated from various types of stock biological materials (e.g., dried activated sludge, lyophilized cultures of organisms from rivers, etc.), with pretested properties.

- Monocultures of various microorganisms cultivated from collection cultures and defined mixed cultures prepared from these mono-cultures.

The preparation of a culture for a toxicity test involves many procedures and operations, from the simplest ones to highly demanding microbiological cultivation techniques and application of automated biostate. These procedures must satisfy the requirements of the particular application of the culture, i.e., small amounts are prepared for inoculation while large volumes are required for monitoring toxicity with a developed culture. The control of the cultivation operations is centered on the species composition and state of the culture and on attainment of reproducible properties of cultures prepared simultaneously or subsequently.

In addition to the substrate and the indicator culture, the toxicity test is defined by the working conditions, especially the temperature (analogous to the real system or standard, $20°C$), the amount of light (affecting the assimilation), the medium motion (usually stirring), and the volume and geometry of the reaction vessels.

12.2.2   Sample Preparation for Toxicity Determination

Three types of sample are common in determination of the toxicity of an aquatic environment:

- real water samples with the original chemical and biological composition
- substances isolated from real water samples, mostly concentrates and sometimes fractions of the total content of the original substances
- substances that enter or may enter waters and whose toxicity for the hydrosphere is to be found (wastes, materials utilized in industry, agriculture, households, etc.).

Prior to the analysis, the samples are pretreated (water and solution samples) or are dissolved for the analysis (solid materials). The water and solution pretreatment involves neutralization, removal of undissolved substances, or homogenization (when solid substances are to take part in the determination) and dilution, depending on the procedure.

The preparation of concentrates and of fractions from samples utilizes evaporation, vacuum evaporation, distillation from various media, extraction, precipitation, and preparative chromatography.

Sampling, sample preservation, and storage are similar to the procedures common in hydroanalytical chemistry.

12.3   The Measuring Technique

Three types of analyzer for the determination of physiological toxicity will be discussed, and these cover the whole field of requirements of water economy and environmental protection.

12.3.1   The Principle of the Type I Analyzer

This analyzer measures the dissolved oxygen concentration using a membrane sensor at a constant temperature and stirring rate in a closed space (incubation bottle) completely filled with the sample. The analyzer measures the oxygen concentration in parallel, in several incubation bottles for the reference solution and for the test sample diluted variously, pretreated, and measured by the selected procedure.

12.3.1.1   Analyzer Design

The analyzer consists of three functional parts: the reaction part (an incubation bottle with an oxygen sensor), auxiliary devices (stirrer, stabilizing bath, and thermostat), and a recorder. An overall view of the instrument is given in Fig. 12.1.

The measuring-reaction part (Fig. 12.2) consists of a 300- to 500-ml incubation bottle provided with a glass thread on a flat ground-glass neck (a commercial infusion bottle) with the sensor attached to the neck. The SKTO3 oxygen membrane sensor (developed by WRI, Prague and manufactured by Chemoprojekt, Satalice, Czechoslovakia) has a silver cathode, a cadmium oxide anode, is filled with 1 M KOH, and is covered with a polyethylene membrane 0.045 mm thick. The sensor signal is 0.55 to 0.70 μA for 1 mg O_2 liter^{-1} at 20°C and 90% of the final signal value is attained within 15 to 20 s.

Auxiliary devices are a six-place electromagnetic stirrer and a stabilizing water bath. The recorder is a six-curve point recorder with a basic sensitivity of 2 mV/scale to which the oxygen membrane sensors are connected through regulating 1-kΩ potentiometers for the sensitivity adjustment and calibration. The recorder scale is calibrated from 0 to 10 mg O_2 liter^{-1}.

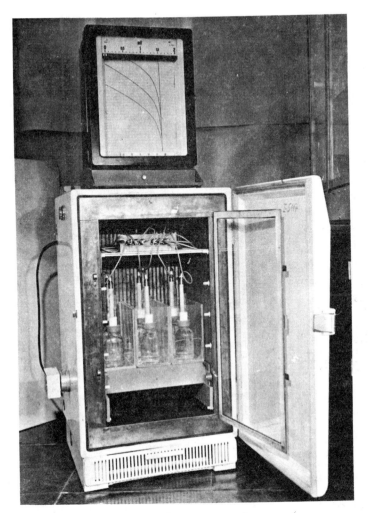

Fig. 12.1 Type I toxicity analyzer.

12.3.1.2   Analyzer Operation

The analyzer preparation, i.e., thermostatting of the sample at a given temperature and calibration of the sensor, should be carried out one day prior to the test. The thermostat is switched on and the stabilizing baths are filled with water. In one bath, a vessel for the calibration with the sensors containing a thermometer is placed above the stirrer magnet, air saturation is begun, the stirring is switched on, and the thermostat is closed. The sensor signals are recorded until they become constant on attainment of the equilibrium saturation of water with atmospheric

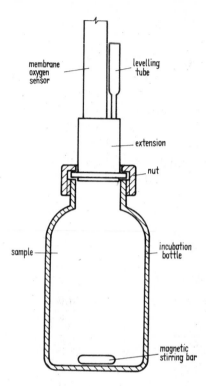

Fig. 12.2 Incubation bottle with the extension and the oxygen sensor.

oxygen and stabilization of the temperature occurs. The calibration vessel temperature and the air pressure are read and the equilibrium oxygen concentration in the water is found in tables; on this basis the sensor signals are adjusted by the regulating potentiometers. The sensors are left in the calibration vessel and the thermostat with the stirrer remain switched on. It is recommended to check and, if necessary, to adjust the calibration prior to the beginning of the determination. The value obtained is recorded as the calibration value at the beginning of the test.

The samples pretreated by a selected procedure, i.e., usually five solutions with various dilutions of the test sample, with an organic substrate and an inoculum added, and the blank sample are thermostatted at the required temperature and saturated with atmospheric oxygen. The solutions are transferred into the incubation bottles, air bubbles on the vessel walls are removed, and stirring rods are placed in the bottles. The sensors are then gradually taken from the calibration vessel and, after drops of water have been removed using filter paper, are inserted into the bottles, care being taken to prevent the formation of air bubbless, and are fixed with the caps. The sample forced out of the bottle should reach about to the middle of the wide part of the tube; the tube is closed with a PTFE perforated stopper. The bottle is placed in the stabilizing bath in the thermostat and another sensor is placed in the next bottle. After placing all the bottles in the thermostat, it is closed and the time of the beginning of the test is marked on the recorder chart. The oxygen concentrations in the bottles are usually the same and correspond to saturation; small differences in the beginning of the experiment, within a maximum of 30 min, are usually caused by not adjusting the solution temperatures exactly.

The oxygen concentration is monitored during the test with sufficient frequency. With experiments lasting several days, the sensor signal is recorded every 72 s with each sample. The water temperature is controlled daily in one of the stabilizing baths and the barometric pressure is recorded.

After the completion of the test, the sensors are taken from the bottles, rinsed, placed in the holder of the calibration vessel containing water thermostatted at the experimental temperature, and saturated with atmospheric oxygen, and the calibration is repeated in the same way as before the experiment, measuring the temperature and the air pressure.

The above procedure is applied to parallel tests with solutions containing various concentrations of the test substance. If the effect on a developed culture is to be monitored, the test substance in solution is added to the incubation bottle containing a medium whose oxygen consumption is recorded, through a hole in the sensor holder, best by a syringe with a long needle or through a capillary tube with simultaneous aspiration of the forced-out liquid (a water pump with tubing provided with a needle extension). The volume of the solution added must not exceed ca. 3% of the bottle volume to avoid disturbing the concentration ratios in the medium. When additions are made to several bottles simultaneously, it is recommended that the same volumes of the variously concentrated solutions or of water be added to each bottle.

12.3.2   The Principle of the Type II Analyzer

This analyzer utilizes measurement of the oxygen concentration by a membrane oxygen sensor at a constant temperature and stirring rate, in the gaseous phase of the incubation bottle partly filled with the sample. The gaseous phase is freed of carbon dioxide and is connected by a capillary with the ambient atmosphere to compensate for the consumption of oxygen by aspirating air into the bottle. The analyzer contains several incubation bottles for a reference solution and variously diluted samples, pretreated and measured by the given procedure.

12.3.2.1   Analyzer Design

This analyzer is virtually identical with the Type I analyzer described above, with small differences permitting the measurement of oxygen in the gaseous phase, the removal of carbon dioxide from the gaseous phase, and the connection of the incubation bottle interior with the ambient atmosphere.

The oxygen sensor is placed in the holder so that its tip reaches about 20 to 30 mm under the holder bottom. A glass-fiber filter paper ring is placed around the sensor and is fixed with an elastic PTFE clamp below the holder. This ring is wetted with ca. 0.2 ml of a 70% solution of potassium hydroxide before the sensor is placed in the bottle containing the sample (trapping of carbon dioxide). The recorder scale is calibrated in percent, up to 110 to 120 divisions.

12.3.2.2   Analyzer Operation

The analyzer preparation for the measurement should be carried out one day before the test in the same way as with the Type I analyzer, but with only a small amount of water in the calibration vessel, so that the sensor tips are in the air. The air temperature inside the thermostat is measured. The signals of the oxygen sensors are adjusted to 100% by the calibration potentiometers.

The samples pretreated by a given procedure, usually five sample solutions with various dilutions, and the blank sample are thermostatted at a required temperature and, if necessary, saturated with atmospheric oxygen. Depending on the range required (see above), a certain amount of the sample is measured into the incubation bottle, the sensor is inserted (with the filter paper ring soaked with the alkali) and fixed, and the stoppers are placed on the tubes. The bottles are then placed in the stabilizing baths over the stirrer magnets and weighed down. The thermostat is closed and the beginning of the test is marked on the chart.

The recording of the control values during the test, the measuring frequency, the calibration, and the recording of the final values are analogous to the procedures described above for the Type I analyzer.

If a developed culture of indicator organisms is used for the toxicity determination, the test substance is added at a selected time through a hole in the sensor holder, using a syringe. The volume of the solution added must not be large to avoid disturbing the concentration ratios in the test medium and producing large changes in the ratio of the volumes of the gaseous and liquid phases in the incubation bottle (up to 3% of the smaller volume).

12.3.3   The Principle of the Type III Analyzer

This analyzer measures the dissolved oxygen concentration with a membrane sensor, at a constant temperature and with a constant stirring rate, in a closed space (measuring vessel), in which a small part of the test sample is periodically isolated, while the main part of the sample is constantly saturated with air in an open space. The analyzer contains one or more measuring units for measuring with the test medium and for analyses of samples pretreated and analyzed by a selected procedure.

12.3.3.1   Analyzer Design

The analyzer consists of four functional parts — the reactor (including an incubation vessel connected by means of a valve to a measuring vessel containing a membrane oxygen sensor), auxiliary devices (a stirrer, stabilizing baths, a thermostat, and an air-saturating apparatus), the control unit, and the recorder.

The reactor (Fig. 12.3) consists of a membrane oxygen sensor (1) fixed in the measuring cell (2) by a PTFE stopper with a nut (3). The measuring cell is cylindrical with a volume of ca. 15 ml and is provided with two outlets. The upper outlet is placed tangentially with respect to the motion of the liquid stirred by the stirring bar (4) of the electromagnetic stirrer. The measuring cell is actually a small pump causing circulation of the medium (a pressure of ca. 5–7 cm of the water column). The incubation vessel (5) is made of glass, is cylindrical with a volume of 200 to 300 ml, and has two outlets. The bottom outlet (6) has a valve. The valve saddle (8) is a rubber ring on a glass rod, connected with the core of an electromagnet (9). The liquid inlet into the incubation vessel (7) as well as the outlet (6) are connected by tubing with the measuring cell. The air saturation tube (10) is connected to a small air pump. The analyzer contains a stabilizing bath for two or three reactors, which is connected to a flow-through laboratory thermostat.

The control unit operates the electromagnetic valve and consists of a time-circuit that produces signals for closing (signal S1) or opening (signal S2) valve 8. The valve can also be opened by signal S3 produced by the terminal switch of the recorder (in procedures where the measuring stage is terminated on the attainment of a certain oxygen concentration) (Fig. 12.6b). The length of the two signals (S1 and S2) can be selected from 0 to 120 min, retaining a constant time interval between the signals. Signal S3 can be adjusted over the whole range of measured oxygen concentrations (0 to 10 mg liter^{-1}).

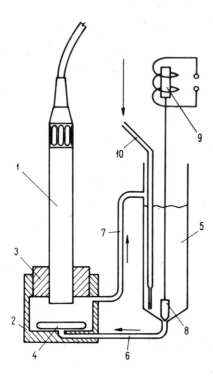

Fig. 12.3 Type III toxicity analyzer (for description, see the text).

12.3.3.2 Analyzer Function

The analyzer operates continuously during the test, in repeated time cycles. Each cycle consists of the measuring and the regeneration stage with the cycle length, from 15 to 120 min, selected depending on the properties of the test medium. For the measuring stage, an interval from 10 to 110 min then remains, as the regeneration stage should last at least 5 and preferably 10 min; longer regeneration times can also be used.

The measuring stage is begun by stopping the medium circulation (signal S1 in Fig. 12.6) in the reactor (the measuring and incubation vessels) and isolating the liquid in the measuring cell from the air supply. The content of the measuring cell is not exchanged and is stirred, and the recorded decrease in the oxygen concentration corresponds to the rate of oxygen consumption in the biological process taking place in the test medium (see the time interval following signal S1). The measuring stage is terminated after a selected time (signal S2 in Fig. 12.6a) or on attainment of a selected oxygen concentration in the measuring vessel (signal S3 in Fig. 12.6b) by resuming the circulation of the medium through the measuring and incubation vessels.

The regeneration stage is started by resuming the circulation, the measuring cell aspirates the medium from the aerated incubation vessel, and a rapid increase in the oxygen concentration is recorded (see Fig. 12.6, after signal S2 or S3). The contents of the measuring cell are rapidly mixed with that of the incubation vessel and the oxygen concentration slowly approaches the equilibrium value. The equilibrium state depends on the rate of oxygen consumption by the medium and on the efficiency of the aeration apparatus, its establishment takes some time, and its attainment is not a necessary condition for beginning the next cycle. The recording during the regeneration stage serves only for checking the analyzer function.

12.3.3.3   Analyzer Operation

The analyzer is prepared one day before the analysis, is cleaned, and is filled with distilled water, and the thermostat and the aeration apparatus are switched on. The intervals of the measuring and regeneration stages are adjusted using the time switch, or the recorder switches are connected in the measurement to a certain oxygen concentration. The switch blocking the connection to the reactor during the regeneration stage and the recorder switch are activated. After the signal stabilization, the equilibrium oxygen concentration values are adjusted with the regulating potentiometer; the calibration is checked at the beginning of the analysis.

Thermostatted samples are mixed with the inoculum and a solution of nutrients, with constant stirring, immediately before filling the reactor, so that the time from the beginning of the action of the toxic substance to the commencement of the first measuring stage is as short as possible.

The reactor contents (water used for the calibration of the oxygen sensor) are removed through a thin tube that is inserted after removal of the magnetic valve and cessation of stirring and aeration into the lowest point in the incubation vessel in the bend of the bottom outlet. In this way, both the incubation vessel and the measuring cell are emptied. The reactor is filled with the test mixture so that the level is 0.5 to 1.0 cm below the upper outlet from the incubation vessel. Stirring is begun at a low speed, all the air bubbles are removed from the measuring cell, and the velocity of stirring is increased. The electromagnetic valve is fitted in place and the aeration is begun. The time switch is adjusted to the regeneration stage. The measurement is thus begun.

After termination of the test, the analyzer is emptied as above and the remainder of the organisms adhering to the walls are removed. After prolonged tests, the oxygen sensor usually has to be dismantled and the measuring cell with the sensor membrane must be cleaned. When used frequently, the analyzer is filled with water during storage.

If a procedure involving direct addition of the toxic substance is used, the analyzer operation is analogous; the reactors are filled with the test medium and about three measuring cycles are performed, in order to check the measuring reproducibility and obtain the basic parameters of the inoculum (the rate of oxygen consumption). Then a small amount of the test substance is added (the dilution of the inoculum should be negligible) during a regeneration stage, 1 to 2 min before the commencement of the final measuring stage. Within this time the solution added is thoroughly mixed with the inoculum, with aeration and medium circulation in the reactor.

12.4   Measuring Results

The results obtained using the Type I analyzer represent the time dependence of the biochemical consumption of oxygen dissolved in a closed environment. A typical recording obtained in a medium with a low substrate concentration (up to ca. 8 mg liter^{-1} BOD$_5$) is given in Fig. 12.4, curve A. After a short adaptation stage, the organic substance is degraded until the oxygen consumption attains a value corresponding to complete decomposition and the oxygen concentration changes very little (decomposition of the microbial culture). The curve corresponds to the first carbonization stage of the degradation. Curve B corresponds to the process in a solution containing a substance that affected the development of the organisms — the initial adaptation phase is longer, the rate of the culture development is slower, and thus the oxygen consumption is smaller. Curve C corresponds to conditions in a toxic medium in which the organisms do not develop and the oxygen concentration virtually remains at the initial value.

The monitoring of the effect of a toxic substance on a culture of indicator organisms in its development stage is demonstrated in Fig. 12.5. Curve A is the blank. Point Z is the instant of the addition of the toxic substance and the subsequent curves illustrate possible effects of the test substance.

The operational range of Type I analyzers lies within the region of saturation with atmospheric oxygen, i.e., from 0 to 9 mg liter^{-1} at 20°C, and can be extended to ca. 30 mg liter^{-1} during saturation with pure oxygen.

The results obtained with Type II analyzers are analogous, but the range is greater and is determined by the ratio of the volumes of the liquid and the gaseous phase in the incubation vessel. It should be borne in mind when estimating the analyzer range that the system is closed, i.e., the amount

Coral Fishes

their care and maintenance

Coral Fishes

their care and
maintenance

T. Ravensdale

GREAT OUTDOORS PUBLISHING CO.
FLORIDA

GREAT OUTDOORS PUBLISHING CO.
4747 28th STREET NORTH
ST. PETERSBURG, FLORIDA 33714

Printed in Great Britain

Contents

Dr. Hans Hass
INTRODUCTION

One of the greatest wonders of Nature – accessible now to an increasing number of admirers – is the truly enchanted land of corals. There are reefs steeply ascending from the deeps, forming walls and towers – so picturesque as if they belonged to a medieval castle which for some magical reason subsided under the sea. This castle is overgrown by thousands of unusual flowers – and thousands of unusual butterflies glide gracefully about among them. Mr. Ravensdale's book tells of these butterflies, and how to bring their beauty to your own home.

How can you make these delicate creatures feel at ease in a transparent box far away from their natural habitat? How should this box – the aquarium – be made, and how should it be decorated? What kind of water do these graceful fish need, what kind of filtration should be used? If different species are combined – will they live in harmony? How can you obtain these fish, how should you feed them? what are their possible diseases . . . Mr. Ravensdale is presenting a great amount of practical advice, and answering these questions.

As I have had the privilege of visiting coral reefs in almost every tropical sea, I have never possessed an aquarium; therefore I can hardly comment on this book, nor give any further advice.

A very interesting problem is the basic question: Why are these fish in fact so colourful? A second, equally interesting one: Why are many of them so very aggressive towards their own kind? As Konrad Lorenz in his recent book "ON AGGRESSION" has explained these two questions have a common answer. To all who grow interested in the more subtle details of their pet's behaviour this book of Konrad Lorenz can also be very recommended.

I shall never forget my first impression of a coral reef and its inhabitants – it was in 1939 at the Island of Curacao in the Dutch West Indies. We had to swim across a rather murky lagoon and felt uneasy. The muddy bottom had faded away and through our diving masks we could not see anything at all but our own hands holding a spear and a frightening empty vastness all around and beneath us. We were afraid of sharks – even more so as we did not really know how far we could see and, consequently, how close they would be should they enter our field of sight. We had seen that at the other side of this lagoon there was a coral reef and into this direction we swam. And then – as if a curtain was drawn aside – we saw this very reef through our masks. It looked like a petrified forest, and the hundreds of fish gliding about through the branches of these trees – elchorn corals – glittered like jewels in every possible hue. This first impression – especially after these frightful minutes of suspense – imprinted itself in our memories for ever. Since that day I have spent thousands of hours observing coral fish and admiring their beauty. – To bring these jewels to your own home may be a quite costly and irksome task. But I should think that such effort is well rewarded.

Foreword

WRITING A BOOK about such a new field as coral fish keeping must surely be putting one's head into a noose. Theoretically all books about marine fish should differ only in text and illustration, for facts are facts – no matter how they are presented. Unfortunately, however, I have yet to read two books agreeing with each other on anything but the most basic fundamentals and the unfortunate hobbyist has the consequent dilemma of which to read and which not. One eminent author often conflicts with another to a considerable degree. I have seen such examples where one well known authority states quite emphatically 'Dim lighting is essential or the aquarium fish will be blinded' whereas another states, with equal zest 'Aquarium fish require powerful lighting or their smaller food dies'. And now we have presented you with yet another work to contend with. We have perhaps covered ourselves by calling it 'Coral Fishes' and not 'Marine Fishes' for the word marine covers such a gigantic field that several volumes would hardly touch the subject and we are interested here only in those creatures one can keep in the home aquarium, namely coral fishes. But however conflicting our views may seem, compared with others, rest assured that this book is based upon visual experiment carried out here, in London, during the year 1966/7 with water, conditions, climate and equipment as per availability. Therefore, apparent contradictions to other authorities do not necessarily mean that either is at fault. The scene can differ so greatly 5000 miles away that two authors writing 'correct facts' about the same subject can 'apparently' violently disagree – whereas it is only the elements that differ.

Tropical fishkeeping in the home dates back to the Chinese civilisation but coral fish keeping has hardly begun, consequently literature peculiar to this rather specialised subject is limited. The hobby is however growing in leaps and bounds and the time is not too far distant when the hobbyist will be keeping the bizarre salt water species alongside his more common freshwater collection.

It seems incredible .that, in this modern jet age, where man is spending countless billions on space study and unearthly phenomena, he has hardly a notion of what lies beneath the oceans of his own planet. The coral seas, let alone the great ocean depths, have hardly been penetrated at all – we cannot even identify most of our own sea creatures. Breeding, or even sexing, is, as yet, still a mystery where most coral fishes are concerned.

This book is, we hope, one of many to come in these early days in the fascinating study of an entirely new and remarkably interesting branch of Zoology as yet relatively untouched by man. Whatever else you may achieve, the keeping of coral fishes in your home could help you become one of the pioneers of an entirely new science. Remember, even the scientists – with all their resources, only know perhaps a little more than you, the aquarist, and whatever you may learn during your practical coral fish studies should be passed on to those concerned. This is one field in which the man in the street can teach the experts a thing or two. The Ichthyologists responsible for listing, naming and cataloguing of the countless thousands of marine species have an almost impossible task unless help is given by the people who can study, first hand, living fish habits – the aquarists. The Ichthyologist normally deals only with formalin treated corpses, and as we all know, a fish can dramatically change colour several times in as many minutes when alive, let alone dead!

There are a considerable number of so called different species which are in fact one and the same. In coral fishes the colour, shape and habits can differ so greatly as the specimen grows to maturity that several classifications can be given to the same fish! The popular *Pomacanthus Imperator* changes through life to such a degree that even now, a juvenile is often thought to be a different species from its adult form.

So remember, things you witness in your own home aquarium may well be phenomena unknown to Science – surely a more absorbing hobby could not be devised.

The first coral fish enthusiasts were undoubtedly the skin divers, for only a few feet below the surface of any barrier reef lies one of nature's greatest wonders – a breathtaking panorama of unbelievable beauty and splendour. Colours so startling, shapes so weird, and forms so fantastic that the enhanced diver would hardly have been human had he not wanted to capture some of the wonder he had witnessed,

14

by bringing home some of the glory of the deep. And so the coral fish hobby began; soon we began seeing these wonderful creatures – so different from our freshwater specimens – here in our European Zoos and public Aquaria. The inevitable followed, and now we are attempting to keep them in our own homes. As interest rises so prices get lower, and although coral fishes are not cheap, they are no longer beyond the pocket of the average aquarist.

Coral fishes are, without doubt, the most colourful creatures ever created. From the amusing Wrasse to the weird Sargassum all, in time to come, will be a commonplace sight in our aquariums. Their colours must be seen to be believed, and once a beautiful marine set-up has been viewed by the enterprising enthusiast, he finds it difficult to remain satisfied with his own drab freshwater specimens. Marine fish-keeping is by no means a beginner's game, but any average aquarist should have no troubles that are not already associated with fishkeeping, provided he sticks to the basic rules. Marine fish are as expensive as they are colourful and for this reason it may be better to start simple and learn the easy way. Don't hurry off to buy the finest equipment available, experiment first – by the time you tire of the simpler species, you will have learned enough to help you with the more expensive specimens.

Finally, although an attempt has been made to be as up-to-date as possible, in the short time it will take this draft to be edited, checked, corrected and printed it will reach you – the reader – long after new specimens, which arrive daily, have been added – that's how fast the growth of this new and extremely interesting hobby is.

1
The Coral Sea

IN ORDER TO keep animals of any description it is first necessary to know something about the environment from which they come and then attempt to imitate it in everything except natural enemies. An attempt to simulate a coral sea with its tropical climate, tides and vast size, would obviously be a feat beyond comprehension, let alone possibility, and so we created a miniature sea – the aquarium. With this simple container we must now do our utmost to make the fish we intend keeping, feel as much at home as is within our power.

A coral sea is the wildest, least cultivated and most vicious place to be found anywhere on earth. A vast, dangerous, frightening wilderness completely lacking in organisation or community. There are no policemen or doctors, and as in all such environments, the strong grow stronger whilst the weak live in constant fear of their lives, and not without just reason. The lesser fish venturing too far from its home invites instant death and, in consequence, the coral fishes' sense of self preservation is highly developed. The slightest upset of this delicate condition may well prove fatal.

Unquestionably, the African jungles are mere playgrounds when compared to the sea. Disease runs riot and, with no help forthcoming from man, as on dry land, epidemics flourish.

When considering these conditions, removing animals from the sea and placing them into such vastly different surroundings as the aquarium, we are expecting rather a lot in hoping that they will survive at all. The shock sustained through this tremendous upheaval in the life of a timid specimen is a paramount cause in the high death rate of coral fishes when compared to freshwater specimens.

How can we even hope to begin imitating the surroundings in which coral fishes have existed? True we can remove a considerable number of their large and obvious enemies but what of the smaller ones and, most important of all, *which of the latter are in fact enemies?*

The crystal clear, beautifully kept, hospital clean aquariums one usually associates with marine fish look fine, but just how important *is* such cleanliness and what have we removed by super filtration that *should have been left where it was?*

Our aim should be to compensate for whatever a coral fish is deprived of when removed from its natural habitat.

When we remove the obvious enemies such as large fish, have we also removed the natural fear, inbred since time itself, which plays such an important part in the well-being of many timid specimens? Many coral fish can, without doubt, become very tame with their new owners and perhaps forget the wilds to a certain extent. But this usually applies only to the larger fish such as *Platax* or *Balistes* where natural existence would not have been one of abject terror anyway.

Most of us have, at some time or other, seen glorious colour films of tropical coral reefs; fabulously clear underwater scenes, photographed in crystal clear water but these must surely be the only type of scene one *can* photograph anyway. Filming in murky or dark water is impossible therefore we *can* only see underwater scenes taken in perfectly clear, light conditions. This does not mean that *all* coral fishes live in these conditions, it simply means that these are the only ones in which we *can* observe them.

The mere size alone of the average coral fish eye indicates their almost nocturnal existence. Most coral fishes therefore exist in conditions where eyesight plays only a secondary part in their lives. Transferring a fish with extra large eyes into an overbright aquarium can obviously lead to trouble. Indication of the obvious sensitivity peculiar to the eye of a coral fish is the fact that Popeye (exophthalmy) is relatively more common among saltwater than freshwater fish.

Darkness we cannot completely simulate without impeding the whole object of our intentions but dark corners and caves must be of paramount consideration when preparing an aquarium. We then remove the burden of decision from the aquarist and place it into the hands of the fish. Should the light prove too intense for any particular specimen he will no doubt seek protection in the murkier depths which you should generously provide.

Another subject we have yet to study is pressure. Whether or not this is in fact a problem we do not know. Divers are first taught to surface slowly from great depths or their respiratory system will be damaged, and yet many eminent ichthyologists state, quite categorically, that pressure changes do not affect fish. Perhaps not, but even if it does, there is little that we can do about it. On the other hand it is common knowledge that any fish will grow considerably larger when placed in a very large aquarium. Is it in fact the greater amount of water that is responsible for this increased growth rate – or the greater *pressure?*

To return to the point however, it is reasonable to assume that the bulk of available coral fishes have been caught in comparatively light, shallow, waters or the collector would not have been able to catch them at all, for most coral fishes are caught in a hand net, one at a time (thus the reason for the higher cost of marines).

Another thought worth considering in an effort to understand the changes a coral fish suffers in captivity, is the fact that sea fish, unlike freshwater, have never been subjected to close contact with man. Most freshwater fish have existed through the ages in very close contact with human beings – in rice paddies, shallow pools and small rivers. They have eaten his refuse, swum beside him and been captured, eaten, or even bred by him. Sea fish however are probably the only animals left on earth that have been scarcely touched by the inquisitive hand of men. If you were to gather all the ships in the world and place them, refuse and all, into the Pacific, they would scarcely make a splash. Even the mighty Amazon would hardly be felt.

And so one would ask yet again; how on earth are we to provide conditions even remotely similar to those beneath the sea? This is the object of this book – to help the aquarist achieve the *best* possible conditions in an aquarium, and create, in his home – a miniature coral reef.

2
The Aquarium

NOW THAT WE have some idea of the natural habitat of a coral fish, our job is to provide a home large enough for comfort and with as few enemies as possible. The larger the home, the less chance of unavoidable enemies taking hold and getting the upper hand due to low resistance factors in fish cramped for space. The selection and preparation of your aquarium is the most important decision you will have to make, so make it wisely and start on the right foot.

In order to limit the amount of 'known' undesirable factors in the aquarium we want the least 'unknown' ones possible and every item, however remotely capable of introducing detriment, should be isolated. This includes the frame and jointing compound of the aquarium itself. An all glass container must therefore be the obvious ultimate but water volume is of equal importance so, until the all glass 50 gallon aquarium is produced, we must satisfy ourselves with framed containers and eliminate all possibilities of contamination systematically.

Select the largest and finest quality aquarium you can afford or accommodate – the capacity of which should not be less than 10 gallons. Some stainless steel models are recommended for use with marine water but do ensure that only those *guaranteed* as such are used, and treat the glazing compound with suspicion. There *are* one or two glazing compounds guaranteed impervious to salt water but they are rare.

An all plastic aquarium is an ideal container for salt water fish but as these are usually of low capacity they are of little use for anything but the very smallest specimens.

Nylon coated angle-iron framed aquariums are by far the best for salt water does not attack the sheathing even after a considerable time. The putty however should be regarded as toxic and not allowed free contact with the water at all. An existing angle-iron framed aquarium can easily be converted for salt water use by simply removing the glass, having the frame dipped in nylon and replacing the glass with non-corrosive putty (which should *still* be isolated from the water).

Plastic coated aquariums are perfectly alright provided the sheathing is handled with care, for unlike nylon, it will peel off the framework if damaged.

Asbestos is not an undesirable material for aquarium construction. Indeed quite large inexpensive units can be easily made by simply cutting a 'window' out of the side of a domestic water supply tank and fitting glass into the aperture. The Germans use asbestos units extensively and most European Zoos utilise these very tough aquariums for their smaller marine displays.

Should you be fortunate enough to be able to have a say in the manufacture of your aquarium, have the glass extended beyond the top edge of the frame. This will alleviate any chance at all of water coming into direct contact with the frame – especially if the frame is further protected by nylon.

Having eventually decided upon the aquarium best suited both to your needs, and the fishes' health you must now treat and condition it. We have already decided to consider glazing compound detrimental, so now we treat it accordingly. To begin with, do not purchase any aquarium which has badly fitted glass and remember to allow for a greater putty exposure when it is filled with water. If you buy an aquarium with a large gap where glass meets glass to begin with, the gap will be even greater when the water pressure forces the glass further into the putty.

Fill your new tank with water and allow it to stand for several days before attempting to seal off the putty. Once the glass has 'settled in' (This could take months if the glazing compound is excessive) the water should be drained off with a syphon. *Never* tilt an aquarium in order to empty it or leaks will surely follow. Dry off the interior and remove all excess putty with a knife. The inevitable seam left behind should now be sealed with a liquid plastic, black asphaltum varnish, aqua-sealer or one of the new wonder sealers such as *Dow Corning aquarium sealer*. The latter incidentally, can be used to build an all-glass aquarium, for it is powerful enough to hold a fifty gallon *frameless* aquarium together!

Undoubtedly the aquarium manufacturers will soon follow up the meteoric rise in popularity of the saltwater fish keeping hobby with a purely marine aquarium, sealed with flexible plastic channelling instead of putty but, until then, treat your present one with meticulous care.

Blue Koran Angel *Pomacanthus Semicirculatus*

Once the tank has been sealed and allowed to set, your attention should be turned to the accessories which will be undoubtedly fitted both inside and above. You will obviously wish to illuminate the unit and a hood is the recommended storage place for lighting equipment. The hood also serves other purposes such as preventing surface dirt forming, deterring fish from leaping out of the aquarium (many coral fish are prodigeous jumpers) and, most of all, it will prevent excessive water evaporation. Do not however be over zealous in fitting a hood too closely. Ensure that air is not cut off from the water surface, for saltwater contains 20 per cent. less oxygen than freshwater so it is imperative that no artificial impediments are introduced.

The aquarium cover should not be manufactured from unprotected metals for condensation forming on the inside constantly drips back into the water. The canopy can be protected by a coating of nylon, epoxy resin or polyurethane paint but better still is the addition of a sheet of glass separating the canopy from the water, not forgetting of course to allow an air access gap. This system is by far the superior one, for even a nylon coated hood cannot allow for such dangers as brass capped bulbs.

Your tank is now ready for sterilisation and all intended interior equipment such as heaters, thermostats, internal filters, decorations, etc. should be dealt with at the same time. This does not include such decoration as coral or floor covering which is dealt with separately in a later chapter. Heaters or thermostats with pure rubber end fittings are inferior to plastic construction for the rubber breaks down in salt water and soon becomes slimy, and unserviceable. After a thorough individual cleaning of these accessories, put them into the aquarium and fill with fresh water. A disinfecting agent such as Diseasolve, methylene blue, or potassium permanganate should be added at double the indicated strength (usually given by the manufacturers). After temporarily wiring your heating system, switch on to full power and allow the temperature to reach 120°F. As most aquarium thermometers do not read beyond 100°F, the instrument should not be included in this operation or damage will ensue. Allow the water to stand at this temperature overnight, dropping it sharply the following morning by adding plastic bags filled with ice. Maintain this low temperature for a further 24 hours. This process can be repeated in full, using salt water, for the more meticulous. There aren't many microbes capable of withstanding this sort of treatment.

Now your attention may turn to the heating equipment. Having previously pointed out the undesirable characteristics of rubber we must now replace such items as rubber thermometer end caps and heater holder stickers with the easily available nylon types as used exclusively on Eheim units. We must then set about the concealment of heaters and thermostats. It may be advisable at this stage to fit a safety unit by the simple method of putting a heater at each end of the aquarium, operated by individual thermostats. Should one thermostat burn out, there is at least a chance that the other will prevent a complete catastrophe. However, it has always been my personal experience that the thermostat contacts weld together on these unfortunate occasions and the heat builds up instead of down but, even so, the secondary system will not come into operation in this case and so will not affect or intensify the heat. It will however prove vital in the case of a burst heater or broken wire.

Concealment of these ugly necessities is difficult for there are no plants to utilise in the marine aquarium. The heater and thermostat should be parallel and close to the bottom edge of the rear glass and this makes the job of hiding them easy for large pieces of coral will no doubt be added, and these will cover where necessary. The wires however are much more difficult to hide as are the surface tubes of an undergravel filter, should one be used. A good method of internal equipment concealment is the one adopted by the Hornimans Museum Aquarium in London. A sheet of tinted pliable plastic is curved around the three non-viewing sides of the aquarium. The two rear corners of the tank are thus left free for unsightly equipment. Needless to say, perforations must be made in the lower part of the screen to allow water circulation.

One other method of concealment can be utilised where a 'freewater' filter is in use. The heater can be put into the filter and not the tank! This also ensures that the heat is evenly distributed all the time.

Care should be taken not to cover any heater or thermostat with gravel, crushed coral or any other similar floor covering. One inch above the bottom of the aquarium is ideal. A smothered heater will not only work poorly, but also its life will be shortened.

Lastly, should you decide upon an external thermostat of the type requiring a steel clip to hold it in position, you should either run a piece of air tubing over the whole length of the clip, or dispense with it altogether and glue the thermostat directly onto the glass.

3
The Water

WATER IS UNDOUBTEDLY the most important element in the life of marine fishes, but even this problem has been simplified for us by the chemists, who have developed artificial compounds which are quite safe to use, and generally superior to natural sea water due to its lack of natural bacteria. They are in fact totally free of the countless myriads of plankton and nectons which natural water embodies. 'Live' water may be more natural when attempting to recreate 'homely' conditions but it would be a fantastic coincidence if the Caribbean held exactly the same microscopic life as the British, highly trafficated, coastal water so, assuming you don't want to go to Bermuda for your water, we must settle for Meersalt, Tropicmarin or some other well known brand of artificial preparation.

It is sometimes possible to purchase clean genuine tropical salt water from a zoo or public aquaria. The London Zoo for example will willingly supply their salt water (which is brought in by tanker from the Bay of Biscay) at a cost of little over one shilling per gallon to genuine clubs or organisations. But remember; you must be consistent with water, it is dangerous to use 'live' water one day and synthetic the next. Remember also that artificial water is usually 'buffered' to the correct PH – an important consideration and, even more important, it is adjustable. Density can be changed at will with artificial water but only a salinity change is practicable with genuine sea water. If you live near the sea and are sure of the cleanliness of your water or can even collect it out to sea a mile or so, then by all means use it, but do make sure that you thoroughly filter every drop you use. Change it often and don't even allow it to smell, for a bad odour can only mean decomposition, because cold water plankton cannot live long under tropical conditions in the aquarium and their death soon causes the water to foul.

For the more industrious aquarist who would like to manufacture

his own salt water the following formulae are recommended:
The 'Kramer and Wiedemann' formula (To be mixed with 100 litres of water).

  0.5 grams potassium iodide (K.1)
  2765 grams Sodium chloride (NaCl)
  692 grams Magnesium sulphate crystals ($MgSO_47H_2O$)
  551 grams Magnesium chloride crystals ($MgCl_26H_2O$)
  145 grams Calcium chloride crystals ($CaCl_2\ 6H_2O$)
  65 grams Potassium chloride (KCl)
  25 grams Sodium carbonate ($NaHCO_3$)
  10 grams Sodium nitrate ($NaNO_3$) (not essential)
  0.5 grams Sodium phosphate ($Na_2\ HPO_4$)
  1.5 grams Strontium chloride ($SrCl_2$)
  10 grams Sodium bromide (NaBr)

The potassium chloride should not be added until afterwards when it should be introduced in a dissolved form. The sodium nitrate can be excluded altogether.

The Berlin Aquarium formula (To be mixed with 100 cubic ·metres of water).

  2816 kg Sodium chloride (NaCl)
  65 kg Potassium chloride (KCL)
  550 kg Magnesium chloride crystals ($MgCl_2+6H_2O$)
  692 kg Magnesium sulphate crystals ($MgSO_4+7H_2O$)
  25 kg Sodium bicarbonate ($NaHCO_3$)
  122 kg Calcium chloride ($CaCl_2$)
  100 g Iodine potassium crystals (KJ)
  100 g Sodium bromide crystals (NaBr)

There are of course many other formulae but these two have been tested and recommended by many leading authorities. The obvious omission here is the one used by the fabulous 'Exotarium' in the Frankfurt Zoo but, as this is extremely complicated we do not recommend anyone with less qualifications than a chemist to attempt it. A letter to the curator will however be answered if you are interested. The density of your salt water is of paramount importance. A specific gravity reading can range from as little as 1.016 or as high as 1.030. These figures should not be confused with salinity content but simply mean that the water weighs e.g. 1.030 more than pure distilled water when salts, etc. have been added. Salinity measurements however are expressed by content – or parts per thousand. Water having a salinity of, for example, 34.5 per thou. would have a specific gravity of 1.027.

Temperature can affect water density so beware of water taken from a cold sea which is heated for the tropical coral fish aquarium. Density readings are more accurate when taken at a temperature of 60°F.

Determining the correct specific gravity for any given community tank is not an easy task. The difference in nature itself can vary greatly in a mere one mile stretch of water. I personally checked the density of water used in the marine aquariums of four different European countries and found each one to differ. In Antwerp the readings were as high as 1.032 whereas the Frankfurt experts preferred 1.025. I have personal tanks containing fish that are perfectly happy at a specific gravity reading of 1.020 which incidentally is the average of sea water surrounding the British Isles. But I must confess that the coral fishes I saw in the Antwerp Zoo, so ably run by Mr. Van den bergh were the finest I have seen anywhere in the world.

The smaller coral fishes unquestionably prefer a heavier density whilst the larger specimens such as Platax and Angels are happier around the 1.020 mark.

The specific gravity you eventually decide upon must therefore be determined by the fish themselves (a point worth considering with many decisions). The magnificent coloration of many coral fishes is decided to an extent by the intensity of salt in the water for even the slightest density will be far in excess of the salt within a fish's body. With this information alone it is quite possible to determine an acceptable water density for any given specimen. It need hardly be pointed out that fish differing greatly in salinity preference should not be mixed.

Start with a specific gravity of 1.025 – a simple hydrometer will indicate this – and raise or lower where necessary by judging the coloration of your fishes. An anaemic colour indicates a possible under-rating of salt whereas overintensity accompanied by constant trips to the surface for air shows plainly a need for the addition of freshwater. Do not make any change in density too quickly and, most important, remember that salt takes some time to dissolve so do not take readings until enough time has elapsed for this (up to twenty-four hours).

By all means check the specific gravity of the water your specimen arrives in but *never* accept this figure as being necessarily accurate.

The final and most important factor to consider when changing salinity is that additions of salt (or temperature increases) should be accompanied by an air supply increase for oxygen is consumed faster at higher temperatures and decreased in quantity by the addition of salt.

The temperature of your aquarium water will depend upon the variety of fishes kept but an average figure should be near 75°F. This leaves margin for increase should an epidemic break out, allow reasonable oxygen consumption, and prevent excessive water loss through too high an evaporation rate. Raising the temperature may of course induce new arrivals to feed – an important point – since fish consume larger quantities of food at higher temperatures.

Water loss by evaporation can, in theory, be compensated for by the addition of pure freshwater for only pure distilled water is lost thus but, in practice, a certain amount of 'salt crusting' may occur around the filter and on aquarium edges and this must be compensated for. As some of the trace elements, so carefully added by the chemists, must obviously leave the water in addition to pure salt, it would be senseless to attempt regaining the specific gravity by the mere addition of salt alone. There is no way in which the lost trace elements can be measured so the obvious answer can only be a complete water change. The period between water changes will depend entirely upon the number of fish per gallon in your aquarium but, provided the recommendations in this book are adhered to, a partial (20 per cent.) water change once a month and a complete one every six months should eliminate trace element loss factors.

Another undesirable element which frequently occurs in the aqua·rium is surface scum. This should be removed periodically by drawing sheets of Kleenex over the surface of the water at high speed. Newspaper should not be used for it is usually manufactured from low grade paper, is contaminated with ink and is rarely very clean. Kleenex however is spotless, convenient to use and, due to its high absorbtion quality, suction is created as the water floods into the fabric causing a minor 'vacuum cleaner' action which sucks up the scum.

Soft water is preferable when mixing your solution so, wherever possible use clean rain water. Tap water is perfectly alright provided it does not exceed fifteen degrees of hardness. Distilled water can be used for adjusting purposes.

Now we come to the most important aspect of all, water acidity. You will no doubt be adding coral to the aquarium for decorative purposes. Coral, due to its calcium properties, will help prevent an acid build up but rarely sufficiently enough because calcium is not particularly soluble. Therefore we must accurately measure the acidity or alkalinity of the water – *after* our synthetic salt preparation has been added. There is no point in doing it before as most commercial salts are buffered already. A saltwater PH test kit is essential – the freshwater type being completely unsuitable in this instance.

PH value is a means of expressing the concentration of Hydrogen ions in a given volume of water. It is not constant however, varying according to the weather, atmosphere and temperature. The PH value of your water will determine to a great extent the types and number of organisms capable of existence in your aquarium. It is therefore most important that the pH value of your water is correct and constant. Experience has shown the best pH value to be 8.3. A difference of more than 0.5 either way should not be tolerated. Like density, acidity changes must be made slowly therefore the best way to introduce chemical pH adjusters is via the filter. Chemicals added in this way will obviously be distributed through the aquarium at a slower rate than direct treatment.

It is rarely necessary to lower the pH value but, should you find it necessary, add Sodium biphosphate until 8.3 is attained. More likely will be the necessity to raise the pH value and this can be done in a most astounding way – by adding pure washing soda! To explain; numerous reputable authors state quite plainly that Sodium bicarbonate should be added to water that is too acidic. There are even test kits on the market with Sodium bicarbonate supplied for this very purpose. Sodium bicarbonate will unquestionably function as directed in *fresh* water but *never* use it in salt water. It simply will not work so disregard any such instructions whether in a book or on a product. The chemical in fact required is Sodium carbonate and this, believe it or not is exactly what washing soda is and no ill effects will ensue provided it is added to the water slowly. A correct PH value of 8.3 can easily be attained with this chemical.

Algae, although having a preference for a PH value of 10, will flourish in any well-lit aquarium; it can even be accelerated by the use of 'Growlux' lighting. Although unsightly, algae is essential to the well being of many coral fishes. It supplements food (as shown in a later chapter), provides oxygen and helps prevent nitrate production but beware of algae destroyed by cleaning (glass scraping, etc.). Dead algae is highly undesirable to the marine aquarium. Lastly, do not use water for fish that is too 'new'. Salt water, whether synthetic or not, should be allowed to stand under vigorous aeration for several days before the introduction of fish. This will give any excess sterilization chemicals chance to leave the water in the case of synthetic preparations and time for impurities to be filtered from sea water.

4
Filtration

NOW THAT YOUR WATER has been cleared of impurities, carefully treated and meticulously prepared for the acceptance of fish, your labours must turn to the continuation of this condition – the obvious answer to which is filtration. This can be achieved in a number of ways, the simplest being a subgravel filter. The use of this device will depend on the amount of floor covering you intend using. A great depth of this medium is not desirable for it soon turns into a home for bacteria, not all of which are harmless. Therefore, provided you use a subgravel filter which does not require a great deal to hold it down there are no objections. The type of gravel however is restricted to a grade large enough to be able to resist filter penetration yet not so large as to become host to pieces of uneaten food which may then become trapped at the bottom of the aquarium. Sand is much too fine for use with most subgravel filters so crushed coral, quartz or gravel graded to one-eighth of an inch should be used.

The general objection to subgravel filtration is the fact that refuse is not removed from the tank but simply pulled down below the gravel, where, (unlike freshwater where the plant feeds upon it) it promptly rots and causes gas pockets which can be dangerous. This seems poor logic when one considers that hardly any filter system will remove refuse from the aquarium circulation anyway. It simply collects the dirt together into one place where it can be conveniently removed – the refuse is still in circulation until then. In actual fact, only the very finest of dirt is moved at all by the subgravel filter and this is why water filtered in this way always looks cleaner. It is the fine suspended water particles that mar that clean crystal effect not the larger dirt such as fish droppings. These should be removed by syphon in any case, no matter what filter is in use. My personal objection to the *sole* use of a subgravel filter is the considerable effort required to clean it. The whole tank has to be disturbed to such an extent that the fine rubble collected in the first place is instantly spread throughout the water. The use of a subgravel filter as a *secondary* filter however receives full blessing for then we are getting the best of both worlds; crystal clear water plus external filtration. The most obvious point in favour of the subgravel filter is its cost, which is very little, and, for this reason, it is well worth considering as a supplement to a good outside filter.

Inside filters other than subgravel, are not recommended, for they merely act as gathering points for noxious gases and are notorious food concealment 'homes'.

A first-class outside air driven filter such as the 'Bubble Up' or Halvin 'Champagne' can be thoroughly recommended for smaller aquariums. Aeration is not necessary with these types for an air stone is incorporated into the design. The filter medium however should be selected with great care. Nylon floss, if changed frequently, will collect a considerable amount of debris in a very short time but do not be misled into believing that your aquarium is clean because the filter is dirty. The dirt in the filter has not been removed from circulation at all – it is still in the aquarium, the only difference being that it has been collected into one spot. Never use fibre or spun glass. The minute glass particles soon enter the aquarium where they are swallowed by the fish or introduced into the gills where inflammation soon breaks out. Gill epidemics can often be traced to fibre glass so do not use it.

Activated carbon, as used in fresh water aquariums can cause heavy changes in the pH value of your water, so in preference to this, use one of the resin exchange compounds. They are rather expensive but well worth the extra cost. A new type of carbon with the ammonia removed is being used in Germany and, with the moderate personal success I have experienced with this substance, I am sure it will soon be generally available. Carbon of any description should always be soaked in salt water, preferably in the dark, before use and then thoroughly washed in fresh water.

The external filter which is not motorised can only function within the limits of the air pump operating it, so never use one of doubtful quality. A good vibrator type is inexpensive and usually efficient, being quite capable of operating an air stone in addition if required. Should the room in which your aquarium stands not have a reasonably pure atmosphere (Heavy kitchen smells or smoky air can prove fatal to fish) fit the pump outside on a window sill. There is little sense in drawing in air from a smoke laden atmosphere and pumping it straight into a tank. Do not be afraid of turbulence for the barrier reefs suffer a considerable pounding most of the time and no sea is ever still.

Fine air bubbles are the best form of aeration but beware of using a soluble air stone (most of them are) or, better still, use a wooden block in the place of stone. A small piece of silver birch with an air tube fitted into a drilled hole will generate the finest of air bubbles and there is no fear of adverse affects on the water. The only instance where larger bubbles are preferable is when pure oxygen is being used in the form of cylinder aeration. Small bubbles of pure oxygen are lethal to coral fishes.

When deciding the necessary size of an external filter in relationship to the aquarium, allow enough volume of filter to accept 1 oz. of carbon per gallon of water. A ten gallon tank will therefore require a filter large enough to accept 10 ounces of carbon plus the nylon wool. The gallonage of an aquarium being calculated by multiplying its cubic capacity in feet by 6¼.

External filters, especially power driven types, should always be protected from overflow accidents such as blockages by the fitting of an extra return pipe. A non return valve can also be added to the air supply line – essential if the pump is placed at a lower level than the aquarium.

The ideal filter for a marine tank is of course the modern electric power filter. These come in various shapes and sizes but those not reliant upon a syphon system are generally preferable. A power filter relying solely upon a syphoned supply can cease to function if the water level drops as little as ½ inch and they are apt to be rather noisy in operation.

Due to the extremely fast rate of water· movement in high power filters, algae tends to collect faster. The colour green tends to slow down this reaction so filters with a water consumption greater than forty gallons per hour should be either coloured green or shaded. Tubing can be purchased in this colour and should be used to replace the normal transparent type. The heavier filters such as Eheim are coloured green throughout and are extremely well suited for marine use.

Due to the fast turnover of water in high speed pumps or filters, the medium contained inside should be changed regularly. The greater the flow rate of a filter the more frequent a medium change should be.

The most desirable aspect of a power filter is its ability to draw 'old' water from one corner of the tank and, after filtering it, return it to another location altogether. The Eheim filter returns water through tiny holes drilled in a tube which i suspended above the water level. This causes a considerable turbulence which aerates the water fully.

Care should be taken however that these bubbles are not too fine when in great numbers for great masses of 'overfine' air bubbles can actually get into the fishes' blood system. They are then taken through the arteries until they reach the point where artery becomes vein – the capillaries and, in the more minutes ones, such as those of the eye, blockages can occur and even develop into popeye (Exophthalmy).

Another filtration system suitable for the home aquarium is one that is still at an infancy stage. It operates on the principle of gravity. A normal outside filter can be used along with a very simple air pump but without the use of carbon or nylon wool. The filter is allowed to fill by a very slow air lift. This unclean water is fed into the lower half of the filter and the return is taken from the top. Turbulence is not permitted, the whole principle being that the sediment collected by the intake tube is allowed to fall to the bottom of the filter by sheer gravity – there being no fish in the filter to keep the particles suspended as in the aquarium. Sediment settling in the bottom of the filter will not be returned to the tank via the outlet, provided water movement is restricted to such an extent that syphoning is not induced. A good method of preventing too great a current is to add a little brine shrimp to the filter. If the brine shrimp is sucked into the aquarium then the filter is being operated too fast and adjustment should be made accordingly. Fine particles, too light to drop to the bottom of the filter or 'Sediment remover' may be eaten by the brine shrimp, thus assisting the job of filtration. Should this system eventually be noticed by the manufacturers a 'sediment remover' will undoubtedly become available fitted with baffle plates and incorporating removal trays but, until then, you may wish to experiment yourselves. Try new ideas – after all, this is a new hobby.

There are of course many filters not mentioned that are easily adaptable for marine aquarium use but they are far too numerous to list. The most remarkable aspect of this fact is that not one of them deals with the removal of liquids! Most aquarists go to great lengths not to overfeed for fear of uneaten foods fouling the tank, spend hours removing droppings, corpses and algae, spend considerable sums on dip tubes, sediment removers, electric vacuum cleaners and filters – all designed to keep solid waste products out of the aquarium, but what about the waste fluids? A coral fish passes considerably greater amounts of waste fluids than solids. It can, on observation, be actually seen urinating. Unlike freshwater fish (which have a greater salt concentration in its system than in the water it lives in) the coral fish has to 'absorb' great quantities of salt water, by drinking, in order for their gills' to be able to extract the pure water, an element vital to all forms of life. This means that coral fishes consume great quantities of water daily and, as the bulk of this is excreted into the aquarium, a saturation point must inevitably be reached. It is not absorbed by plant growth as in the fresh water aquarium, nor is it removed by evaporation so the urine content builds up until it becomes necessary to change the water completely.

As there are, as yet, no products available which remove these undesirable fluids (They would not have been passed from the fishes' system at all had they been wanted) the more enterprising reader may wish to construct one of his own. In order to do this he needs only two old inside filters and an understanding of the basic principles of fluid separation. Without being too technical, we know that water is a reasonably hard fluid therefore, if a vigorous stirring is given to a bucket of water, very little will happen but the addition of a softening chemical and a repeat of the process will result in a dense foam forming on the surface. Remove the foam and continue the beating until no more foam occurs. You will then be left with the original hard water.

Urine is a very soft fluid and will foam readily. Mixed with water and given the 'stir treatment' it can be removed in the method previously shown.

A simple device for the removal of urine from the marine aquarium can be made from two old inside filters in the following way:

The two filters are placed on top of each other. The top one will need an extension fitted to the air stone access hole to raise the level of fluid allowed to stand in it. The bottom filter should be fixed into a corner of the aquarium with its top one inch above the surface of the water. The air stone, which is fed by air line to the bottom of the filter, should be adjusted to give its maximum output. Water is drawn through the holes (drilled all round the filter) into the chamber, where it is subjected to the violent agitation caused by the air stone. This causes the water to foam and consequently rise through the air stone access hole of the second filter, which should be placed directly on top of the first one. If a two inch length of tubing is fed into the bottom filter through the top air stone access hole, the foam will be forced up through this and into the top filter. As this filter is not in contact with the heated water there exists a slight temperature change which immediately cases the foam bubbles to burst and consequently liquify in the top chamber. As they enter this chamber through a tube two inches higher than the bottom level they can only run down the wall where accumu-lation soon converts it into pools and, eventually a dark yellow liquid fills the top chambers. This liquid should be emptied before it reaches the two inch level. The result of this removal of urine will be happier fish and cleaner water, which is white and not the usual yellow. A thoroughly filtered tank.

An Ozone reactor tube, when turned up to give a high air output, will do the above job quite well but ozone and reactor tubes are discussed fully in a later chapter.

5
Decorating the Aquarium

DECORATING A MARINE AQUARIUM is quite different to decorating a freshwater set up mainly because plants, which are the basis of design in fresh water, are very rare in salt water. Plants are in fact quite out of place in the coral fish set up for only one or two species such as Sea Horses, Trunkfish and Sargassum are found in the marine 'grasses'. In fact most marine 'plant' life is not a plant in the true sense of the word at all but merely algae and, as we have already discovered that algae prefers a PH of 10, we have this problem to overcome right from the start.

Most algae is comparatively rootless, the 'plants' are found either loosely held in sand or adhering tightly onto rocks. Removal from the rock usually results in death so do take the rock as well if you are collecting wild specimens. Should this be the case do not expect 'plants' taken from a cold sea to live in a tropical or even room heated aquarium for periods longer than a few weeks at the most. Beach 'plants' should be rejected on the grounds that they are used to conditions which cannot exist in the aquarium. Shallow water 'plants' are almost as impossible for they are subjected to great lengths of intense light and ultra-violet rays almost constantly from the sun. They are furthermore allowed long periods in the dry when the tide ebbs and flows. To reproduce even the light alone would mean the use of powerful Growlux and other strip lighting for long hours each day. This lighting would of course not only accelerate upper plant life but lower as well and the result would soon lead to dense algae growing at random throughout the aquarium and on the glass. Chemicals used to impede the progress of lower algae forms would eventually lead to the destruction of all the 'plant'.

Plants which *are* rooted to the sea bed should also be rejected if they cannot be removed without damage (a difficult job) for it is difficult enough to keep healthy 'plant' alive let alone damaged ones.

There are one or two 'plants' offered for sale in shops specialising in marine aquaria and these, although difficult, have been found quite hardy once the transfer has been overcome.

Penicillus Capitatus is a nice little 'plant'. looking somewhat like an upturned palm tree or shaving brush. It requires of course considerable light before settling down. *Udotea Flabellum* is another tough 'plant' and looks very much like a miniature sea fan. *Grassfolia* comes in two forms; the small fern like creeping 'plant', which soon covers the bottom of your aquarium once it has been established, and the giant variety which needs very little light at all. The latter is one of the hardiest of all marine 'plants' whilst the former is an excellent food for herbivorous fishes. For this reason it may be difficult to grow in a community tank. Giant *Grassfolia*, apart from being strong, has its uses as a refuge for sea horses and other 'plant' living animals.

All these 'plants', although possible for the expert, are not advisable for the novice. His efforts should first be restricted to the task of keeping his fishes alive. Once this has been achieved he may wish to extend his studies to plants and algae, about which so little is known. The more enthusiastic aquarist having reached this stage and wishing to try 'plant' should begin with a second tank and succeed in isolation before combining fish and plant together. After going to a great deal of trouble setting up a near 'hospital clean' aquarium it would hardly seem logical to introduce such an unknown factor as 'plant'. Until you can discover the best conditions your 'plant' prefer you must expect 'plant' fatalities, and dead 'plant' in the *community* aquarium can only lead to foul water   a condition unacceptable to most fish. So keep your 'plants' in a separate container; discover, by experiment, the conditions they prefer, and compare them with the existing conditions in your fish aquarium. When the two sets of conditions are compatible you may begin a transfer.

Another different feature between fresh and marine water aquariums is the amount of floor covering used. It is quite common in a fresh water display to see great bankings of gravel piled up into rifts and with large rocks scattered around the bottom but, in the marine aquarium, we want only the very shallowest of floor covering and as little water displacement as possible. Water volume is essential to coral fishes and great chunks of rock, although beautiful to look at, displace a considerable amount of our prime requisite – water. A twenty gallon aquarium can be dwarfed in water capacity by the addition of unnecessary rocks to such an extent it will only hold ten gallons or so.

35

The lesson is plain, do not overcrowd your aquarium with too much decoration at the expense of water volume.

The first decision you are faced with is the floor covering. Silver sand is quite common and, provided it is kept clean and not allowed to blacken, will look quite attractive. It also serves as a 'trouble beacon' and will inform you of imminent pollution by turning black. A sharp darkening of silver sand is a sure indication that pollution is beginning. Care should however be taken when using this as a floor covering medium when a subgravel filter is being utilised. Silver sand is so fine it will soon filter into the subgravel unit and possibly cause poisonous gas pockets, thus accelerating pollution. Sand also produces an incredibly vile smell when polluted.

Quartz is another extremely good floor covering medium and is usually obtainable in a grade best suited to the subgravel filter – one eighth of an inch. It is almost non soluble and does not interfere with such vital factors as pH value or water hardness. Sand is perhaps a little more use to the fishes in the aquarium for many, such as Wrasse, like to bury themselves at night. Their body shape makes this feat easy when the sub sand is fine. Marine eels can also have free access to the tank bottom without damage to their skins.

Crushed coral is, in my view, the very best possible floor covering for use in a marine aquarium. Its calcium properties alone make it an extremely desirable addition to the water, for this helps to maintain a steady pH and prevent an acid build up. Furthermore it is preferred by many of those fishes which like to 'chew' on coral – again the Wrasses will be kept happier whereas they will seldom 'chew' on Quartz or sand. A layer of crushed coral, not more than half an inch deep (unless a sub gravel filter is being used) will suit the needs of most of your coral fishes and it is a simple job to syphon the whole of it from the bottom of the aquarium into a bucket when cleaning time arrives. It should be thoroughly washed, soaked in a mild solution of bleach (two tablespoons of bleach to one bucket of water) for one hour and then rinsed for at least half an hour with freshwater. This rinsing should be maintained until the coral no longer smells of bleach. It may further be cleaned by a soaking in salt water for as long as possible before returning it to the aquarium.

A few rocks may be added if you wish, but do place them in such a way that food particles or dead fish cannot be trapped out of sight where they can pollute the water, and make sure that the rocks are igneous and cannot dissolve into the water. Fused glass can be used

36

Wimple or Bannerfish *Heniochus Acuminatus*

Common Clown *Amphiprion Percula*

in place of rock and, once an algae growth has coated it, the ruse cannot be detected.

Slate is often seen in a marine aquarium but, if you really must use it at all, take care that there are no iron deposits streaked through the layers – or any other undesirable elements.

Coral is an attractive substitute for rocks and certainly of more' use. It can quite easily be set up to simulate a miniature coral reef and, being calceous, it will prove helpful in maintaining a correct acidity balance. Coral comes in all shapes and sizes but can unfortunately become a good breeding ground for unwanted bacteria. Should uneaten food be allowed to penetrate the delicate fingers of the coral, foulness will be the inevitable result so do not feed directly above 'fingered' coral, but over a bare patch of sand. The food can then be removed after the fishes have finished with it. Do not be over zealous in cleaning 'finger' coral after its initial purification for many of the long nosed Chaetodons such as Chelmon Rostratus will spend many hours picking at the crevices of the coral in search of the minute animals so vital to their existence with their long snouts specially designed by nature for just this purpose.

The coral we are discussing now is of course dead (a later chapter deals with live coral) and is easily obtainable in marine specialist shops. It is however very often sold as 'ready for the aquarium'. This statement should be completely disregarded and all coral treated as highly contaminated. Unclean coral does not necessarily smell although the existance of an odour obviously indicates decay of some description. All coral should first be placed in a strong bath of bleach (one cup full of bleach per gallon of water) for several hours and then boiled in a non metallic container for several more hours. It should then be soaked and rinsed in freshwater for two or three days at least and finally soaked in salt water for a further day. The fastidious aquarist (we should all be such) may also use a mild solution of hydrochloric acid for removing the dead animals adhering to the outer layer of coral but great care should be taken not to completely dissolve all the coral – especially organ pipe coral which is very delicate.

Shells should be regarded with the greatest of suspicion as they are rarely free from some form of dead animal which once inhabited it, and this, if not removed, could prove fatal to the aquarium inhabitants. Shells may also upset the balance of chemicals we are so careful to place in the water and often act as host to unwanted parasites which seek refuge around the very tiniest of curly bends in the structure of the shell. Any shell you intend placing into your aquarium should be thoroughly examined for corpses. I once found a small octopus in a giant queen helmet shell, so treat shells with even more care in cleaning, than coral. Soft shells should be avoided altogether. Clam shells make excellent hiding places for fish and are relatively harmless provided they are treated well before immersion in the aquarium. As previously stated, hiding places *must* be provided for all coral fishes and now is the time to prepare their 'homes' All coral or rockwork should be arranged into 'bridges' and caves of all sizes. Nooks and crannies will inevitably result where your fishes may go for apparent safety. A coral fish, when first placed into an aquarium, will immediately head for the nearest hiding place proceeding with great caution to examine it for existing occupants before venturing inside. Finding none there he will most probably make this his headquarters unless another more attractive 'house' is discovered when his first fears are overcome enough to allow further examination of the aquarium. Once a 'territory' has been established it is protected and defended with great ferocity by the owner and any attempt to 'invade' this area by other inmates of the tank will result in the damage or death of one or both fishes involved. Without such hiding places the timid coral fishes must eventually succumb to either starvation, brought about by fear, or exhaustion.

Most coral is white and attempts to change this colour artificially may result in disaster so accept this fact from the beginning and never attempt to stain – or purchase stained – coral. It soon bceomes 'plated' with a film of algae which would hide the colour anyway. One exception is Organ-pipe coral which is bright red and can be placed in such a manner as to break up the monotony of white if required. It is also very easy to carve and holes can be cut into the coral to provide refuge for the smaller fishes. Do not allow a housing shortage in your aquarium – fights and squabbles will be the only outcome and nervous disorders are the usual preliminaries to organic ones.

Organ-pipe coral, like shells, should be given an extra cleaning for it is extremely porous and can contain much undesirable life.

Available in some gift stores are a considerable variety of fired pottery or china shells. Many of these are remarkably lifelike and can be used in preference to 'real' shells. They are non-porous and therefore completely free from living organisms – they tend to remain that way too.

Sea Fans can add a touch of plant effect but, as these are composed of many dead animals strung together over the years, they should be boiled in order to remove the flesh. The skeletons of these minute creatures will remain intact and resemble a dried tree leaf or fern when placed into the water. They are most useful objects with which to conceal such equipment as wiring and heaters etc.

You need not worry unduly about the sharpness of some of these decorative objects such as razor sharp coral for, unlike fresh water fishes, which will cut themselves to ribbons on a piece of sharp edged coral, coral fishes can rush through these jagged needles at remarkable speeds without once touching it. A coral fish in the wilds is somewhat like Brer Rabbit in his briar patch and would be most difficult to remove from a bed of sharp coral.

Plastic decoration, in the form of plant, can be added to the salt water aquarium provided it has been soaked in water for at least forty eight hours. Care should be taken to examine a prospective 'plant' for metal wires which may be used to hold it together. A large aquarium can be provided with apparent greenery by the simple expedient of a false back. The plastic – or even genuine – plants can be placed *behind* the aquarium and, provided the rear glass is not allowed a coating of algae, they will appear to be in the water and give an effect of great depth in the aquarium.

There are of course countless other forms of decoration but, before attempting a pet scheme, remember the golden rules; the least unknown elements in your aquarium, the better chance of winning the fight for survival of your coral fishes so keep risks down to a minimum. Before you add *anything* to the tank ask yourself if it is really necessary.

6
The Community Aquarium

THE INTRODUCTION OF more than one species of fish into an aquarium changes the plural to fishes. A tank with several species of fishes then becomes a community aquarium. This community must spend the rest of its life together and therefore we want as few variations in character as possible amongst the inhabitants. Harmony is of utmost importance so make your selection of fishes that are to live together with great care and forethought. Your selection must at first be determined by space for this factor alone will decide the number and size of fishes which can be safely placed into the aquarium. Many coral fishes grow to sizes far in excess of fresh water specimens and this means that a normal aquarium usually housing numerous fresh water specimens may not even be large enough to accomodate one marine fish. Taking into consideration the vast amount of water per fish in the sea, in theory two fish per tank – regardless of size – is one too many so do not overcrowd under any circumstances. It is very difficult to lay down hard and fast rules for a lot will depend upon filtration and types of fishes kept but a general permissible size ratio can be calculated by allowing two gallons of water for each fish or two inches of fish per gallon of water – whichever gives most volume of water to the fish. These figures must of course refer to adult fish and not purchase size fish and allowing for the fact that fish will seldom reach a natural size in a restricted aquarium. A sailfish for example can grow from one inch to six feet in a matter of six short months so allow for this when calculating. The more fish you put into the aquarium – the lesser their adult size will be. Having decided from the start not to overcrowd the water it would seem foolhardy to then buy fish which will shortly outgrow the safe level of balance so carefully calculated.

Once we have removed from our list those fishes which we regard as too big for the set up we pass on to the next most important consideration – temperament. We intend keeping a happy, community of fishes and therefore all possibly ill-tempered inmates must not be placed on the consideration list – or otherwise removed from the aquarium if this condition becomes apparent at a later stage.

It would be impractical to ignore all specimens which show aggressiveness for we would be left with very few fish at all but the degree of violence must be limited. For example, six small Sargassum fish left together in an aquarium overnight would probably result in only one, very large, fish in the morning.

Most marine fishes spend a great deal of time squabbling and chasing each other around the aquarium but we must decide where to draw the line and, for this reason we have added a safety factor to each fish listed in the index of fishes. This factor refers only to those fishes we have personally kept and, although as accurate as our experiences show, cannot be held as hard and fast, for coral fishes very often do the exact opposite of expectations. I once fed a guppy to a large Scorpion fish. The Guppy immediately proceeded to set about the giant with gusto taking great pieces out of his fin tips. The Scorpion, greatly perturbed at his supper's antics opened his jaws to their fullest extent and fled. The lesson is clear; a fish classified as 'quiet' may well awake on the wrong side of its bed and promptly set about his wife but this is rare and generally speaking the habits described apply.

A considerable amount of aggression amongst the smaller coral fishes such as Damsels does not always result in actual damage at all and the bullied fish rarely suffers more than a nipped fin but should damage exceed this almost playful fighting the offender must be 'arrested' and put into a place of confinement. The length of time he spends in 'prison' will depend on his submissiveness, for excessive detainment may result in death through shock so keep a close eye on 'prisoners'. A detainment area is easily made from a nylon net or even a proper breeding trap (net type only to allow free water circulation). Clip the net onto the side of the aquarium and place the offender into it. After an hour or two he should have been tamed enough to be allowed back with the other fish. Detainment usually results in a less boisterous attitude towards other fishes but if this fails after several attempts the incurable must be removed for the safety of the community.

Going to the other extreme where a fish is over timid, can be just as dangerous. A too timid fish cannot be regarded as suitable for a community aquarium. It is always last to the food and very often goes without altogether, eventually ending its life through malnutrition.

It spends most of its days in a corner where no pleasure is gained from keeping it so there is little point in having it at all if your reasons for keeping it are visual. Furthermore it is more likely to die in a dark corner where you will be unable to spot it and his body can be more lethal dead than a vicious fish when alive. His over-timidity can even induce otherwise passive fish to commit acts of agression.

Unlike fresh water tropicals, where one tends to purchase fish in pairs, marines should at first only be acquired singly, especially if equal in size. If you really must have pairs and want to try your hand at breeding, then examine the fishes for sale in a plain aquarium. If you see two fish contantly nipping at each other with obvious aggression there is a good chance that these are males whereas the more sedate fish may just as possibly be a female. Just because two fish fight however this does not mean that they *must* be males so make use of your own judgment as well for there is little other chance of sexing most of the coral fishes without resort to surgery. With this great difficulty in sexing marine fishes there is very little chance of breeding them to any but the most experienced aquarist and only one or two people in the world have managed this, so there seems little point in buying pairs at all. I have made many personal excursions down under the sea in my frog suit in an effort to study breeding and have even bought a boat equipped especially for the job, yet my success has been limited to a few spawnings – none of which resulted in successful hatchings – so the average aquarist has little chance in his front room. Do not however let this completely deter you from any attempt whatsoever for you may well be the first to succeed where all others have failed – someone must be the first!

In the coral fishes' natural habitat there is far more space and many more hiding places than you can possibly provide artificially consequently many natural enemies are forced to reside next door to each other and feuds are inevitable. In order to limit these feuds as much as possible it is necessary to understand the reason for them. The main reason for aggression in most wild animals whether under the sea or in the African jungle is fear. Each animal believes the other to be a threat to its own existence and therefore a show of strength is called for. The violent Damsel fish is simply showing its neighbours how strong it is and how unwise they would be to attack him. This 'bluff' technique has been used effectively by all manner of animals since life began and can still be seen in the lion tamer's whip and the wild birds' "display" (where the bird puffs itself up to prodigious proportions in an effort to 'bluff off' the invader.) The solution to this bluff aggression, which so often leads to more serious fights, is not to combine fish of equal size and family. A large Angel fish for example does not need to impress a smaller one for the latter is fully aware of its size deficiency and will keep well away from the larger specimen. Eventually both will realise the absence of danger and act accordingly – rather like the sparrow feeding at the pigeon's side, cheekily but with an eye on quick retreat.

Another compatability desirable in the community aquarium is for food. All the inmates should be able to exist on similar foods. There is little sense in having six specimens – all of whom require a different diet. Keep to a selection of foods which all your fishes can enjoy and thrive on. Cardinals, Scorpions, Snappers, Groupers and Frog fish for example rarely accept any form of dead foods and it is not always practicable to hand feed them. On the other hand Surgeons require great quantities of algae so, once again, keep a happy balance and beware of mixing non-compatible fish haphazardly. There are of course many other reasons for not mixing certain fishes, the obvious being predators (fishes of prey) and small fishes. The answer is to study each specimen you intend keeping before the purchase and ascertain both its requirements and undesirable features. For example some Cow fish can excrete a poisonous substance lethal not only to the fish it was intended for but to the Cow fish itself. Octopii (sometimes called octopuses) possess this suicidal weapon also. Most Surgeons can tear open a fish twice its own size with its built in 'scalpel'. On the other hand Sea Horses and other slow swimming fishes simply aren't quick enough to get food before it is all eaten by other inhabitants of the community aquarium and so on. The obvious formula for a happy aquarium is a complete examination of all intended purchases before introduction to the tank. Do not simply purchase those fish which catch your eye for too many will do so. Look them up in this book and determine their compatibility to conditions you can provide.

Once you have made your final selection and purchased your fish, do not tip each new arrival into the tank and hope for the best. Keep a sharp eye on each new fish individually for several days and watch its behaviour. Persistent offenders, whether too aggressive or over timid, should be ruthlessly withdrawn before too much damage has been caused. Your community aquarium must be a happy one to be a healthy one.

7
The Brackish Aquarium

WATER IS CONSIDERED brackish if it has a salinity of not less than one part per thousand or more than thirty parts per thousand. This type of water usually occurs where river meets sea and fish which are considered brackish are those which either spend their lives in this river mouth area, come in from the sea to feed there, or come up river in search of food. In view of the considerable variations in water density these fishes are able to withstand, they are much easier to keep alive than many pure salt water fishes. Brackish water fishes are not coral fishes and should therefore be technically absent from this book. We feel however the liberty to be acceptable in view of the fact that much can be learnt about coral fishes and salt water life by keeping some of the brackish water fishes as a start. The aquarist intending to keep coral fishes will enrich his chances of success to a great degree if he tries the comparatively easy brackish water fishes first.

The amount of salt added to your brackish water aquarium at first will depend entirely upon the salt content of the water in which your specimen arrives. This is very often nil and your job is to begin a slow conversion from fresh to salt water stopping where you will. Many fish sold as 'fresh water requiring a little salt' should in fact be kept in salt water completely. They do not live in brackish water areas at all but were simply caught there whilst feeding. One may well ask then 'Why must the change be actuated slowly for the fish must surely head either straight out to sea or up river after feeding and can therefore stand such sudden changes several times a day?' The answer can be a complex one and will depend upon the length of time the 'brackish water' fish has actually been kept in fresh or near fresh water, sometimes for weeks. Therefore a sudden plunge into a much heavier water could prove fatal. The nervous system of a fish subjected to the terrors of captivity and claustrophobic conditions whilst travelling in a plastic bag also suffers to such an extent that physical resistance is almost nil. The fish simply isn't fit enough to cope with *any* sudden changes, least of all one which requires great physical adjustment.

Unfortunately when converting brackish water fish to pure salt water all intended specimens must be dealt with at once. As you will be raising the salt content daily new specimens would have a greater differential to overcome as each day passes so collect all your fish together or at least keep them in their original water, until you have all the fishes you wish to maintain ready for the conversion. Many brackish water fish, such as Scats and Mollies, consume large quantities of algae and for this reason your intended brackish aquarium should be fitted up well in advance of removal day. This allows algae spores to form and thus provide a certain amount of greenery.

Your water should be mixed in a separate container at first and daily topping up made by tapping this concentrate. The reason for doing it this way and not simply by adding the powdered salt is the impossibility of accurate trace element separation in commercial preparations. For example a ten gallon packet of commercial salt would have all the trace elements mixed. If you then divide this into ten gallon segments you may have all the correct quantities *en masse* but certainly not in their correct proportions. You must therefore mix the whole packet at once but in perhaps a two gallon container. This concentrated solution can then be tapped at will and diluted to its correct ratio before introduction to the aquarium.

Do not be in a great hurry to raise the density of your water. The slowest, easiest and most consistent way to convert is to replace the naturally evaporated water with salt water in its pre-concentrated form. With this method you are only replacing pure distilled water with marine water whereas with the more common method of syphoning off an amount each day and replacing with salt water removes a great deal of salt too. Replacement of only evaporated water will not only avoid this inevitable inaccuracy but will eventually result in a complete transfer of waters with the least use of salt. Furthermore, as very little more than one eighth of an inch of water will evaporate each day, the change isn't too fast or erratic.

As your salt intensity increases you will realise how much better the fishes are for the change. Colours are accentuated and body shapes soon begin to improve plus a far better general condition resulting in a much more desirable fish. The difference between a fresh water mono' (which rarely lives long in any event) and a salt water one is remarkable and you are more likely to keep one when full strength

45

salt is used.

Once your brackish aquarium has been finally converted to a pure marine one, if you go all the way there is no reason at all why the inmates cannot be transferred to the community coral fish aquarium if you wish but do consider the dispositions of intended transfers. On the other hand there is no reason at all why you should not keep your brackish aquarium brackish all the time. and keep the water density at around the 1.010 mark. New brackish acquisitions in a similar density may then be introduced at any time, but please – never keep brackish fish in pure fresh water as so many aquarists (unsuccessfully) do.

The following fishes are all tolerant enough to be able to exist in salt, brackish or fresh water;

Family *Scatophagidae*; Scatophagus Argus, Scatophagus Tetra-canthrus Micrognathus Strigatus, Selenotoca multifasciata, Scato-phagus Ornatus

Family *Theraponidae*; Therapon Jarpua

Family *Monodactylidae*; Monodactylus Argenteus, Monodactylus Falciformis, Monodactylus Sebae

Family *Toxotidae*; Toxotes Jaculator

there are of course many other fishes which will stand considerable salt variations such as the common Molly, Ambassis Lala, Stickleback, Malayan tiger many toothcarp and Cichlids but all fishes which we know to survive in brackish or salt water are listed as so in the index. These fishes should never be purchased whilst in pure fresh water if they have spent a long period therein. Pipe fishes for example will rarely live longer than two or three weeks in freshwater and you may be too late to save them.

This list of course is far from complete and great pleasure and satisfaction can be gained from trying out your own experiments. I have even kept Guppies for weeks in pure brine!

8
The Invertebrate Aquarium

A WELL DESIGNED and populated invertebrate aquarium can rarely be surpassed for sheer beauty of the most bizarre nature. Strange creatures and weird shapes so startling in colour and unbelievable in design that it is difficult to believe that they are animals at all. An invertebrate is an animal having no spinal cord or column and there is an abundance of such creatures to be found under the sea be it cold or tropical. You need not spend a fortune to own a lovely invertebrate aquarium for many of these creatures can be found on the beaches of your own sea-side resorts and a little foraging in the shallow rock pools left by the tide will be rewarded by hundreds of different specimens. For the aquarist who wishes to build up a collection of 'local species' in an aquarium a brace of plastic buckets, a long handled net and a set of strong finger nails is all he will need. Throw in some fine weather and all is set for a trip to your local coast. If you are in time to be able to catch the tide flowing you will undoubtedly be able to catch more specimens by following the tide out. Most of the more lively creatures caught out in pools soon die due to the temperature increase in a small volume of water. You will find many forms of life trapped in these pools but do not make the common error of overcrowding your bucket. If you spend several hours on the beach the specimens you have caught should be transferred into a second bucket filled with fresh sea water every half hour. Those which have already died should be discovered and disposed of. Most of the smaller crabs and shrimps will be too frightened to cause damage in the bucket to other occupants and will be more interested in hiding themselves, but remember that overcrowding your bucket will almost certainly result in a gooey mess at the bottom by the time you reach home. Beware also of large animals such as adult crabs for their weight alone may cause fatal damage to the smaller animals.

Anemones and crabs

The obvious first choice for your invertebrate tank is anemones (Actinaria). These highly colourful and most interesting creatures are a delight to behold when displayed and a collection of them take on the appearance of a fabulous under water garden. Sizes range from tiny half inch 'minis' to the giant Stoichactis which can grow to a size exceeding four feet in diameter. Most of the common cold water anemones found around the English coastline rarely exceed four inches in size and range in colour from deep brown to bright green with the beautiful Strawberry anemone (Actinia Equina) in the middle. Their colours, which are rather anaemic and semi-transparent hardly compare for beauty with the vivid colours of tropical specimens but their availability must make them a firm favourite with the amateur aquarist in England.

The anemone is perhaps the easiest of shallow water creatures to find and capture for they are almost completely immobile and are usually found adhering to the undersides of large rocks or on the lower edges of sea walls. They can remain out of water for several hours, usually for the duration of the tide, and combat dehydration by filling up with water as the tide begins to flow. The anemone is completely contracted whilst full of water with all its tentacles withdrawn into a sheath and over zealous efforts to remove them from their anchorages will result in a jet of water being emitted with considerable force from the pin pointed mouth. This fine stream of water can be very irritant to the skin and is often accurately shot directly into the eye of the collector – not intentionally of course but nevertheless extremely painful so once you begin the removal of an anemone from its base do not be tempted to watch the proceedings but rely only upon your sense of touch whilst pressure is being applied.

Anemones are quite easy to spot when under water and opened to their fullest extent but rather resemble a dried fig when exposed to the air and 'tentacleless'. The strawberry or beadlet anemone looks remarkably like a real strawberry when closed.

Those anemones which are attached to shells or odd sea rubble should be ignored for we do not want wild shells in the aquarium as they can act host to a multitude of unwanted microbes and one can often be tempted to bring the whole shell home complete with anemone sooner than go through the ordeal or removing it. This does not apply if your experiments are conducted with the unknown in mind for fur covered shells often produce the most interesting forms of life.

The adhesive powers of the anemone are astounding being even more powerful than its own flesh and is actually attained by a series of tiny suckers operated by transparent muscles stemming from the mouth. Due to this fact – adhesion being more powerful than substance – great care and patience should be exercised removing the anemone from its moorings or damage will result. This is where the fingernails come in. Gently prise the anemone from its anchorage by sliding your nail under the skirt taking great care not to split the skin. Half fill your bucket from a clean pool and place your first specimen carefully on the bottom where it will promptly curl up like a hedge hog and will not be harmed by small crabs or such in the same container.

Should you decide upon a crab or two make sure they are of the green variety and not more than one inch in size – red crabs should not be collected communely for they will attack all and sundry especially each other. These wild crabs will incidentally exist quite happily in a fresh water tropical aquarium provided they are converted slowly. They are extremely good bottom cleaners. Spider crabs (Maiag Squinado) can grow extremely large.

Once all your specimens have been collected you must await the return of the tide before filling your second bucket with fresh clean water. If however you intend keeping life other than anemones and crabs, such as shrimp or fishes, do not use genuine sea water but commercial synthetic preparations for this is far cleaner and totally free from plankton which soon dies in the aquarium.

A piece of flat stone or similar objects must be provided in the aquarium for the anemone to live on and, although the anemone will eventually chose a spot of its own, it should be placed directly onto the rock by hand – the majority of anemones, the Portuguese Man-O-War being the painful exception, are unable to penetrate human skin with their sting so do not be afraid to handle one.

Feeding an anemone is simple; tubiflex, small guppies or pieces of fresh shrimp etc. will be accepted by the seething cauldron of tentacles, which contantly wave around in search of food, and will be thrust slowly into the mouth for consideration. Unfortunately this applies to all foods regardless and the anemone will not have decided to eat it until it has been already eaten! Food which the anemone then decides to be undesirable will be rejected – even up to twenty-four hours later so take great care to examine for rejected food and do not feed it again. The very greatest care should also be taken not to over feed an anemone and never feed one at all if its tentacles are not waving around in search of food. It can live for weeks without any food and rarely needs feeding more than twice weekly.

In its natural state the anemone stings its victims to death with a jolt rather like an electric shock as soon as contact is made with the tentacles. The tentacles immediately begin to close and an almost magnetic force holds the intended food in place. The tentacles then pass the food along the line until it is forced into the mouth. Unappreciated food will be regurgitated along with the natural wastes in the form of a slime coated ball which is ejected at intervals from the mouth. These refuse packs, neatly wrapped and almost perfectly spherical, should be disposed of immediately for they obviously contain many undesirable elements. I recently dissected one of these disposal bags to find a perfect tropical red crab – complete with limbs, hair and all!

Many wild anemones will produce young soon after capture and, as they are viviparous (livebearers), the young are not difficult to raise. A few hours before actual birth commences the animal begins a series of contortions. Towards the end of these puffings, swellings and gyrations the mouth contracts to a pin point and a miniature of the parent only one eighth of an inch in size and differing only in quantity of tentacles is forced out. The young is still attached to the mouth of the parent by a thin thread. The young drifts away from the parent and soon attaches itself to an anchorage point. Several minutes later another baby follows, also attached to the parent and on the same length of thread. This process is repeated until soon there are a number of tiny offshoots all strung across the floor of the aquarium and linked to each other by the thin thread. An hour or so later this 'umbilical cord' (which serves in nature only to prevent the young from being swept away by currents and not as a food or oxygen passage) disintegrates. By this time all the young will have attached themselves to a temporary mooring from which they will move as they grow in size.

Growth is rapid for the first month and young are soon able to cope with the same food as the parent although in smaller quantities. Growth from this stage onwards slows down but this may be solely due to the restriction of aquarium confinement.

Anemones and small crabs will live together provided the crabs are kept well supplied with food for they are immune to the sting cells of anemones and will steal their food straight from the actual tentacles. Cold water shore crabs should further be supplied with an area of 'dry land' if only by means of a rock or two projecting above the water line and all crabs should have hiding places provided for the moulting period or they may be eaten by each other at this very weak time. Most crabs are also quarrelsome.

Consideration should also be given to the types of fish kept with anemones for, against general opinion, even Clown fish are not safe with some. We know that the Clown fish has been kept successfully with anemones other than its natural ones but this most certainly does not apply to all anemones. The English coastal anemones for example will sting any fish to death upon contact – including the Clown fish. On the other hand most Scats or Monos when placed into an anemone tank will promptly eat them all. There are however many other exceptions to the rule – there usually are – and these can become quite amusing at times. The antics of some crabs such as the Hermit Crab, which spends its life in a shell, are hilarious at times. One particular species endeavours to gain the favour of an anemone or two by tickling them until the hysterical anemones climb onto the back of the crab's shell where it cannot be reached by the probing pincers. The crab then sallies forth with its partners in crime, locates a fish and crawls cunningly underneath it whereupon the anemone promptly stings the fish to death. The villains then share their boodle. Another species of crab, when endangered, actually uproots an anemone from a nearby rock and tosses it straight into the face of the aggressor, obviously well aware of the lethal character of their sting.

Although most of the cold water beach anemones rarely live longer than a few years the giant tropical types such as Stoicactis often live to a hundred years or more.

Although one often sees the anemone in company with the Clown fish I do not personally agree with any invertebrates other, than perhaps a small Hermit crab for cleaning duties in company with fish in a community aquarium. Invertebrates are far happier and less trouble when kept alone where their often dangerous mucous emissions are suffered only by themselves. This does not mean that an invertebrate aquarium may be permissibly dirty, on the contrary, an unfiltered invertebrate tank will foul much more quickly than a fish tank.

There are many invertebrates in the sea which simply will not survive at all in captivity, and these should not be allowed to endanger other occupants of your invertebrate tank so, once again the golden rule; do not entertain the unknown at any time in the *community* aquarium, keep them in a separate container until you are satisfied that they are harmless.

Scallops

One of the most interesting denizens of the sea which is impractical to keep in the aquarium is the Scallop. As with most plankton feeders their natural requirements exceed those which we are able to provide. The Scallop is well worth a mention here even if only in order to understand some of the difficulties you are liable to meet with whilst experimenting. The Scallop has been one of the ancient symbols of man ever since the first artists sketched on the walls of their outhouses. Shells have always fascinated man since he first discovered their cutting edge possibilities and no shell has since received more attention than the Scallop – you probably pass one, even if only on a petrol advertisement, every day of your lives.

The latin for Scallop is Pecten and this, as every schoolboy knows, means comb and no lady of ancient times would be seen without a hairful of scallop shells.

There are over one hundred and fifty varieties of Scallop ranging in colour from white, red, blue, purple and black so no doubt their ornamental attraction was strong.

The Scallop is an extremely lively creature and can propel itself through water at considerable speed by opening the shell some twenty five degrees and closing it with a slam by actuating the central adductor muscle. This action causes the creature to leap along in ungainly twelve inch bounds. It can also frighten away intruders by raising the velum on one side and closing it with a snap thus causing the Scallop to pivot in a most nerve racking manner. An interesting escape trick is performed by raising the velum on one side only and closing it with great rapidity, an action which causes the Scallop to jerk violently backwards – a startling thing for a shell to do. A Scallop freshly removed from water will actuate all these escape devices at once and beat its shells together until exhaustion overcomes it.

Once the shell has been opened a short white circular muscle can be seen. The muscle is the food referred to as Scallop (pronounced Scollop) on all the best menus and, although the whole of the Scallop, except the shell, is edible only this adductor muscle is usually eaten for the Scallop, unlike most shellfish, remains open when dead and the organs soon decay.

Apart from the adductor muscle an enormous reproductive gland wraps around the shape of the shell under which lies the gills. Food is filtered through these gills and passed to the digestive organs located between the adductor and shell hinge.

roon Clown Fish *Amphiprion Tricinctus*

Clown Trigger *Balistoides Conspicillum*

The Scallop is generally unable to adapt itself to an aquarium existence for it lives mainly on plankton which it collects by drifting along the sea bed in a horizontal position with its shell open just enough to admit water high in bacteria content at this level, a condition hardly available in the aquarium – we hope. This feeding process is aided by high speed gill beating which separates food from water.

The most remarkable aspect of the Scallop is the number of eyes which surround the inner edges of the shell. These tiny organs glow rather like a cat's eye for they incorporate the same mirror system as the feline family. The eyes are perfect in every detail and complete in every aspect including a retina. Perhaps the most 'eye catching' of all Scallops is the common blue eyed (Pecten Irradians) which has over one hundred eyes gleaming from between rows of tiny tentacles, similar to those of the anemone, which surround the fringe of the shell. These perfect little eyes grow at the end of stalks which in turn grow at random just where they feel like it and appear irregularly scattered all around the mantle. Groups of eyes combine to share an optic nerve and the set of individual nerves all link up to the main one which is connected directly to the brain. These eyes, which are completely blue, grow wherever a space occurs and eyes removed by either accident or surgery will be replaced by new ones within a month. It seems however that the brain is incapable of making full use of these remarkable eyes for although they are equipped with a double retina, a feature not found elsewhere in the invertebrate world, the Scallop will only react to large bodies passing over them, leaping violently when the shadow reaches them be it a cloud or a shark. It is the tentacles that 'feel' an enemy more often than the eyes which see it.

The Scallop's deadliest enemy is the Starfish which wraps its powerful arms around the shell and forces it open just enough to insert its stomach. The Starfish then feeds on the Scallop, withdraws its stomach and moves on. A Scallop lucky enough to survive a Starfish mauling usually loses a number of eyes in the fray and this may be the reason for such an excessive number. Perhaps one day we will be able to keep one alive long enough to study in the aquarium.

The Starfish

The tropical starfishes vary greatly and sometimes grow far beyond the possibilities of an aquarium but the average Starfish one usually finds in marine specialist aquatic shops is quite safe in the coral fish tank. He does however feed on crustaceans and should therefore only be mixed with those intended as food. The method of feeding adopted by the Starfish is as remarkable as most of its other characteristics. Its powerful arms are quite capable of forcing open the largest of shells and, as explained previously, instead of putting food into the stomach the Starfish puts its stomach into the food. This highly flexible stomach, when filled, is then drawn back into the body. A further remarkable aspect of the Starfish is its marvellous regenerative powers which enable it to reproduce almost any part of its body which becomes damaged. This incredible feat is probably due to the great danger limbs are placed into when thrust into a shell, which may just resent the removal of its interior and nip off the arm. This regenerative power, although remarkably advanced in evolution stages, is not however restricted to the Starfish for many higher animals such as coral fishes are capable of growing completely new eyes and fins, almost overnight, when in good health.

The method of locomotion employed by the Starfish consists of the expansion and contraction of many tiny suckers, similar to the octopus, but on the end of stalks attached to the undersides of the arms.With this method it can propel itself along at a slow but steady speed. It cannot swim.

Starfish collected from coldwater areas, such as Asterians Rubens should never be put into heated tanks. They will seldom survive no matter how slow the change. A Starfish should be quite firm and hard and a softening of the skin is usually followed by death. Starfish found on a dry beach cannot often be revived if they have rested there too long.

Sponges

Sponges can be kept in the invertebrate aquarium and very often arrive there accidentally, being attached to the shell of a hermit crab or via a rock which wasn't cleaned too well. They require a great deal of oxygen, crystal clear water and succumb very quickly to even the slightest injuries. They will not last long in company with anemones either and this makes them perhaps a little less desirable for the anemone will sting a sponge fatally upon the slightest contact. Sponges will therefore be of secondary consideration, preference usually being given to the far more interesting anemone.

Crustaceans

This group embraces such a vast number of species that it is difficult to know where to begin without writing a separate book. The crabs of course could have been added to this section but, due to their close association with anemones, they were dealt with there.

A great deal of crustaceans are used in the coral fish aquarium for food and this indicates the obvious risk entailed in placing the smaller ones into a community aquarium. Once again a separate container is recommended.

The one crustacean most of us have kept at some time or another is the common brine shrimp (Artemia Salina). Strange to relate however is the fact that brine shrimp is not found in the seas at all but in lakes. This does not mean that it will not live in the marine aquarium but, as it is more commonly used as food, it would not last long in a community set up. Brine shrimp eggs can remain dormant for many years provided they are not allowed to get damp. A production of the correct conditions will result in the eggs hatching. The eggs are obtainable all over the world in commercially prepared packs and are much easier to hatch than many aquarists realise. The secret of rearing brine shrimp is not to use too many eggs, a fingernail full of eggs will result in many hundreds of shrimp. Fed correctly the young will soon grow into adults and will produce their own eggs. Males however are rather rare and many hatchings produce only females.

A circular container of non metallic construction is better than a rectangular one for the latter induces great numbers of newly hatched shrimp to gather in the corners, for they never stop swimming, and the bulk of these suffer from overcrowding and oxygen shortage in an otherwise sufficient volume of water. Brine shrimp spend their lives swimming towards light so never allow a variation of light intensity unless you want to catch the shrimp and require them to congregate for netting purposes.

A salt water preparation with a density of 1.025 should be poured into the circular tank (a goldfish bowl is ideal) and placed where it will be subject to long periods of daylight such as on a window ledge. This water will soon turn green and be ready for the eggs. The young should hatch in a day or two and will depend upon temperature for an actual period. The higher the temperature the faster the hatching but do not raise it beyond eighty degrees farenheit. The hatched brine shrimp will soon begin feeding on the algae and will do so for up to one month. They may then be fed with – newly hatched brine shrimp! Dried foods will also be accepted but great care should be taken not to overfeed for this is the main fault in breeding. The healthy brine shrimps should attain the size of around half an inch.

There are of course numerous other shrimps perhaps better equipped for community life than the Brine shrimp and the most attractive of these is the banded coral shrimp (Stenopus Hispidus) which is a contrasting red and white with extremely long feelers. This 'shrimp' is in fact a prawn and can grow to lengths of up to four inches. They spend their lives in sponges and are very good 'fish cleaners'. This habit should be carefully watched in a community aquarium for the fish in question may repay the shrimp's kindness by eating it. Banded coral shrimps should not be kept in pairs in captivity for they are notorious fighters and one, the other or both may be killed. They are quite easy to sex, if you want to try a pair, for the female sports a green belly. The young can be raised in the same manner as brine shrimp except that the parents must not be allowed to access them or they will regard them as food and act accordingly. The adults should be separated after mating.

The Anemone shrimp is another colourful creature which will add interest to your invertebrate aquarium. As one would suppose his name comes from a habit of marching around amongst the tentacles of sea anemones assisted by his very powerful thick set legs which do not appear to be drawn by the adhesive powers of the anemone skin.

The Mantis shrimp is another well built animal and is extremely powerful. As its name implies, it can be an extremely vicious one. Its claws are razor sharp and are quite capable of gashing a careless hand so make use of a strong net when moving is necessary. It should be provided with a 'home' and kept alone for although it may well ignore shrimp of another species it will almost certainly attack another Mantis shrimp.

The Sixteen gun salute shrimp, the dancing shrimp and others can all add to the colour and interest of an invertebrate tank but do remember that many shrimps enjoy battling and keep an eye out for corpses which should be immediately removed. Do not mistake 'phantoms' for corpses and imagine them to result in less specimens but remove them just the same. Neither should you be alarmed to apparently find more shrimps (or crabs) one day than you started with for many shrimps are completely transparent and when, like the crab, it discards its shell periodically, a perfect phantom like replica, complete with eyes, feelers and all, is left in the most lifelike positions.

These 'used' models should be removed when discovered and all shrimps or crabs undergoing a 'new suit' fitting should be carefully watched for invasion as the unfortunate creatures are completely defenceless at this time.

Lobsters

Lobsters are amongst the larger of crustaceans and should not be kept with other animals. A lobster taken from the sea will probably settle down quite quickly but beware of its pincers and only hold one in such a position that the arms cannot reach your hands. The easiest way to obtain a fresh lobster is to pay a visit to your local fish market early one morning. There are usually plenty of live lobsters there but make sure you only purchase one which is reasonably lively. Do not place a lobster obtained in this manner directly into water or it will drown! Gently splash the animal from top to bottom until it is soaking wet inside its shell as well as out. Only when it is completely saturated should you place it in water. Feeding should not be too frequent and should be with fresh clean brine foods. A lobster seldom lives for longer than one year in an aquarium and will not reach this age if large hiding places are not provided in the form perhaps of a few giant clam shells. Do not overbrighten a lobster aquarium but keep the lighting rather subdued and from the top in preference to the sides.

The Octopus

The Octopus is of the Mollusc group known as *Ephalopods* and is an animal well known to be equipped with eight arms. These arms are studded with powerful suckers which are used both to propel the octopus along flat surfaces and to take a firm grip upon food. The method of locomotion used for swimming consists of opening the 'bat wing' type arms, which are webbed with membrane, and closing them again umbrella like. This causes the octopus to rocket along at considerable speed in great bounds.

The octopus must surely be the most fearsome nightmare of all sea creatures and few who have seen a large specimen are likely to want to fondle one. Nevertheless, for all its ungainly presentation and ugliness, the octopus is an extremely intelligent and sensitive animal. We will not dwell long on the subject, firstly because very little is known about the octopus, secondly because they are very difficult to obtain and lastly because they seldom live for more than a few months in captivity, even with such expert facilities and attention as are available in public zoos.

The octopus is just an enormous bundle of nerves every one of which reacts to even the slightest upset or movement. They prefer to live in dark holes and black water. When removed from these conditions and placed into a bright nerve-racking aquarium with nowhere to hide for they are very shy, they are usually frightened to death – literally.

The dwarf octopus is by far the hardiest and certainly the most practical one to attempt keeping for the octopus is very strong and able to remove the heaviest of aquarium covers. I have actually seen an octopus lift off a cover glass, climb out of the tank and scuttle laboriously down stairs, into the roadway and across one hundred yards of beach to the sea.

The octopus is quite capable of changing its colour several times in as many minutes and this is a useful trouble beacon, for the more frequent these changes the less happy he is.

Trouble in feeding the octopus is seldom experienced for he is a predator and will accept most forms of shellfish which he cracks open with his powerful horny beak. He will also catch fishes.

More than one octopus in the same aquarium is not advised for an unexpected movement by either one may cause the other to release its poisonous 'ink'. The octopus is extremely sensitive to light and should not therefore be photographed by flash. The lights in his aquarium should also be subdued. An octopus must not be kept in the community aquarium at any time for the slightest upset may cause it to eject a poison which will kill everything in the aquarium, including itself. As the octopus is a cavity dwelling animal it should be provided with a cave or two in which to hide. An exposed octopus will seldom live longer than a few days.

Snails

It is widely believed that snails are fresh water creatures only but this is of course erroneous and many are to be found in the sea.

The predator snails such as the murex or whelk will live longer in the aquarium than most and will become good scavengers, helping to keep the bottom clean. Beware of such snails as the Conus which sports a built-in hypodermic, the needle of which is usually inserted into an unsuspecting fish late at night. The resulting injection of poison usually proves fatal to the fish.

Many snails, however, especially those found around the colder shores, will not live in temperatures higher than those found in even a normal room so you may well prefer to keep only the tropical specimens. Cockles and winkles, which require vast quantities of plankton, are completely impractical.

Worms

An excursion to the mud flat type beaches at low tide with a pitch fork will usually result in the collection of many types of worm. Some of these worms are prodigious in size and astounding in colour. The number and type of worms available is uncountable and yet, from such a great selection, only the delicate Fan worm is, in my opinion, worth bothering about. Worms which spend their lives under the gravel and out of sight are both pointless and dangerous. You will never know if they are alive or dead and they are valueless from the observation point of view – and highly dangerous as bacteria food.

The Fan worm or Tube worm (Sabellidae Family) is however a delightful and strange creature which one can study closely in the aquarium. So fair is this bizarre flower of the deep that laymen often refuse to believe that these highly coloured feather dusters are in fact worms at all.

The Tube worm builds its own horny case in which to hide then they bury themselves, with the tube, in the sand with only one end protruding. When all is quiet they push out a mass of delicate tentacles which search ever endingly for food such as plankton and other small animals. Tube worms will not live long in artificial water for it is too sparse in the vital organisms necessary for the worms existence. Sea water should be used and frequent changes are advised.

Live Coral

Live coral has been regarded for some time with great suspicion and criticism. Even now one rarely sees a piece of coral which is truly living. I recently made a tour of many European zoos and not once did I see live coral, although I saw many difficult plants. For this reason you may well decide not to bother with it at all but, should you ever have the opportunity to study a piece, I am sure you will make an effort to acquire it.

A piece of coral is a complete colony of tiny living and dead animals complete with their homes. Coral takes many shapes and sizes and can grow to the most enormous extent. The Giant Australian Barrier reef is of course built of solid coral.

These minute animals are completely incapable of moving and spend their lives firmly attached to the dead bodies of their parents. The parents are rewarded for giving birth by being smothered to death by their young which, in turn, die in the same manner. Consequently, although a large piece of coral may contain many thousands of these animals, only those at the surface of the mass are

59

alive. These animals are called polyps and their sole existence is reliant upon their ability to catch and have access to the very smallest of nectons.

Although live coral will exist only in the very cleanest of water this water must be natural, full of plankton, well filtered, fast moving and frequently changed.

Coral is best viewed when feeding, usually at night. The colours enrichen and the tiny polyps open rather like eye lids. Their method of feeding is solely dependent upon the hope that food will swim directly into the 'eye lids' and a live condition cannot be maintained if food is not generously provided. In daylight they rather resemble tiny butter-cups.

The hardiest of corals likely to live in aquaria is the Gorgonian and this should be treated as though a valuable fish. In fact if you regard all live animals, be they snails or fan worms, as such you will enjoy a happy and beautiful invertebrate aquarium.

9
Obtaining Specimens

THERE ARE MANY ways to purchase fish, the most obvious being from a dealer specialising in marine fishes and equipment. It is of course possible to obtain an import licence, contact a collector abroad and order direct but, unfortunately one seldom gets what one asks for in such cases. Although the dealer which has his fishes on show may not be able to offer such a variety at any one time at least you are able to see with your own eyes the condition the fishes are in and, for this reason, I shall only deal here with this type of vendor.

Marine fishes are very often scarce and your local dealer may at times have to wait weeks before he can change his stocks and consequently one may often be tempted to purchase fish which are available rather than wait for those which are not there at the time. This leads to collecting fish which are not necessarily wanted. An enthusiast may trek half way across the country to purchase a specimen only to find that it has already been sold or has died. Sooner than go home empty handed the disappointed aquarist may purchase an unexpected fish about which he knows nothing. This is fatal and very often the beginning of his troubles. You are attempting to set up a Utopian community of coral fishes and should be just as meticulous about choice of inhabitants as you were about every other aspect of the operation. This can only be done by making your selection of fishes at home with no pretty colours or fast talking salesmen to distract you. Never accept second best or alternatives and, once you have made a selection, do not change it without forethought.

When making up your list of fishes there are many points which you must consider; will they mix? Where do they come from? In what conditions have they been in? Can you simulate these conditions? Will you be able to feed them correctly? Do you know *how* to feed them? And so on. The make up of this list and selection of a possible community may be assisted by the study of the Index of fishes at the end of this book.

Never purchase a fish simply because of its eye catching appeal, it may well turn into a headache within days of purchase by either gobbling up all your existing fish or simply dying through lack of intelligence on your part about that particular specimen. Never purchase a fish.which shows even the slightest suspicion of unhappiness. Most coral fishes are extremely active, incurably greedy and persistently nosey, constantly darting around the aquarium either seeking intelligence of the disturbance you are causing with your inquisition, or refuge from your gigantic monster-like proportions. An enormous face peering into one's window *should* have some sort of effect on the nerves so beware of indifference or corner sulking. As usual however there are exceptions and one or two specimens are completely unmoved by anything at all. It takes a great deal, except food, to send a lion fish scurrying. Your job is to determine the difference between complacency, due to the nearness of death, and indifference due to a lack of fear for the symptoms are similar. Ensure also that your prospective purchase is feeding, for complacency at meal times is a sure sign of distress, not of course forgetting to make allowance for nocturnal feeders.

To begin with, when looking through a display of sale fishes, make sure that the vendor's tanks are clean, correctly filtered, not too new and protected against such obvious dangers as metals – even if only by the salesman's use of an unprotected wire framed catching net. A dealer not prepared to take the precaution of using an all nylon net may well be lax in many other ways. Dealers with coral fishes in ordinary non-coated metal aquariums, other than stainless steel, should be avoided at all costs. You do not want specimens which have been contaminated by possible metal poisoning and this condition may take several weeks to become apparent. A marine fish aquarium, whether in a shop, fish house or home, should be clean and spacious, so do not accept fish which have been kept in conditions you would not keep them in yourselves without preparing for the consequences. Do not however mistake an algae coated aquarium for a dirty one. The presence of algae is not necessarily indicative of dirty conditions, on the contrary it shows the tank has not just been set up and that the fish have possibly already overcome the first dangers of captivity. Care should be exercised in recognising the difference between brown algae and rust, which is rather similar in appearance.

Care should be taken not to purchase fish which have lived in a water density differing greatly from your own. Assuming the specific gravity of your water to be between 1.020 and 1.025 then discover the density of the water in which your prospective purchase lives and compare it with your own. If your water differs by anything more than .003 then, provided the vendor is correct (not always necessarily), you must change the density of your water at least twenty four hours before the purchase. This of course cannot be done if you already have fishes in your aquarium and the newcomer must be subjected to a change itself in the quarantine tank. Salts added to the water must be given at least twenty four hours to dissolve or a false reading will result. Once the purchase has been made insist on two large plastic bags (in case one bursts) and plenty of water. You may be better off by taking your own transit container complete with insulation to prevent chilling. Should you have a quarantine tank – and you should – use Methylene Blue, Diseasolve or Pottassium Permangate as a disinfecting agent and maintain strong aeration for at least a week. It would be a pity to spoil all the work you have put into the community aquarium and possibly lose expensive fish, for the sake of impatience (the biggest Ichthyological fault of all) and a small quarantine tank will cost little more than the price of one fish. A quarantine tank should be of all glass construction and shaded from strong lights. It should also be placed in a situation least liable to permit any sudden external movements. Water density changes as well as hospital and quarantine facilities can also be conducted in such a useful accessory.

In an emergency, such as when the quarantine tank is already occupied, a plastic bucket may be used but only for the very shortest periods. If this situation arises make sure that the bucket is white and not highly coloured for coral fishes do not like strong colours and will succumb quickly if enclosed in a highly coloured container. The strong unnatural intensity of any one particular colour completely upsets the fish to such an extent that it probably dies from sheer shock.

Shock, as discussed in chapter eleven, plays a great part in the life or death of a newly captured coral fish. A hospital casualty, suffering from any form of accident, is first treated for shock, the wounds taking secondary consideration. It is often shock that kills where the wounds would not have done. No operation is ever performed without prime consideration first being given to the shock factor in the patient. The lesson is plain, handle your new purchase with the utmost of care at all times, he is probably suffering from shock.

Before having your new fish actually removed from the sale tank take a rough respiration count by watching the mouth opening and closing and, after it has settled into your own aquarium, take it again. The first sign of distress in any animal is a change of breathing. A doctor will always take his patient's respiration count *first*, no matter what the ailment. You will soon learn to recognise a fish which is not breathing correctly.

Once you reach home with the new fish, the container in which it has travelled should not be opened in a brightly lit room. A sudden change from the darkness of a closed box to the brightness of daylight will only further the state of shock. Place the plastic bag into the quarantine tank after first checking that the temperatures do not differ greatly and allow a few minutes for the two waters to become exactly even. If the densities are within .001 of each other then a few pin pricks in the plastic bag will allow the two waters to combine slowly. The punctured bag will eventually sink and the fish will be able to leave it without the indignity of being tipped out with a splash as is so often the case. The plastic bag should be quietly removed once the fish is free. Should there be a difference of water density of up to .003 then it should be preferably stronger in the aquarium and not the bag for a weaker solution can be provided in minutes by dilution whereas stronger one can only be accurately made overnight. In order to have the happier state of affairs it is far wiser to avoid buying fish first and then preparing the correct conditions, better to prepare the aquarium and then buy the fish!

10
Feeding Coral Fishes

THE FIRST DEMAND your new coral fish will make upon you, is one for food. At least that is the hope for a hungry fish is more likely to be a healthy one. The most important point to remember is that whatever food you give your fish it will almost certainly be different to its normal diet. Fish are not particularly intelligent animals in fact they may even be too stupid to realise that you are feeding them at all. They just see you as an enormous monster who enjoys throwing things at them every now and then. There are many coral fishes which will remain unaware that you are attempting to feed them for some time and even when it finally sinks in, the food may not be to their taste and they will prefer malnutrition to eating it.

Coral fishes eat far more food in comparison to most fresh water specimens so once you find foods which are acceptable to your fishes do not be afraid to make use of it. There is a difference between over feeding and starving the fish – a common occurrence in marines – and the coral fish, unlike the fresh water one, has no plant life to fall back on when hungry.

Study each new specimen from arrival and during his stay in the quarantine tank for food tastes. He should not be taken from this tank until he is eating. The physical layout of a fish's body should give a fair indication of feeding habits so study this aspect too. If it only has a very small mouth such as many Butterflies have then it will require small foods such as plankton or algae. A predator will sport an ear to ear mouth whereas a pointed nosed fish such as the Chelmon Rostratus has obviously been provided by nature with such a weird structure with a purpose in mind, in this case to enable it to poke into the closely formed coral fingers which conceal such a myriad of fine life.

If your new specimen refuses all the regular dried foods then he will probably be tempted by one of the live foods. If even this delicacy is refused then it either requires algae or is sick. It is often a simple case of trial and error but you will soon learn, from simple anatomy, the type of foods your new charge is liable to accept.

Conditions under the sea are seldom compatible with a well lit banqueting hall and many fishes of the size we intend keeping find it necessary to seek foods in semi or total darkness. They may even find the jungle too frightening to venture into unless it is dark after all, if the smaller fishes are able to find their dinners easier in bright conditions then so are the predators. Therefore, as always, nature has provided coral fishes with an aid to food location – sound or, perhaps more accurately, a sense of feeling through underwater vibrations. This sense of 'feel' is very often the only way in which a fish can locate food at all, especially in the deeper areas where light is unable to penetrate. Vision is not always limited at depth in the sea by light but by the sometimes poor visual capabilities of the fish itself. Unlike man fishes are at birth equipped with monocular vision. This simply means that, instead of simultaneous impulses being passed to the brain in order for that component to sum up a situation, the nerves of the eye pass separate pictures and to the same side of the brain as the eye, not, as in man, to the opposite side and without the luxury of binocular view. Therefore the fishes' brain has two pictures to cope with at once – one for each side of the fish, except of course those fishes with both eyes on top of the head such as flat fish. This design of the fishes' optical system may well be the reason why a fish will often rush in the completely opposite direction to logic – into a net for example. This handicap – be it one – is compensated for by this extraordinary ability to be able to 'feel' vibrations, created possibly by food, at distances far beyond the visual capabilities of any optical system. A shark was recently placed into a special aquarium in Florida with impulse reactors attached to its heart. Reaction was recorded when other fishes were introduced into the same tank but a damaged fish, when also added, caused considerable excitement for distances up to three hundred yards! This indicates that not only was the shark aware of the other fishes but quite capable of recognising one in distress – or easy food.

Many tropical marine fishes are difficult to induce to feed at first and live foods are very often the only way in which to set them off but care should be taken not to continue the pampering with live stock for too long, or the fussy fish will never accept dried foods at all and eventually suffer from a vitamin deficiency due to the fact that fresh water fishes, as food, cannot contain the same protein as sea animals and it is most unlikely that you will be able to provide the latter. A prod on the nose with a piece of fresh shrimp often results in a quick snap and dried feeding has begun. There are of course many coral fishes which will *never* accept dead foods and these should occasionally be treated to a live shrimp or two which one can easily obtain with a net, a trip to the sea side and a little patience.

A good way to teach a new fish how to eat the strange foods you are supplying is to provide him with a good teacher. This old hand's frantic rush for food at meal times often induces a hungry fish to eat. Remember, *all* coral fishes are wild and too much emphasis cannot be placed upon the experiences which they suffer. With such a disruption of their lives and systems it is hardly likely that they will arrive with healthy appetites.

The usual exceptions arise once more and the Sergeant Major or Beau Gregory seem quite unaffected by all they suffer, at least their appetites stay unperturbed. A pinch of food sprinkled in the aquarium usually produces the desired results and a movement outside the tank often brings one or the other rushing up to the surface. These type of fish make very good teachers and both are hardy enough to warrant inclusion on your list of possibles as 'first fish'. The myth that Clown fish are the easiest coral fishes to keep of all should not be regarded as fact for many of the Clowns can be most difficult. To say however that the Clown is the easiest coral fish to catch diseases may be somewhat nearer the truth for most diseases of the salt aquarium are usually first noticed upon the Clown. Some Clowns are not even good feeders and should not be regarded as teachers. There are many Damsels quicker.

Damsels on the whole are among the easiest coral fish to feed and there is seldom a problem with live food beginnings. They are however prone to quarreling and this may even deter the timid specimen from feeding at all for they will have enough fears to overcome without the added fear of being chased away from the food as many overbearing Damsels will tend to do. For the larger fishes where only live foods will be accepted Mollies make good food and should be given preference to the usual guppy. The Molly will be entirely unaffected by the water density and will live for days in salt water whereas a guppy seldom lives the night out. A nylon breeding trap can even be fitted into the corner of your aquarium with one or two pregnant females inside it. The young, when born, will pass through the net and into the main tank to be eaten. This may seem callous but is no different to using daphnia or brine shrimp.

For the persistent non feeder perhaps the best live food to begin with is brine shrimp and once the ice has been broken your difficulties will be restricted to the job of keeping up with appetites as most coral fishes are gluttons.

Algae

An essential addition to the coral fish diet is green algae. Many marine fishes will in fact only eat algae and some, such as Surgeons, will soon die without it. Brown algae is not usually acceptable so, unless your aquarium is situated in a position where it will receive at least eight hours of direct sunlight each day, you must cultivate algae in another container such as a goldfish bowl set up in a window box. If you are using brine shrimp as a food then you will probably have a separate tank set up for the growth of eggs and this tank will be ideal not only to provide algae for the fish to eat but to encourage the growth of brine shrimp, which in turn like algae. If your algae starts off brown do not immediately assume it to be of no use for there is a good chance that it will eventually turn green given enough light. Algae that turns brown after it has been green however should be regarded as lifeless and removed from the tank. Algae which grows in excess of your requirements should not be simply scraped off (from the front glass for example) with a razor blade, for it may die and help pollution at a later date. A good car sponge is far better but fit it onto the end of a wooden dowel in preference to using the hands. Hands are seldom free from bacteria and even freshly washed hands may harbour lethal soap beneath the finger nails. The sponge scraper will absorb the algae into its pores instead of allowing it to drop to the bottom of the aquarium. Incidentally, a household sponge left in the corner of the tank will act as a magnet for algae and attract it to the porous fibres of the sponge where it may be periodically removed and washed.

Algae can be stimulated in growth by the fitting of strip or, better still, Growlux lighting. Never be afraid of algae encouragement for it is an essential part of the diet of most sea animals and sometimes the only acceptable food at all, especially to the small mouthed coral fishes. Algae also produces a considerable amount of oxygen during the hours of daylight, it can often be seen covered with masses of tiny oxygen bubbles which rise occasionally to the surface. Lastly, as we have already seen, algae prefers a pH of around 10 so do not forget that

n Fish *Pterois Radiata*

Power Brown Tang *Acanthurus Glaucopeirus*

community tank algae is being retarded by the fact that the pH of this water is (or should be) only 8.3.

Brine Shrimp (*Artemia Salina*)

Brine shrimp has been used as food, usually for fry, in the fresh water aquarium for many years and in practically every set of instructions I have seen the advice given is "Feed the brine shrimp when newly hatched". The reason no doubt for this instruction is because small fish can obviously only cope with the shrimp during its early stages of life when it is a suitable size. We are not dealing with fry in the coral fish aquarium but with fishes quite capable of eating well-grown brine shrimp. If you are going to follow these instructions to the letter you may just as well give your fish a glass of water because the newly hatched brine shrimp contains very little else. Perhaps at best you are teaching the fishes to eat and even expanding their digestion tracts (fatally, should any unhatched eggs pass accidentally into it) but you are certainly not feeding them. The solution is to *feed the brine shrimp* before using it as food and thus turn it into a nutricious food instead of a laxative. Brine shrimp, during its early stages of life will feed on algae. Even a few hours after birth it may have consumed enough algae to be of food value but it is most certainly not within minutes of hatching. A small bowl placed near a strong light source will suffice for a hatching tank but make sure it is round in construction and not rectangular (see chapter eight). The hatched (and well fed) brine shrimp can be rounded up by the simple expedient of shining a light onto a small area of the tank glass. The stupid animals will soon collect around this point and can be gathered, without their shells or stray unhatched eggs, with either a brine shrimp net (easily made from an old nylon stocking toe) or, better still, an ear syringe. It is far easier to examine the shrimps for unwanted egg or egg shells in a syringe and these 'intruders' should be removed for they are quite capable of causing digestive troubles if fed to the fishes. Should a separate tank be impractical or difficult to arrange then the brine shrimp eggs can be hatched in the community tank with the assistance of a dark brown container such as a medicine bottle. The eggs are placed into this bottle and prevented from floating out into the aquarium by leaving enough air to carry them just above the neck of the bottle. Those which sink in the bottle will of course be below the neck. Precautions should be taken to ensure that no fishes in the actual aquarium are small enough to be able to enter the bottle or they may either get stuck due to their inability to turn or choke themselves to death on the unhatched eggs. As the eggs hatch the live shrimp will naturally swim towards the light coming from the aquarium and thus out of the bottle. Only living shrimp will be able to gain access to the aquarium for the only way out of the bottle is to swim. Eggs are therefore not introduced into the aquarium at all and, as this danger is a real one, you may prefer this method even though the shrimp may be of less food value than ones which have been given time to feed, for the direct hatching system does not give the shrimp much chance to do this.

Shellfish and Mollusca

Mussels, lobster, crab, prawn and shrimp are all accepted by the majority of coral fishes. All these foods should be cooked, cleaned and thoroughly strained. All bones, shells and horny particles should be removed and the flesh cut into manageable pieces unless of course you prefer the large chunk method of feeding. Either way the 'cured' fresh food can be frozen and kept in the refrigerator for weeks on end and fed when required. The frozen food market catering for human needs has produced many frozen foods which one can obtain in the grocers but these should only be used if fresh shellfish is unobtainable. One never knows in this competitive world of chemistry which preservative or colouring agents have been employed and, as most products claim a 'taste of its own' they cannot all be completely natural as fresh fish would be. Many aquarists actually prefer to buy tinned lobster, crab or such and freeze it for use at will but these aquarists are usually the unsuccessful ones for feeding tinned foods to coral fishes, although common and convenient can only lead to eventual metal poisoning. Tinned foods should therefore never be used under any circumstance.

The size of pieces used for feeding will depend entirely upon the size of your fishes and their habits. You will soon learn which fish like the larger chunks and which the smaller provided you present them with a choice. It is sometimes better to feed all the fishes in the aquarium with one very large piece of flesh and let them all nibble on it. In this way you will soon see when they have all finished with it and it can be removed. This method is not practical however if timid specimens are included alongside such rowdy feeders as Sergeant Majors. The timid fish usually manage to nip out of his hiding place, snatch a small piece of food from the floor and hurry it back home for consumption. He is hardly liable to barge in with the crowd and squabble for a share if only one piece is communially provided. If the large food system is acceptable to all your fishes then the best method of introducing it is by means of a length of nylon thread. The chunk of food is tied firmly onto the end of the thread, suspended in the tank within three inches of the sub sand and removed as soon as the fishes show signs of satisfaction. If small pieces are fed then care must be taken to remove the untouched fragments which inevitably find themselves unseen and untouched upon the floor of the aquarium for they will pollute the water in no time at all. These pieces must be removed carefully to avoid frightening the fish and for this reason the large food method should be attempted first. There is little alarm caused by the removal of uneaten food on an almost invisible thread and foreign bodies are not introduced every feeding time when removing unwanted foods.

Tubifex

This excellent fresh water food leaves much to be desired in the salt water aquarium. It is seldom really clean and contact between worm and salt water usually results in almost instantaneous death. If the Tubifex worm is not eaten before it reaches the bottom of the aquarium it will probably not be eaten at all for it soon turns white and looks most 'unfoodlike' so only feed the tubifex worm singly, not as in fresh water fish feeding, by the ball. Clown fish and most Trunk fishes are very fond of Tubifex worms and so are the sea anemones but excesses should be avoided. Tubifex can be kept alive for long periods provided it is left in running water. The simplest way to keep Tubifex in running water without inconvenience is to put it into a porous container such as an old plastic tea strainer and place it into the lavatory system. Every time the flush is operated the tubifex is provided with a complete water change. It should also be fed with potato peelings if kept for long periods.

Mealworms

Mealworms are extremely nutricious and seem to spend their lives being eaten by some form of animal or other. Very few birds, lizards or toads etc. will refuse a nice juicy mealworm – they are aptly named too. And now this unfortunate creature finds itself yet again gracing the pages of a food chapter, this time for coral fishes. There is an old zoological saying "If it won't eat, give it a mealworm", and this applies to coral fishes. The mealworm is quite large and cannot survive long in salt water and so, for the smaller fishes, there is little lost by cutting it up with a razor blade and using it as a dry food. Very few large fish will refuse mealworms but they seem to prefer them in a defunct condition, pulling and pecking at them if alive. You should not of course cut the worm up until you wish to feed it for the worms will live for a long while if provided with a little lettuce and dry conditions.

White worms

These very easy-to-breed worms are by far the best possible live food for coral fishes other than actual live fish. They will also live in the salt water aquarium for at least twenty four hours thus reducing water pollution possibilities. A breeding box can easily be made up out of wood and left out in the shed for use at any time. Unlike their close cousins, the Grindal worm, they need no special heating. The box should be shallow and with a tightly fitting lid, not of course forgetting air access holes. A mixture of peat and loam to a depth of two inches, should be put into the box and a wad of white worm, which is obtainable in most pet stores, placed in the centre. The compound can be easily replaced at any time by the simple method of exposing the surface to a bright light. This has exactly the reverse effect to brine shrimp and the worms quickly make their way to the bottom of the box in order to avoid the light. The top soil can then be removed and replaced with fresh.

White worms should be fed on potato, porridge or in fact any of the soft breakfast cereals.

White worms are hermaphrodites and contain both male and female sexual organs in the one body. Two specimens, therefore, having both sexes each, lie side by side and join together by means of a secreted mucous. The eggs resulting from a 'match' are spawned in cocoons and hatch in a very short time. Provided plenty of worms are left and the compound not overdrawn upon, the culture will go on for years providing your fishes with as much food as you require, but new blood should be added occasionally to keep the strain strong.

Adult worms can be separated from the earth by the method employed to assemble brine shrimp but with the opposite effect – the worm moves away from the light. Most aquarists however prefer to pick out the worms with a pair of tweezers and feed them individually. This is not always good practice, especially with coral fishes, as there is always one fish far better than the others when it comes to retrieving food and he will inevitably grab every single worm leaving none at all for the rest. The best method that I have come across is to pick out the worms with tweezers and put them into a plastic cup. When sufficient worms have been collected, the cup is filled with water. This not only rinses off particles of dirt but causes the worms to collect together in a solid ball, the same as Tubifex. This ball of worms can then be placed into the aquarium where it is more easily seen by the fishes than are single worms.

Daphnia

Once again we have a food which, although excellent for fresh watre fishes, is of little value to marine fishes due to its inability to exist in salt water. It dies upon contact with tropical salt water, or shortly afterwards, and is usually ignored in a defunct state. Daphnia is however an extremely good laxative and attempts to feed it should be made. Many of the coral fishes seem to enjoy cracking the shells and spitting it out. The dead husks should be removed.

Animal and Fish Flesh

Scraped heart, liver and beef make excellent additions to the coral fish diet and should be offered occasionally to those fishes willing to accept it. An excess of heart or liver should be avoided but beef may be fed liberally. Beef is by far the best of animal foods to use as it is highly nutritious and has excellent preservative qualities – salt beef! Fish meats can be fed to a certain extent but under no circumstances should oily fish such as Halibut or cod be used. There is also a view held by many experts that fresh water fish flesh if not entirely healthy can cause many of the coral fish diseases (see Chapter 11)

Green Foods

Greenery, other than algae, is an essential part of the diet necessary to keep many coral fishes alive. Lettuce or cabbage leaves should be occasionally dangled into the water but never leave it there over night. It should also be added in large pieces and never cut up into small segments which may become concealed behind ornamentation and rot. The disturbance undoubtedly caused by the foraging for leftovers simply isn't worth it. Sea weed, provided it is green and thoroughly clean, can be suspended in the aquarium for a short while but never for long periods.

Dried Foods

There are many commercially prepared dried foods specially designed for marine fishes and they are usually devoured greedily by most coral fishes. The freeze dried variety of fish foods recently introduced, such as Longlife Wonder Worms, Freeze-Dried Liver, and Freeze-Dried Shrimp Gels (Adult Brine-shrimp) are very good, especially the latter, but consideration should be given to the fact that it is dehydrated and consequently most concentrated. The pieces of food should therefore be considered as much larger than they in fact are. Small chunks of freeze dried food can be either stuck to the inside of the aquarium glass, where the fishes can nibble at them, or soaked before feeding. It should never be placed loosely into the tank to float or it may be devoured in one gulp causing acute indigestion and possible death to the fish.

Flake foods such as "BiOrell are very carefully prepared by experts and contain most of the protein essential to the coral fish diet but remember that it is unnatural for any coral fish to feed from the surface even if they will in the aquarium. There isn't a sea in the world calm enough to allow food to float. Furthermore such species as the Batfish or Moorish Idol would find it necessary to either tip upsidedown or to raise half their bodies completely out of the water to feed.

Many of the commercial vegetable foods such as Mollyfare are of great value in completing a diet but should not be considered as a complete one. Foods, no matter what they are, should be varied.

The food requirements for individual species are listed in the Fish index and prospective additions to your coral fish aquarium should first be studied for food demands in this section. You may for example acquire a strange new fish whose diet consists solely of the occupants of your aquarium, a fact you will become painfully aware of when it is far too late.

Many coral fishes are carnivorous and living fishes must be provided for those which are incurably so. Occasionally, with extreme patience, one may be coaxed into accepting dead food by being constantly prodded on the snout with dainty morsels of food. Eventually the enraged fish may display his annoyance by snapping at the food and thereby discover the edible possibilities of its tormentor but this requires a great deal of time and patience and the food he may eventually accept cannot possibly compare with his natural diet.

Fry can easily be raised in a small breeding tank and many fresh water fish can produce prodigious amounts of young every month. The common Guppy or sword fish will keep you well supplied with live food and these will usually live long enough in salt water to fulfil their purpose. There is very little difference between raising fry as food or white worm as food if you are squeamish and the former is far easier. Sea Horses or pipe fish will seldom eat anything but small fry so you must make a choice first, before buying the fish.

Many aquarists make the mistake of feeding their 'runts' and outcasts to coral fishes. It is easy to say 'I don't like the look of that fish – feed it to the marines!' This must surely be the finest possible way to introduce disease into the coral fish community. *Never* feed a fresh water fish to your marines if it shows even remote signs of illness, and don't use fishes which are too big to be eaten. Most coral fish predators swallow their prey in one gulp, they seldom tear food to pieces or bite chunks out of it – just one great gulp. It is also wise to stick to one type of fish for food as well. If you feed all types of fish at random don't be surprised if your predators swallow a prized new specimen the minute you put it into the tank. After all the hungry fish isn't to know he just ate a day's pay if you haven't trained him to one kind of fish. A predator fed for example solely upon female guppies will quite often refuse a male when confronted with one.

When feeding your coral fishes do ensure that they are not constantly forced to eat the same dried foods every day. Unlike the predator, the dried food consumer becomes bored with no change. Freshwater fishes are also less liable to refuse foods through boredom but a coral fish will undoubtedly refuse foods if he has been kept on them too long and even starve himself to death sooner than eat. Vary foods as much as you are able to for, apart from boredom, the fish may be lacking a vital food and a change in diet every so often, although unscientifically, may accidentally provide that ommission.

11
Diseases

DISEASES IN THE coral fish aquarium is, at the moment, a subject about which very little is really known. This is remarkable fact when one considers that at least half the fish imported into any one Country are infected with one form of disease or another. Furthermore there cannot be an enthusiast in the world who has kept coral fishes for even the shortest periods and not suffered trouble of some sort or other. From the very greenest newcomer to the finest public aquariums, all have experienced disease in many forms. The remarkable fact is that with so many sick specimens to examine and so many enthusiasts working long hours in an effort to cure coral fish diseases, why are we all still, to a great extent, in the dark? We cannot even recognise most of the diseases let alone cure them!

Do not be too discouraged by these facts, for now that marine fish keeping is becoming more popular, a breakthrough is just commencing and the time is not too far distant when the hobbyist will be able to obtain specific disease cures in his local aquatic store.

Disease is one of the greatest challenges of all in the keeping of coral fishes and the finest weapon you have is prevention. Provided that all the aforementioned points are strictly adhered to then your diseases should end, one way or another, in the quarantine tank. This, in theory, seems logical but, as usual, is completely untrue in practice. This does not of course mean that there is no point in obeying all the instructions but simply indicates the unexpected behaviour of animals about which we are not entirely knowledgeable.

Recommendations and intelligence presented in this chapter can therefore only be applied as a guide and is only intended as such. Your job, as a coral fish enthusiast, is to apply your experience to mine and perhaps between us your diseases will be lessened.

Quarantine

All objects, be they fish, invertebrates or plants must be subjected to the same rigorous quarantine conditions. The quarantine period of your fish will be the most important time of their lives so perform it strictly, well, and with patience. If anything is going to show up in a fish, let it be in the quarantine tank and not in the community aquarium where all your stock can become infected. *Never* remove a fish from the quarantine tank until you are absolutely certain it is fit. Quarantine periods should not be fixed or restricted to any particular period but a minimum of seven days is the very least time that a specimen should be detained. This does not mean that a specimen kept in isolation for a period is then safe to be added to the main tank but that any ailment should have shown itself by then. The quarantine tank is also a place to study a specimen and try, from observation, to determine the requirements he demands. A mild disinfecting agent should be added to the water and I have found Diseaseolve, Methylene Blue or Pottassium Permagnate to be the most useful mixtures for this purpose. New fish should not be added to the quarantine tank whilst it is occupied, for the object of the quarantining would be defeated.

Water density changes should be made in the quarantine tank by reading the weight of the water in which the new specimen arrives, adjusting the quarantine tank water to match and slowly performing a change, if necessary, to match the community aquarium. This should not exceed a difference of more than .001 per day.

A familiar guide to the diagnosis of trouble in fresh water fishes is the condition of fins, the dorsal in particular. If this organ sags or shows signs of tiredness, then a fresh water fish is regarded with suspicion. On the other hand, a fish holding a firm, upright dorsal fin is regarded as looking 'quite healthy'. This is particularly so with goldfish. My personal observations of disease in the marine aquarium however have resulted in my forming the opinion that this case is completely reversed in most coral fishes. A frightened coral fish will always raise its dorsal fin and this very often transforms it into a very potent weapon. The Stickleback is a fine example of this as is the Surgeon. Furthermore a troubled or diseased fish will also raise the fin erect. Many other moods are expressed by the dorsal fin of a coral fish and it can often be seen flickering up and down, during feeding for example. The fish's mood can often be read by this 'near semaphore' but it is the permanently upright dorsal fin which we are concerned with here. The coral fish normally carries this fin folded and with complete disregard to the symptons of unhealthy fresh water fishes. I have found that in many cases, where a salt water fish carries its dorsal fin erect, an illness breaks out shortly afterwards.

In short, unlike the fresh water tropical fish, a coral fish, when upset, will 'put its back up'— perhaps this is where the expression came from.

Shock

This is unquestionably the prime killer of many marine fishes and many diseases or mal functions stem solely from it. We have already seen the terrible upheavals and vast changes a salt water fish must undergo before it reaches your kindly hands and it never fails to amaze me that the coral fish survives them at all. It surely indicates a tremendous will to live but, as so often with wild animals, and none are wilder than sea fishes, once the upheavals are over and the fight ends, shock strikes.

During the few terrible moments of a high speed automobile accident, when every nerve screams out in protest, one never for an instant suffers shock. The heart and brain are far too occupied with actual events to have time for it. Shock only sets in when the car has stopped and all is calm again. It may seem strange that the heart should wait until the danger has passed before protesting but it is a fact and delayed shock can be fatal. For this reason – delayed shock – it is the unfortunate aquarist who has to overcome the resulting effects and not the person who caught the fish.

There is little one can do to actually cure shock but the furtherence of it should be impeded by the provision of appropriate conditions. The quarantine tank, of all places, must be generously provided with hiding places where the unhappy fish may rest and catch his breath. It should also be left with subdued lighting which should be gradually increased during the quarantine period. Disturbances should be limited only to essentials such as feeding.

Poisoning

Second only to shock and perhaps malnutrition comes poisoning, hard on the heels of which follows pollution. We are therefore including both poisoning and pollution in the same category even though both can be caused by entirely different actions. Poisoning may be caused by numerous external factors whereas pollution is usually brought about internally. Either can, and usually does, cause the other. Natural metabolisms can create poisons and such obvious mistakes as allowing uneaten food to rot, possibly due to concealment, in the aquarium can result in pollution by causing Hydrosulphuric acid. Excess or unwanted foods, corpses, etc., should never be left in the water and the design of your aquarium should not be such as to interfere with your vision. You should be able to see the food at all times and thus be able to know when all has not been consumed for removal of the left overs. Healthy coral fishes will seldom allow a morsel of food ever to reach the bottom of the tank at all.

Many marine animals make use of poisonous cells either to obtain food or as a means of defence. A number of unexpected smaller animals fall into the former class and considerable damage may be caused by such surprising creatures as snails!

More obvious poisonous agents such as metals or chemicals should need no explanations other than those in Chapter 2 but beware of the common poison many aquarists overlook – chlorine. Most drinking water is purified at the water works by the addition of chlorine. In America 'the chlorine content of drinking water is usually twice that used in Great Britain. Water suspected of being high in chlorine content should be filtered through activated charcoal or carbon and allowed to stand for several days. It is worth noting that chlorine will leave water faster at a high temperature.

Another form of poisoning not often realised is contaminated air introduced via the air pump. If for example a fly spray or other insect repellent is used in the same room as an aquarium the air pump should be switched off. Air heavy in cigarette smoke should also be refused admission to the water by being drawn into the air pump. Forty milligrams of nicotine per gallon of water is quite enough to reduce your fishes to corpses. Metals and glazing compounds can be poisonous and are discussed in Chapter 2. The most poisonous metals are: Zinc, Brass, Copper, Iron and Steel in that order. Metal poisoning is also accelerated in soft, acid water.

Coral fish poisoning, although quite common, is seldom detected or diagnosed, usually because the lax aquarist has so many other possible causes of death at hand that he cannot possibly know which one to pick. The usual cry is 'I don't know why it should die, there didn't *seem* anything wrong. I tried everything'. Poisoning is not easy to detect but the cause of it usually is and this should be looked for before all else. Poisoning symptoms can often be mistaken for several of the disease symptoms and consists of restlessness accompanied by abnormal movements such as sharp bursts of speed in no particular direction. Following this is a state of breathlessness and an inability to stay upright. Prior to the final gasp the stricken fish changes colour. There is no cure for chronic poisoning and a case should be followed by a complete water change, the location and removal of the poisoning elements.

Oodinium

This is a disease of the gills usually referred to among fresh water aquarists as 'velvet'. It is an epidemic disease brought about by a parasitic flagellate. These tiny parasites cannot usually be seen in the place where most harm is done, the gills, and is usually diagnosed long after the disease has become quite advanced. The disease, by then, affects all parts of the body. Oodinium is easily recognised at this stage for the fish loses its colour and becomes peppered in fine spots which take on a yellow tinge. At this stage the fish is obviously unwell and refusing food. The Oodinium flagellates insert blood-disturbing substances into the skin of the fish. This severe irritation to such delicate parts of the fish as gills causes a swelling which soon interferes with breathing. Following this stage a poor skin pallor becomes noticeable and sluggish movements, combined with a lack of appetite, should confirm the disease. Oodinium is often confused with poisoning. An interesting fact about oodinium is that it occurs more frequently with West Indian fishes. The Clown fish is particularly susceptible and, in view of this I carried out a series of experiments both in the aquarium and under the sea. The Oodinium discovery was incidental but nevertheless interesting. The Clown fishes used in the experiments showed an extreme disregard for all that has been written about them as did other specimens studied. It has been firmly believed for some time that Clown fish, and Clown fish only, are immune to the sting of a poisonous anemone. It has even been suggested that Clown fishes actually emit a substance which is recognised by the anemone and permission is given for the Clown fish to enter the tentacles with immunity. This I found to be complete nonsense. No two wild animals would ever form such an association unless the profit was mutual. The Rhino only suffers the cleaner birds on his back because of their usefulness and the shark seldom eats the sucker fish which are kind enough to keep him free from parasites. My observations show that the anemone is feeding on the skin of the Clown fish and usually with greater enthusiasm when the fish is infected with Oodinium! The Clown fish however is most certainly *not* the only fish an anemone will refuse to sting for I have observed many other small Damsels performing the same 'dance with death' and with the same immunity. The tentacles made no attempt whatsoever to close on the guest but merely gently 'fingered' the skin of the fish removing, almost vacuum cleaner like, the tiny parasites attached to the fish's skin. When the last flagellate was removed the fish usually left the anemone only to return if the disease reappeared. The Clown fish was an exception however, in that once a suitable anemone was found it would actually do battle in order to retain the privilege of the anemone sanction. A frightened Clown fish will also seek refuge from an enemy in an anemone and seems well aware of its safety once buried deep in the tentacles. My personal examinations have shown that the probable reason why Clown fishes are more often seen in the company of anemones is that the Clown fish is particularly susceptible to Oodinium. The observations resulting in the above opinions refer only to the tropical (Stoicactus) anemone. Cold water anemones such as the Beadlet stung to death *any* fish which came into contact with it, including the Clown.

The cures offered for Oodinium vary but the ones I found to be most successful were Copper Sulphate, Methylene Blue or Diseaseolve. The latter is an easy remedy to try first for it does not cause discolouration and stain the coral and is unlikely to affect the fish adversely should it not prove successful in curing the disease. Methylene Blue is perhaps more effective but it will unfortunately stain coral or sub sand. The 'blue' should be added to the aquarium via the filter (as should all chemicals) so as to avoid too sudden a change for the fishes. The amount of 'blue' used will depend entirely upon the severity of the disease and can be added to a strength which will darken the water enough to render the fishes impossible to be seen if necessary. Filters and lights should be turned off and the temperature should be gradually lowered a full five degrees. The oodinium flagelates are unable to survive without light or heat. Do not be concerned about the discolouration of your aquarium water for it will fade away after a while. Should it fade away too quickly and not be detectable after a few days then add more methylene blue for Oodinium has a ten-day cycle and unhatched eggs are unaffected by the treatment. Once a parasite has reached maturity it leaves the fish and drops to the bottom of the aquarium. It then splits up into eight, sixteen, thirty two, sixty four, etc., parts. These parts become spores and must find a host within twenty four hours to die. Water therefore which has contained Oodinium contaminated fish should be safe for re-use if it has been allowed two or three weeks to stand.

Copper Sulphate can be the marine aquarist's best friend but it should not be forgotten that copper is not only toxic to many parasites but, in sufficient quantity, just as deadly to fish. Copper is a poison and should always be treated as such. A little can cure whilst a little more can kill. A maximum concentration acceptable to most coral fishes is 20 milligrams per gallon and even this strength should only be resorted to when all else has failed. Some coral fishes can accept more copper concentration than others so begin with at most 10 milligrams per gallon strength. This can be increased to the maximum strength when the situation is desperate but ensure that it is so first. Have you made a complete water change? Have you raised or lowered the water density? Varied the temperature? Your job when presented with any form of illness in the aquarium is to produce conditions insufferable to the cause of the trouble but with as little possible disturbance to the invalid. The biggest nuisance to the germ, combined with the least nuisance to the fish is your ultimate. Delicate microbes are usually quite unequipped to cope with such a drastic water density change as .003 so, if your specific gravity is 1.025, change it to 1.022 or 1.027. Do not use this method unless you are quite sure your fish is able to suffer the change, remember that he is sick.

Most chemists will prepare a solution of copper sulphate to the strength you wish. Simply give him the gallonage of your aquarium and ask for a solution to give you 10 milligrams per gallon in that water. You may then dilute this at will without the necessity for constant calculations and should try your first dosage at half strength first before resorting to a more powerful dose. Always try the weak solution first, do not give a large dose simply because the fish looks very badly infected. The effect copper has on a fish is to cause considerable irritation to the dermal cells of the skin. A mucous is secreted through these cells and the resulting movement of fluids through the skin from the inside, instead as usual from the other way round, causes the parasite to loosen its hold and drift off into the water where it is then promptly poisoned by the copper.

It is most important to note that many invertebrates, especially the anemone, are completely unable to stand even the slightest amount of copper so never allow these animals to come into contact with it. Remove them if they have been in contact with the disease and treat them as new specimens by once again putting them through the whole quarantine routine.

Another remedy used and recommended by some coral fish experts is Acriflavine but although reports are good I have never personally experienced success with it and cannot therefore verify cures, but all treatments are worth a try, what is good for the goose, etc., does not always work in coral fish keeping. My water may differ considerably to the hobbyist one mile up the road.

Benedenia

This wretched ailment is brought on by parasites rather similar to the Oodinium flagellate but, in a much larger form. It also places a greater demand upon the fish. Benedenia is easily seen, usually upon the transparent areas such as fins, but the best place of all to locate and recognise this ailment is on the eye where it becomes magnified. Should any fish suddenly decide not to eat then inspection of the eyes will very often result in the discovery of some form of parasite or other. Lion fishes and large butterflies seem more prone to catching Benedenia and once a hold has been grasped on a fish the whole tank soon becomes infected. The parasite seems to affect the fish to such an extent that irritation of the skin soon becomes unbearable. A fish is completely unable to touch its own body in any way. It has no arms to scratch with and can reach no part of its own body at any time. It can therefore only relieve severe irritation by 'scratching' on external objects and this it does by diving into the coral or rubbing up against sharp objects. Scratching of course never alleviates aggravation, if anything it adds to it, but this never deters the fish from doing so. This 'scratching' can be just as dangerous as the parasite itself if prolonged take action if you observe a fish diving into the subsand. If an external parasite is not visible then an internal one probably exists. A temperature drop should be enforced immediately and all lights should be dimmed or turned off. A more drastic cure can be enforced by the use of copper sulphate as for Oodinium. Copper sulphate incidentally can kill algae so treatments of this nature should be performed in the quarantine tank. A better method of treating Benedenia is the use of Ozone, which is discussed later in this chapter.

Ichthyophonus

Unlike Oodinium and Benedenia, which are external parasitic diseases, Ichthyophonus is caused by an internal parasite which grows in the form of a fungus. It is often caused by unclean foods and is relatively common among many of the Damsel fish. It can also be caused by an algae spore. It is recognisable usually by the size and condition of the belly of the fish and looks somewhat like Dropsy. The belly becomes bloated and the scales are unable to lie flat. The area can even become so extended that blood can be seen. Predators fed with unhealthy fresh water fishes are liable to suffer this infection as are in fact any fish fed with food that is not thoroughly clean. Streptomycin used in conjunction with penicillin can effect a cure and should be added at a strength of 250 milligrams of each per gallon. The British wonder drug, Penbritten, may also secure health.

Ichthyopthirius

This disease, often referred to as 'salt water Ick' or 'white spot', is quite common in the fresh water aquarium. It is also a disease which the fresh water aquarist takes in his stride, knowing full well it is easily cured. This attitude should not be permitted where coral fishes are concerned and 'ick' should rate high as a killing disease for it is precisely that. 'Ick' is as easy to recognise in coral fishes as it is in fresh water and is identifiable by the large white spots which appear firstly on the fins of the stricken fishes. They are considerably larger than Oodinium spots and are white instead of yellow. The white spots should be noticed on the fins and not when they have already reached the body. Coral fishes very often have a white spot on a fin, usually the caudal, and, provided this spot does not become larger or turn into two spots then it can be ignored for the fish will probably keep this mark for the rest of its life. 'Ick' is an extremely contagious disease and will run through an aquarium like wildfire, if allowed to, killing all your stock overnight. Like Oodinium this disease parasite drops from the skin of the fish when mature, the only difference being one of method of reproduction. Instead of splitting into groups of 8, 16, 32, etc. several divisions are made from which many other sub-divisions are formed. Therefore the descendants from one single adult may well run into thousands. The young can live for much longer periods without a host than Oodinium and can stand much lower temperatures. In this case therefore the temperature is raised, as much as 10°F. The 'Ick' parasite bores its way into the epidermis of the fish and then slides along *above* the dermis. In this way it is protected on each side by a layer of skin. The holes which they create in this manner often house several inmates. When these 'housing areas' become over populated the skin may burst and further disorders may be introduced such as an abscess or fungus. A heavily populated aquarium will suffer a far more serious attack than a sparse one so don't overcrowd your tank. Copper sulphate, quinine or Acriflavine may be used as tonics and lighting should be reduced. A fourteen day period should be the minimum time allowed for a cure. Should quinine be used at all then the whole aquarium water will have to be changed when the disease is cured.

Tomato Clown *Amphiprion Ephippium*

Undulated Trigger *Balistapus Undulatus*

Lymphocystis

These are mineral masses which occur on the skin of the fish and look rather like cigarette ash particles. The patches are approximately five times larger than the white spots seen in 'Ick'. This form of infection is often taken to be boils and is often found on newly imported specimens. Pictures on page 151 show a severe infection of Lymphocystis and the further damage caused by surgery. In this case the surgery was successful and the healing of all wounds remarkably fast but, unfortunately the fish died at a later date. Surgery is usually the only way to remove the 'boils' and often results in death but, as the fish will die in any case, nothing is lost by the operation and, even if only experience is gained, the time is not a complete waste. Two incisions are necessary, one to slice under the area and a vertical one to remove the growth. Bathing the wound in a dilute iodine solution should follow the operation, instruments of course must receive a thorough sterilisation.

Exopthalmus (Popeye)

Coral fishes seem far more prone to catching, or should I say suffering from, Popeye than fresh water fishes and this is most probably due to a lack of diet understanding. Malnutrition and poor quality foods usually lead to a weak blood and consequently a poor circulation to the smaller blood vessels such as those feeding the eye. The remedy for Popeye caused through incorrect foods is obvious – a new diet. There are of course other ways of causing Popeye but these can be eliminated from the diagnosis list if your aquarium conditions do not provide those causes. For example excessive aeration with over fine air bubbles can cause Popeye. The bubbles, if tiny enough, can get into the blood stream and be unable to complete a full cycle and, where artery becomes vein (the capillaries) in the more delicate parts of the body such as the eye, a blockage can be caused. This blockage will eventually burst and the damage pass on to the eye. Alternatively the eye may suffer through a lack of blood or even the unburst blockage itself. Although Popeye which is caused by the former mal-function may be cured by the correction of diet little can be done about the second except to replace the offending air stones with ones causing larger bubbles. Another condition liable to promote Popeye is foul water. When water is allowed to become dirty, diseases are easily contracted and Popeye is often added. In all cases of Popeye the temperature must be reduced and lights turned off. The less eye strain at this time the better.

Saprolegnia (Fungus)

This infection often follows one of the aforementioned diseases but can also occur as a result of damage caused by a fight. Chilling can also bring about an attack so take care when dropping the temperature to cure ailments. It may seem very strange but, although fungus may be caused by a chill it can often be cured by a temperature drop! Fungus does not always appear on coral fishes in the same form as fresh water ones. It usually takes the form of white patchy splotches on the body of the infected fish. It seldom grows out into great 'cotton wool' growths that almost light up. A water change usually cures this infection but a more drastic cure can be effected by a salinity dilution. Fungus does not often spread throughout a tank but all fishes having contact with the infection should be watched carefully for signs of distress.

Argulus

This is an extremely common malady peculiar to the Sea Horse and very few arrive from the sea without it. The parasite can be seen quite plainly with the naked eye and the method adopted by many aquarists is to pick off the parasites one at a time with a pair of tweezers. The parasites are quite large enough for this to be done and, as the sea horse has an external skeleton, damage to the skin is not likely.

Wounds

Cuts and other damage caused perhaps by battles among fishes are not strictly diseases but they can quite easily be the beginning of one if left unheeded. Severely damaged or injured fish should be separated from the community and given complete peace and quiet in the quarantine tank. No further treatment is necessary but a close watch should be kept for the appearance of infections such as fungus. Marine fishes show a remarkable ability to restore damaged parts and can grow new fins almost overnight provided they are left to do so and are in otherwise good health. There are many animals in the sea quite capable of growing a complete new digestive system for example in days.

Ozone

This is a new word to aquarists having only been available to them in an aquatic medium for some five years. Strange though it may seem Ozone has been used to purify world wide drinking water systems for over fifty years.

Without being too technical Ozone is perhaps best explained in its natural form. We all know the nature fresh smell that pervades the air after a thunder storm. This smell is Ozone which has managed to reach the Earth. Gigantic flashes of lightning sear across the sky burning oxygen particles at an incredible rate. This purified and ultra fine air, heavy in ozone, created by the high voltage flashes, is carried down through the atmosphere without impediment and our noses are able to detect it. Technically speaking Ozone is an extremely active form of oxygen gas produced from the action of the Sun's ultra violet rays at the edge of the Earth's atmosphere. As the gas passes through each atmospheric level the Ozone is de-concentrated until, at the Earth's surface, it occurs only to the extent of .05 parts per million. This de-concentration is retarded by the electrical effects of a thunder storm and more Ozone reaches the Earth. Many doctors recommend Alpine holidays for invalids, especially those suffering from lung infections, because the air has a greater content of Ozone and is therefore much purer. The Queen of England has Ozone installed at Windsor Castle and many hospitals make use of it. The London Zoo has also installed Ozone recently into the Elephant House to reduce smells. Odours are simply minute particles of the object causing the smell and ozone will 'burn up' the offensive bacteria.

Ozone has been adapted to aquarium use by means of a small box into which air is pumped from an ordinary vibrator pump (although the manufacturers prefer a piston type pump). An outlet is provided and air fed into the aquarium from this. These units are becoming more and more popular among coral fish enthusiasts and on a recent trip around the Continental aquarias I did not once see a marine tank which was not using Ozone. Ozone can be used both to prevent outbreaks of disease and to cure those which already exist. Ozone will kill all bacteria, viruses and spores and is the most powerful oxidising agent known to man. It can aid the healing of injuries caused through fights by preventing infection and fungus, once begun, can be checked by a flow of Ozone. Live food sterilisation can also be performed with an 'Ozoniser'. Ozone, in a weak form, can be continuously run through the air system of an aquarium and will increase the height of Radox Potential thus creating conditions similar to those in a natural coral sea.

The three Ozonisers obtainable for use by the average aquarist are all made by Messrs. Sander of West Germany and give proportionally ten, twenty five and fifty milligrams of Ozone per hour. The latter is the more useful unit, for it is adjustable and has a much higher output than the others. For the larger coral fish keeper or commercial fishery a further three models are available. These can however only be used in conjunction with a compressor. They produce progressively 150, 500 and 1000 milligrams per hour. The air pressure used to operate these models should not exceed twenty two pounds per square inch. The maximum air output is again progressive and ranges from 500 to 1500 gallons per hour.

All Ozonisers are equipped with water repellant electrodes and are therefore completely unaffected by the humidity of the air, an important consideration in a fish house.

Ozonisers may be used for the following purposes:

(1) *For the clarification of bacteria turbid water.* Water murky through excessive bacteria content will be rendered clear after three hours of ozone use.

(ii) *For live food sterilisation.* This is executed by running 'neat' Ozone through the water containing food with an air stone for ten minutes or so.

(iii) *For disease cures.* Ozone should not be used at full power for long periods without protecting the fishes from the stream. Ozone is poisonous but, when used in moderation, can have the same effect as the 'hair of the dog'. A strong Ozone current used for too long, and without a means of separating it from the fish, can cause severe gill irritation and defeat the purpose of using it at all. Treatment should be restricted to one hour of 'neat' Ozone per day. It should never be used for more than twice this period in any one twenty four hour period in neat form.

(iv) *For continuous use to prevent the outbreak of epidemics.* The continuous use of Ozone must be accompanied by the use of a reactor tube. This not only keeps the Ozone stream from direct contact with the fish but also acts as a skimmer and performs similar functions to those described in the fluid separation on page 43. The fluid separator was in fact devised from an idea presented by the reactor tube.

Ozone Reactor Tube

This ingeneous and simple instrument consists of a tube some two inches in diameter through which water, entering through holes situated half way down the stem, is drawn by means of an air lift, through the tube and back into the aquarium. Whilst the water is being passed through this tube it is oxygenated by Ozonised air. The burnt bacteria is lifted into the tube by means of the air bubbles which rise to the surface and are passed into the skimming cup.

The reactor tube must be fitted into the aquarium so that the water inlet holes are one inch below the surface. Air bubbles supplied by the pump and passed through the Ozoniser unit must be fine and cover the whole width of the tube. Skimming however is only a secondary function of the tube, its main use being as a reactor. Oxygen shortage caused by the fishes and their foods, can be reduced by the use of this tube in conjunction with Ozone. Also the formation of bacteria requires organically overcharged water and this condition cannot exist at all when Ozone is present. This makes the outbreak of an epidemic extremely difficult.

The method I use to calculate Ozone strength in conjunction with the reactor tube in the healthy coral fish aquarium is to apply one half a milligram to each gallon of water. Distress signals are usually quickly followed by an increase to one milligram per gallon and, failing a quick recovery, chemicals are then resorted to, but not until then. The Ozone can be increased beyond these figures and even until the fishes show signs of discomfort not in uniform with the disease in question.

More detailed information about Ozone (O^3) will be supplied by the manufacturers upon application.

The need to study this chapter again at a later date will be limited if you apply all your resources in an effort to prevent diseases occurring.

12
Maintenance of the Aquarium

A WELL KEPT aquarium has a far better chance of remaining healthy than one which is left to its own resources. Treat your aquarium as though it were a new car. You would not run a new car for months on end without a service would you? The first service a new car expects is an adjustment of all moving parts which have settled down after being 'run in'. Oil will require changing, points will need resetting and many other similar adjustments will have to be made. The aquarium will need a similar service after it has been 'run in', within perhaps a month of creation date. The only differences being that instead of cleaning the carburretor you clean the filter, instead of changing the oils you change the filter mediums, etc. Air flows and pumps are adjusted, the water level is regained and so on. As the aquarium ages it will require further services just like your car would so why not keep a log book of changes, renewals, new fishes, successes and failures, etc? You are not likely to forget a mistake once made and are able to see at a glance whether or not you are repeating yourself in vain on many points. Servicing also becomes regular and methodical. There are many old timers who will say 'Let well alone' but this saying, although very often sensible, should never be applied to coral fishes. Do not assume that just because all is well you would be foolish to touch anything, or when something does happen, it will happen with a bang. A car would not last long if regular services were not maintained and neither will a salt water aquarium.

You have put a considerable amount of care and effort into setting up the most Utopian community possible and now your job is to maintain that condition. It is a mistake to assume that only pure distilled water for example is lost from the aquarium through condensation. Although pure water is the only loss into the actual atmosphere 'crusts' of salt and trace elements are removed from the tank water during the process of evaporation. The water which leaves the main body contains many of the elements that we would prefer to leave behind. These pin points of water collect together on the rim and on the cover glass and form large droplets which either drip back into the tank or disintegrate into the atmosphere. Only actual pure water does in fact leave the droplet but the remaining salts do just that – remain, and on the cover glass not in the aquarium where we want them. They stay on the cover edge or glass undisturbed until another droplet is formed on top of it and so on until the whole area becomes quite thick in trace element and salt crusting. If we were to continue topping up our aquarium with fresh water in the belief that we are replacing all that is lost we would eventually reach the point where the aquarium water was seriously lacking in vital elements. For a certain length of time, even up to a year, this loss could be compensated for by the addition of pure salt, added when the hydrometer indicated the need for it, as is usually very often the case, but there isn't an instrument in the world that can tell you which elements have been lost or deteriorated. The answer must therefore be a complete water change at regular intervals. The duration of the period between intervals will depend entirely upon the water loss rate. An aquarium which loses water heavily by condensation will require a complete change more often than one which loses very little and which contains fewer fishes. As this time factor is almost impossible to calculate you may be better off providing say three partial changes followed by a complete one on the fourth. For example, every six weeks replace one gallon of aquarium water with a new mixture and, on the fourth change, or after six months, replace fifty per cent. of the water altogether. At the end of the year change the whole aquarium water. These figures apply to aquariums of up to twenty gallons in capacity, those larger should not be calculated by simply doubling the periods but by trebling for the larger aquarium will require less changes. All will depend upon the size of your aquarium and the amount of fishes therein. Also it will depend upon how clean you keep it and whether it falls into the Ford or Rolls category.

Uneaten foods, no matter how meticulous you are in cleaning up after meals, will eventually foul the water to a certain extent but this is not always necessarily detrimental for much of the bacteria is essential to the fishes health but this must be limited and partial water changes alone will not alleviate this. The sub sand or floor covering must also be removed and cleaned periodically. If the floor covering is not too deep, and it should not be, it can be syphoned off with the water when you are conducting a water change. A half inch bore tube will suffice for the syphoning and the gravel being syphoned .into the bucket can be netted during or after the operation. A thorough and absolute cleaning is vital and no gravel should ever be returned to the aquarium if it smells even faintly. Take great care not to disturb the fishes more than is absolutely necessary whilst cleaning the aquarium and do not allow chills to occur by topping up the tank with cold water.

Filter medium must be changed frequently. A high water consumption power filter will require new nylon wool *every week* and a filter operated by an air pump recharged fortnightly. As with the motor car, a complete overhaul is called for every so often and the longest gap between such operations will depend once again upon the size and occupants of the aquarium. A tank having less volume than twenty gallons will require a complete overhaul every six months whereas a large capacity unit can go on for up to two years provided the periodical services have been maintained. For a complete and correct overhaul a temporary home is necessary for the inmates. This second unit should be identical in temperature and water density provided with places for the terrified fishes to hide and be completely without artificial lighting. Do not attempt to catch fish from the tank you intend cleaning without first removing all net impediments such as coral and rock work. If you have generously provided all the hiding places discussed so often in this book then the fishes will make use of them – and hide. Coral fishes show a remarkable turn of speed and acrobatic agility and there is no substitute quite like coral able to aid the panic stricken coral fish in full flight. Trying to catch fishes in a tank full of coral will only lead to frustration on your part and possible injuries to the fish. Remove *all* obstacles from the tank and treat them first. The time taken in cleaning the coral will give the frightened fishes a breather and allow them time to become accustomed to the further shock of being netted and transferred into a strange place. It will also lessen the period they will spend in the new tank if the coral is ready to be returned before the fishes are even caught. Do not keep the fishes in a strange tank for any longer than is necessary. The coral should be treated exactly as though it were new and cleaned in the same way. The tank should be emptied, cleaned and thoroughly sterilized. New water added to it after the change should be given plenty of time to settle and be strongly aerated. Corals should be replaced in the same order as they were originally. You may prefer a different decor but the fishes will not appreciate it and squabbles may result when more than one fish disputes ownership of a cave that was once two. Fish disturbed by a tank overhaul should be well fed before the event and not fed after it for at least twenty–four hours. Their systems will be upset and perhaps unable to cope with food. Lights should also be left off for a few days after the catastrophe for catastrophe it is to the fishes. Lastly, be gentle when servicing the aquarium, do not prod fish with a stick to move then from a corner and avoid sharp contrasts of any kind. A well serviced aquarium, like the motor car, will result in a long lived and reliable one.

13
Index of Coral Fishes

AN INDEX OF coral fishes could, were space available, run into many hundreds of fishes but space is not the reason for omissions here. The obvious absentees such as Moray eels, Groupers, Sea Bass, etc., are missing for a very good reason, they are not in my view aquarium fishes and I fail to see any point at all in wasting time on fishes which will outgrow the aquarium in a matter of weeks. Some of these fishes can exceed six feet in length and this to me is not an aquarium fish. The other reason for omission is lack of knowledge. There are many books which cover every single specimen of a class discussed and this cannot be honest for it is impossible to study every single specimen. I have therefore tried in this book only to discuss in detail my own observations and findings. In cases where I have not first-hand knowledge of the fish in question I have simply given specifications and not observations. A further reason for certain omissions is the lack of availability for there seems no point in writing about fishes which are near impossible to obtain therefore every fish in this book can be found on the lists of professional aquarists or simply picked up on the beach.

In a number of cases there will be found certain disagreements with other authors but this does not necessarily make either writer wrong, it simply verifies the fact that conditions change greatly in different parts of the world and that, what is fact in America, may well be fiction in Europe. This index therefore should only be used as a guide and applied to your personal experiences with an open mind. Do not be afraid to experiment, you have plenty of scope in which to do so in the world of coral fish keeping.

Finally I have chosen to classify fishes in order of their Latin names and although perhaps a little difficult for the reader at first, will be invaluable once one becomes accustomed to it for there are so many mistakes made through the use of incorrect common names that they have become useless with mal-use. I therefore strongly recommend

94

that you try to forget these common names once and for all for they are of no use any more. There is at the end of this book a list of names commonly used and these may be used as a quick reference wherever necessary. If you make use of this you will soon get to know the correct scientific terms which are of far greater use to you.

Acanthuridae
(Surgeons and Tangs)

SURGEON FISHES HAVE been most aptly named for all specimens carry a pair of built in 'scalpels'. These razor sharp protudences are usually folded safely out of the way and are located just forward of the caudal peduncle. Fear or anger in the Surgeon fish is usually displayed by a stiffening of all the fins and a 'dance' similar to that of a breeding male *Betta Splendens* (Siamese Fighting Fish) and these extremely sharp dangerous 'knives' are thrust out like stilletoes ready for action. The angry Surgeon fish then sidles alongside the reason for his anger, usually a larger fish, and whips his tail across the flank of his adversary cutting it to the bone. Two or three of these vicious attacks can reduce the unfortunate victim to a corpse.

An established aquarium Surgeon fish will usually ignore smaller fishes being introduced into his area provided there is no demand made upon his actual territory. They are chased away by the Surgeon fish and seldom return, but woe betide a large fish which looks a possible threat whether it is or not. The Surgeon fish will attack any new fish which is as large or larger than he, and therefore all large specimens you intend keeping with Surgeon fishes should be placed into the aquarium *before* the Surgeon fish.

The 'knives' of the Surgeon fish are not his only weapons for his dorsal fin is also extremely sharp and can inflict a nasty gash sometimes with a poison thrown in as well. Fingers should never be placed too near the water of an aquarium which houses a large Surgeon fish or they may well be lopped off! Surgeon fishes are not always the ideal fish for private aquarists for they grow quite large and require reasonably large aquariums. When purchasing one consideration should be given to its growth rate and adequate space allotted for its adult state and not its purchase size – a common mistake. A further detriment for the private aquarist is the fact that most Surgeon fish lose their fabulous colours as they age and what was once a highly coloured little beauty may eventually turn into a dull monster which prevents you from keeping other more interesting specimens. Another special requirement for the Surgeon fish which may prevent the keeping of many other fishes in the same tank is the need for a high temperature. Surgeon fishes require a temperature sometimes exceeding eighty degrees F. Furthermore they require vast quantities of algae and must be provided either with this or an alternative such as spinach or lettuce. In view of the quantities absorbed by the Surgeon fish it is not surprising to find that it also excretes a considerable amount and a clean tank is almost impossible to keep. Without this vital diet of green foods the Surgeon fish will almost certainly die. Because of their large appetite and coupled with the fact that it is an extremely small-scaled fish, it is very prone to the disease Ichthyophonus or even Oodinium.

Once the Surgeon fish is accustomed to his new home and inmates he usually keeps to himself and seldom bothers other fishes, provided he is left to his own devices, for he has a very peaceful disposition. However new fishes added to his aquarium are regarded with the greatest suspicion and should be carefully watched until the Surgeon fish has examined them and decides they are not a threat to his existence. The Surgeon fish does quite well in a well-stocked community aquarium and is, in the wilds, a shoal fish.

Surgeon fishes are to be found in most of the warmer seas swimming in large groups. They are unusual swimmers making use only of their pectoral fins but a turn of speed can be switched on when necessary.

Acanthurus Dussemieri

Although this fish is known as the Hawaiian Surgeon it can in fact be found almost anywhere in the Pacific. This is the first Surgeon fish that I ever kept and I soon learned that it is definitely not a small aquarium fish for it can exceed two foot in length. It is quite a bright fish when young being all brown with a very high yellow dorsal fin which is seldom seen unless the fish is excited. These colours soon fade and, within a year it is a slate grey with an almost white tail and there is little of interest in the fish at this time.

Acanthurus Leucopareius

This is a very small Surgeon fish reaching a mere six inches in the aquarium but it is nevertheless quite capable of inflicting nasty wounds when aroused. It is not a striking fish being a dirty brown with a yellow line running down the body just behind the eye and a yellow spot in front of the caudal fin.

Acanthurus Leucosternon

This is the most stunning of all the Surgeon fishes and can be picked out in an aquarium from a great distance. It has a powerful light blue body set off with a brilliant yellow dorsal fin and a contrasting black face. Once settled down in the aquarium he will live quite well and will keep these dramatic colours for some years, a truly beautiful fish and a credit to any aquarium. It can reach a size of twelve inches in the Indo pacific from where it comes but seldom exceeds eight inches in the aquarium.

Acanthurus Lineatus

A less circular fish than most Surgeons, being drawn out at the tail and rather long in appearance this fish is not the best of community dwellers and can grow to quite a large size. It is nevertheless a striking fish being white-bellied with a sharp yellow and white horizontally lined pattern all over its body. It is unfortunate that such a marvellously coloured fish should be such a size as to disallow introduction to the home aquarium.

Acanthurus Nigricans

Although this fish is commonly known as the Black Surgeon fish it is in fact almost completely purple and the red spots which adorn his body when young, leave the fish altogether as it reaches maturity. Surprisingly it is not a very popular fish but this may be due to lack of availability rather than choice, for it is a small Surgeon which never grows more than six inches in the aquarium. It can however reach a length of twelve inches in the Indo Pacific regions from whence it originates. It is a fish of moods which can be determined by the colour of his skin for he is an expert at changing colour to suit his needs and feelings.

Acanthurus Sandivicensis

This fish is one of the rare cases where a popular name is less confusing than the scientific for there seems to be some doubt as to the actual name of this fish and it is sometimes known as *Acanthurus Trigosteus*. The common name however is the Convict Tang and about this there can be no mistake. This possibility of coral fishes being mis-named or twice named is certainly not uncommon for as we learn more about this intirely new (to man) form of Zoology and are able to examine the tremendous colour and pattern changes of coral fishes in the aquarium these mistakes are being located and rectified. The Convict Tang found in Hawaiian waters is known as the *Acanthurus Sandivicensis* whereas the Indian Ocean specimens are usually known as *Acanthurus Trigosteus*. Both however are identical in colouring and have a silver body with five sharp brown bands on either side of the body giving the fish its common name. It can grow to ten inches but is more common in an adult form at a size nearer six inches.

Acanthurus Xanthopterus

This is yet another fish which would be a delight were it to remain small and not lose its fabulous colouration but, once again we have a fish not really suitable for a long term aquarium. It has the face of a horse and is usually very bony looking around the jaws with most of the facial bones showing through. As this fish grows it becomes very elongated and becomes a pale green with yellow fins. It is found almost anywhere between America and the African coast. Size varies but the Antwerp Zoo Aquarium has a specimen of almost two feet in length. It lives communially and is no trouble.

Ctenochaetus Strigosus

This is another large Surgeon fish and a rather rare one being seldom seen in Europe at all. It is a rather washed out blue with hundreds of purple undulations running horizontally throughout the body and fins. It is not usually a community fish and must be kept alone or with very large fishes.

Paracanthurus Theuthis

This fish is usually referred to as the Morpho Butterfly fish although why I cannot imagine for it is most certainly not a Butterfly but a Surgeon fish. This is one example of how confusion can occur through not making full use of correct scientific names. It has however been described as the bluest blue in the world and this I most certainly agree to for it has to be seen to be believed. It is an incredibly vivid blue and will no doubt cause a sensation in any aquarium. It is a rather small fish for a Surgeon and is fairly timid and although I have never seen one larger than four inches it is reputed to reach a size of nine inches. The yellow markings on the caudal fin spreads out as the fish matures and the blue gradually makes way for it, becoming less each month so a small specimen would be a preferable choice for it will remain colourful much longer and you will get greater pleasure from it.

Zebrasoma Flavescens

Another Hawaiian fish commonly known as the Yellow Tang. If the last fish shows off blue to its best then this fish certainly does the same for yellow and usually offsets most other fish in a community aquarium. The bright yellow body is only broken in colour by the white dashes just forward of the caudal peduncle where the 'knives' lay. The colour intensity seems to lessen the farther from Hawaii that the specimen originates and only those specimens caught in the Hawaiian area have the true magnificence of full colouring. The Flavescens seldom exceed six inches in length and are rather difficult fish to keep for any length of time. Very thin specimens should not be purchased for they never seem to live at all.

Zebrasoma Veliferum

Ranging from Africa to Hawaii this fish is aptly named the Sailfin Tang for it has an enormous dorsal fin, sometimes equal in size and shape to the whole body of the fish. Its anal fin is no less magnificent and is not very much smaller. Thick black bars almost hide the dark grey body. It is not a very hardy fish when young but once established may well exceed eighteen inches.

Zebrasoma Xanthurus

Another rather dull fish and one which is subject to malnutrition if not carefully tended. It is dark brown to purple in colour and usually fades off to a dirty washed out grey with age. Not the very best of aquarium fishes.

Antennariidæ
(Frog fish and Angler fishes)

THESE ARE THE true angler fishes and are extremely well named due to the fact that all specimens have a built in antennae which is in fact an extension of the spine, which hangs out above their heads like a fishing rod, complete with baited line. This extraordinary and un-believable 'extra limb' is actually used as an angler uses his rod and line and is surprisingly successful for catching fishes. Apart from their ability to catch smaller fishes the Angler fishes are masters in the art of disguise and, with their built in camouflage look less like fishes than bicycles do, with great dermal flaps hanging all over their bodies like chunks of floating sea weed. The unsuspecting prey, who cannot be blamed for not recognising the danger or enemy, is lured to the bait which is wriggled enticingly in front of them and then they are engulfed with remarkable speed. The size of this prey does not seem to affect the predator for he can usually cope with a fish of almost his own size. Angler fishes are not community dwellers except with fishes of a much larger size and they are quite capable of eating each other so, if you wish to keep one at all you must do just that – keep one! Living foods are essential to the Angler fish and seldom will anything else be accepted so do not obtain one unless you are in a position to supply live fishes as a food every day.

The members of the *Antennariidæ family* can be regarded as certainly the weirdest creatures on earth. Living nightmares of unbelievable shapes they look more like irregular chunks of rock covered with odd pieces of sea weed than fish. This almost perfect disguise makes it most difficult for collectors to locate them under the sea and they are in consequence quite scarce to the aquarist. They are quite hardy once established in the aquarium and can provide good entertainment being very poor swimmers and preferring to 'walk' around on the floor of the aquarium in a slow unsteady gait. It is sometimes most difficult to convince a layman that it is a fish at all and the method of catching fish used by the Angler fish is very hard to believe. It is unfortunate that they should require a tank to themselves but they can be kept in quite a small container for they are not very agile and are quite happy to sit on a rock without moving all day. They are well worth the trouble incurred.

The Angler fishes can also be known as *Pediculati* and these can be divided into two categories, the smooth skinned *Histrio* and the hard, flappy skinned *Antennarius*. The one disease which Angler fishes seem unable to endure is Ichthyophnus and an attack of this is usually followed by death. It is therefore not wise to ever feed an Angler fish with a live fish which is not completely free from sickness. Never feed with your fresh water runts.

Lophiocharon Horridus

A smooth skinned Frog fish looking perhaps a little less horrid than most of this family. It has a bright yellow irregular body with dashes of brown painted blobs here and there making it quite a startling fish. It seldom grows larger than six inches in the aquarium and must be fed only with living foods. It will eat any fish including its own mother if she is available so do not buy a pair.

Histrio Histrio

This is the famous Sargassum fish which normally lives amongst the great floating sea weed masses. It is so well camouflaged that once in the weed it is almost impossible to find. Due to its natural surroundings, which are impossible to reproduce in the aquarium, it is seldom happy in captivity and does not often last very long. Its skin is covered with odd looking flaps of skin rather like the face of a Scorpion fish but much more exaggerated. It is also quite capable of changing colour at will and will blend into any surrounding colour. It is generally found in most parts of the tropical world, has no scales and grows to a length of six inches.

Apogonidæ

APOGON MEANS LITERALLY 'without barbels' The family is known as the Cardinal fishes and most are obviously red in colour. They are not very large fishes but are extremely fast swimmers. This speed how-

colate Clown *Amphiprion Sebae*

Butterfly Fish *Chaetodon Corallicola*

ever is seldom seen in the aquarium for they tend to stand motionless for hours on end in a mid water position. Found in the Indo Pacific area they are surprisingly not very popular in Europe. They are nocturnal fishes and are therefore much livelier at night. Their natural foods are alive and dried foods are not always accepted but an attempt to feed them will probably be more successful if you are able to do so late in the evening. The extremely large size of the cardinal fishes' mouths suggest the possibility of their being mouth breeders which like Cichlids, incubate the eggs in their mouths but his has yet to be proved.

Amia Brachygramma

This fish is the only Cardinal which is not red being as black as a Molly and somewhat similar in shape and size. It is found in Hawaiian waters and is very much like the *Amia Maculatus* but perhaps a little smaller.

Amia Maculatus

This fish is found in the Carribean and around the Mexican Gulf. With lesser colouring than most of the Cardinal fishes it is a pale red and has two, or sometimes three spots of black on the body. The eyes are also black and give the impression that glasses are being worn.

Apogon Nematopterus

This is the most common of all the Cardinals probably because of its tough constitution. It is an extremely rugged little fish and can take many strains. It ranges in colour from an almost transparent white to a dark yellow with red spots dotted around the body. Having very large eyes it is obviously a nocturnal fish and should therefore be quarantined in a very dark tank, the lighting being gradually increased until the fish is able to live comfortably in the community aquarium. It is an easy fish to recognise, mainly by the two large very upright dorsal fins. This condition is not indicative of unhappiness as it is with many other marine fishes.

Myripristis Chryseres

This is one of the largest fishes in the Cardinal family and one which tends to seek out large shells in which to hide. It should be provided with such hiding places in the aquarium but, as it sometimes exceeds six inches in length, they should be large shells. It is a typical cardinal red and with very large eyes. It is found around Hawaii.

Balistidæ
(Triggers)

THIS IS THE well known Trigger fish family. They are mainly found in the Pacific and are shallow water fishes. They rate high as oddities and have the largest head, for body size, in the fish world. Almost the whole of their body has been devoted to the head and yet, again size for size, they have incredibly small mouths. Do not however mistake this for a lack of biting ability for the Trigger fish has an extremely powerful set of jaws and teeth. These teeth are very sharp and are quite capable of inflicting vicious wounds to both other fishes and the owner's Hands. Many of the Trigger fishes are completely wild and cannot be mixed with any other fishes at all. The small size of the mouth is usually camouflaged by colouration which gives the impression of a much larger orifice indeed and at a distance, even in the aquarium with perfect vision, appear to run from ear to ear. Another strange feature of the Trigger fishes is the reason for their name, a very complex dorsal fin layout. As in many marine fishes the dorsal fin is not merely an aid to swimming but an instrument with a precise function. This appendage gives the Trigger fish its name in an almost literal sense for it is actually a working model trigger mechanism comprised of three spines. The main spine is longer than the others and is very powerful. This is raised into a vertical position and locked fast by the second spine. Pressure on the third spine will cause the trigger mechanism to relax and the trigger is released thus allowing the whole device to fold down. This extraordinary device has several functions but its main one is a protection against removal from bed at night by aliens. The Trigger fish usually goes into a cave or crevice at night to sleep. The dorsal fin is raised and locked into the roof of the cave and a further short spine located in the belly of the fish locks it to the bottom. The fish is now rendered immovable and is locked into a position rather like a top which is pivotted at both the top and bottom. The only way to remove a Trigger fish thus locked is to actuate the trigger mechanism by hand. Trying to catch this fish can be most difficult in the sea for, when alarmed, he will head straight for the nearest coral clump and lock himself tightly into a hole and no amount of unscientific prodding will remove him in an undamaged state. When attacked by another fish the Trigger will again operate his mechanism and the would be predator finds that his apparent supper has suddenly become much large in diameter and certainly more difficult to swallow. The triggered spine can actually jam itself into the roof of the predator's mouth and, even if the Trigger is killed in the struggle then the spine stays erect. Because of the Trigger fish's unusual characteristics it is essential that it is provided with a roofed cave in which to rest at night in the aquarium. Do not of course forget to allow for the growth rate of the fish.

Yet another unusual aspect of the Trigger fish family is the fact that they have no ventral fins whatsoever. Without this aid to swimming the fish is in consequence a rather poor swimmer. Most Trigger fishes will eat anything you care to feed them with. They are very greedy fish and usually gulp down all foods regardless of substance. They will also gulp down careless fingers if given the opportunity so take care when feeding at all times.

Balistapus Aculeatus

This is the most common and certainly the most popular Trigger fish of all and has been given many unscientific names. This is a thoroughly senseless habit and can only eventually confuse the aquarists all over the world. If you are sincerely interested in fishes then at least go to the trouble of learning the Latin names for these never change and we all know exactly which fish is being referred to. However, due to the fact that many aquarists are not aware of any Latin at all and many dealers only supply fishes with common names given I am supplying at the end of this book a translation where necessary. I am also adding common names *together* with the correct scientific names where possible in the text.

The *Balistapus Aculeatus*, although misnamed many times is usually referred to as the Hawaiian Trigger although it is in fact fairly extensive in its travels and can be found almost anywhere between Hawaii and Africa. Its colours are rather delicate and are basically pale blue with streaks of brown and light orange. It also has the typical 'painted mouth' of light orange making the mouth look much larger than in fact it is. Although this fish has been reported as large as twelve inches I have yet to see one which exceeded six inches. It is a reasonably peaceful fish and will live quite happily in a community aquarium, if given sufficient space.

Balistapus Undulatus

This beauty is found in the Red Sea, Polynesia, Hawaii and Japan and is usually imported in the larger sizes, probably because small specimens are extremely hard to catch. It has a considerable colour variation and comes pale green with yellow stripes to deep brown or purple with orange stripes. The markings and shape however render incorrect classification doubtful. It is a very active fish and usually a most agressive one. It is not normally possible to keep this specimen with other fishes at all. The Undulatus can however be most amusing and will amost certainly become very tame with the owner. There is little point in decorating an aquarium too fastidiously as a home for an Undulatus as he often likes to move his 'furniture' around at night and piles up the coral you have so carefully arranged into a corner where he settles down for the night. He can often be seen running around in circles with large lumps of coral in his powerful jaws just like a dog with a slipper. Thermometers etc. should not be left in the tank for he may well take a dislike to one and crunch it up into small pieces. When tame the Undulatus likes to be tickled and will even roll upsidedown for that purpose. Troubles in feeding this terrific glutton are restricted to keeping up with his appetite for he will eat practically anything, preferring hard foods on which he can chew for he seems to like crunching as any dog does. He will grow quite large in captivity probably because he is usually left in a tank alone and has room to grow to his maximum size of twelve inches.

Balistes Bursa

This Trigger fish is often confused with the *Balistapus Aculeatus* for they are very similar in shape, size and colour. It is however a much lighter fish and two, side by side, cannot be mistaken. It also lacks the stripes of the Hawaiian Trigger and has a large brown patch near the tail.

Balistes Vetula

This fish is commonly known as the Queen Trigger fish and is a large member of this family being at least twelve inches in size. It is yellow in colour with attractive blue streaks around the mouth and on the fins. The dorsal and caudal fins are extremely large, the former being extended with long filaments. It is found along the North American East Coast and is consequently more popular in the United States than Europe being a poor traveller. It is an attractive but rather large fish. It is also liable to change colouration at intervals and as the mood takes it.

Balistoides Conspicillum

This incredibly beautiful fish must surely rate as the most prized fish in the world, it is certainly the most beautiful with unbelievably stunning colours. It is more submarine shaped than most of the Triggers and is found throughout the Indo Pacific. It is rather rare and expensive and, being a bad traveller too makes it an expensive proposition for the amateur but it is well worth the money. It is almost impossible to keep the *Conspicillum* in the same aquarium as other fishes for it is most aggressive. The *Conspicillum* is basically black with large pale blue spots and an unbelievable pair of 'painted' lips looking for all the world like lip stick. A valuable but difficult fish and certainly not one for the beginner.

Melichthys Piscus
This is one of the less practical Trigger fishes for the home aquarium and in adult form is often a two foot fish. It is almost completely black upon first sight but closer examination shows that it is in fact deep purple. For a while it will be a great attraction in the aquarium but it soon loses its convenient size and developes an aggressive nature.

Odonus Niger
This is another large fish growing in the wilds to some eighteen inches but aquarium specimens do not usually exceed twelve inches. It is surprisingly peaceful and even more surprisingly green in colour although I have seen these fish change colour during a three day period ranging from turquoise to pale blue. It has a vivid and realistic fake 'mouth' and must appear quite capable of looking after itself to the would-be predator. It does in fact have one of the smallest mouths of all the Trigger fishes but do not let this fool you for its jaws are immensely strong. Once again an Indo Pacific fish.

Canthigasteridæ
(Puffers)
THESE FISHES ARE closely related to the *Tetradontidæ* family and both are graced with the same common name – Puffers. This branch of the family are known as the sharp nosed puffers. They are also rather similar to the Trunk fish family (*Ostraciidæ*) except that their bodies are not sheathed by a protective hard shell. They are however scaleless and able to inflate their bodies to extraordinary proportions when attacked or otherwise alarmed thus making digestion near impossible for the aggressor. The Puffer fishes consume a large quantity of oxygen for the size of their bodies and should be rated as twice their actual size when calculating aquarium space for inmates. Feeding is not difficult for they are very greedy by nature and will accept all foods most of the time, tubifex seeming to be the most relished. Most of the sharp nosed puffers do not exceed three inches in size.

Canthigaster Cinctus
This is one of the more popular long-nosed puffer fishes and is found in most parts of the Pacific Ocean. This is an unusual looking fish

resembling a fresh twig. It is brown in colour and covered with tiny blue spots. It has an elongated trunk-like nose and looks almost tubercular and wasted. It has an extremely small mouth.

Canthigaster Jactato

This is perhaps the best looking fish in the family. It is a startling colour ranging from dark blue to black with numerous white spots flecking the body. The darker the fish the more obvious is the white, and a prettier fish is the result. It is somewhat larger than the others growing to a length of four inches.

Canthigaster Valentini

This is a prettier fish than the last one but is unfortunately rarer. It has a shorter nose, a white body with black snake patterning. It is also a slightly smaller fish being only a mere two inches when fully grown.

Chaetodontidæ
(Butterflies)

THIS IS ONE of the largest of the coral fish families and certainly the most popular. They are commonly known as Butterflies and their colours make it easy to see why. The Chaetodon family is very closely related to the Angel fish family (*Pomacanthridæ*) and many aquarists become confused when both families are placed together in a book as so often they are. The obvious visual differences are that the Angel fishes have more flowing, elongated fins whereas the Butterflies have rather compressed, disc-like bodies with seldom a fin extension at all.

Most of the Butterfly fishes are rather timid harmless little creatures and this timidity can sometimes lead to trouble for a frightened fish does not remain healthy for long. Feeding is not always easy and if a boisterous community surrounds the new Butterfly he may well starve to death for fear of leaving his relatively safe hiding place. Live foods are almost an essential at first as they are very often the only foods that a Butterfly will accept. His tiny mouth indicates his normal coral picking habits where he searches for the smaller organisms. Avoid purchasing specimens which have sunken bellies or dark patches on their bodies for they may be sick and will probably never regain their strength. Due to the fact that most Butterfly fishes have not been equipped by nature with a weapon their only form of defence is disguise. This must have been a problem for mother nature because the many different species also needed a means of class identification and brilliant colouring was a necessity in the murky depths for mating to take place. Considering the vast numbers of Butterfly fishes in the seas it seems obvious that they are quite able to recognise each other. It seems strange then to state that these highly coloured fishes have any disguise at all but further examination of the bright colours of a Butterfly fish will show one almost uniform aspect. Almost every member of the Chaetodon family carry body markings which run through the eye making it a difficult organ to locate in the deep. Furthermore there is very often a false 'eye' painted onto the rear end of the body. Ridiculous though it may sound this makes it very difficult for the would be predator to tell which way his intended dinner is facing. A frightened Butterfly will stand motionless if no cover is available, it may even swim slowly backwards. The predator, when making his move, is usually very surprised indeed when the cunning Butterfly fish rockets off in the opposite direction to logic. Furthermore most predators gulp at a fish from the front, head first, and the Butterfly is so cleverly designed that nine times out of ten he is attacked from the wrong end and is able to shoot off in the opposite direction.

Most Chaetodons like to pick at algae and small animals which inhabit the many crevices of the coral and are therefore not the easiest of fish to keep if patience is not practiced. No Butterfly fish should ever be taken from the quarantine tank before he is feeding and in fact well fattened. They are also extremely timid and must be provided with 'homes', one to each fish for they are territorial fish and will become exhausted without a cave. Most of the fishes in the Butterfly family are small and, coupled with their bright colours, make excellent aquarium attractions.

Chaetodon Auriga

This beauty is found from the Red Sea to the East Indies and up as far as Hawaii. It is a rather long-nosed fish – indicative of rock picking habits, and seldom larger than six inches. It has a white body with black lines criss crossing each other. It has a yellow rear half and eyes which are disguised by a bold black line. There is a much more real looking eye painted onto the rear of the body near the dorsal fin to fool predators. One of the tougher Chaetodons he soon learns to take dried foods but he may have to be started off on live brine shrimp, a few flecks of dried being added extra each time a feeding is made until it is almost all dried. He may at first refuse the pieces which do not move but hunger eventually takes first place and your specimen is soon feeding as you wish it to. The Auriga also likes a piece of live coral as a neighbour but take care that he does not eat it!

Chaetodon Bennetti

Like so many of the Butterfly fishes the Bennetti is yellow and the eye is concealed with a black bar. There is also a decoy black eye painted onto the rear end. It is an East Indian fish and is not particularly popular being rarely imported into Europe. It grows to four inches.

Chaetodon Capistratus

This is a very popular fish in the United States and is imported from the West Indies. Although an attractive fish it is not often seen in Europe. Its common name is the Four Eyed Butterfly Fish and the typical eye concealment is prominently displayed here by an identical copy of the eyes transplanted to either side of the tail – thus the Four Eyed Butterfly. It is silver in colour criss crossed diagonally with numerous faded black lines. It grows to an approximate four inches and does quite well in the aquarium.

Chaetodon Citrinellus

A rather large Butterfly growing sometimes in excess of six inches this fish as a more Angel-fish-shaped body than most Chaetodons. It also suffers severe colour changes during its life. A silver body when young with a sharp black line running vertically through the eye. Other colours and a more Chætodon like shape creeping in with maturity.

Chaetodon Collare

This is a four inch fish coming from around Pakistan. It is a very popular specimen being quite handy in the aquarium and somewhat easier to keep than many of the Butterflies. Its colouration consists of a mixture of shaded browns with a white streak running vertically across the face.

Chaetodon Falcula

Found mainly around New Guinea the Falcula is rather scarce and does not always do well in captivity anyway. It grows to six inches in length but seldom achieves this in captivity. It is a similar fish to the Auriga having a black bar through the nose and a rather long nose. Its basic colour is yellow.

Chaetodon Kleini

A rather expensive fish without, in my view, enough colour or interest to warrant the cost and trouble involved in teaching one to feed for they are seldom in an eating mood when first imported. A pasty dull yellow with a banded eye and an almost discus shape the Kleini seldom grows more than four inches. It comes from the East Indies.

Chaetodon Lunula

This specimen is widely distributed throughout the Pacific. It is my personal favourite Butterfly fish. This fabulous specimen is surprisingly overlooked by many aquarists. It displays the most vivid yellow I have ever seen which is probably more exaggerated by the jet black patterns which traverse the body. The yellow is even more offset by an almost unnatural white 'collar' around the neck making the fish, to my mind, one of the most attractive of all Chætodons. A six-inch fish in the wilds but more often seen in the aquarium at a size of two inches. It is a good feeder and does not skulk around in dark corners. A very attractive addition to any aquarium.

Chaetodon Melanotus

An East Indian fish with typical yellow colouration on the fins but with a silver body slashed with black stripes in a diagonal pattern. It has a vertical black line through the eye and another false 'eye' on the tail. This fish has certainly mastered the art of swimming backwards and will do so when alarmed, rushing off in the opposite direction when attacked. A beautiful little fish growing to about four inches but dried foods may not be accepted for some time and quarantining may be a long job.

Chaetodon Meyeri

This is yet another East Indian fish, a large one, sometimes exceeding six inches in size, and certainly the most striking. It is coloured a pale to deep blue, depending upon age and has vivid black markings similar to some of the Angel fishes. It is unfortunately very delicate and not likely to accept community life for very long. Its chances of survival may be increased if a good hiding place is provided and no over-active fishes introduced to the same tank.

Chaetodon Ocellatus

A West Indian fish which reaches four inches in the wilds it is white and yellow and has a black bar through the eye with the usual extra 'eye' on its tail. It is almost perfectly circular in shape and remains timid all through its life. Usually it only accepts living foods at first.

Chaetodon Octofasciatus

This fish hails from the Phillipines and is one of the smaller Chætodons not often seen as large as its maximum three inches. It is silver bodied with yellow fins. Needless to say there are eight vertical bars running through the body one of which conceals the eye. There is another false 'eye' provided for the benefit of enemies. These fish seem much happier in shoals single specimens preferring to skulk around the darker corners of the aquarium.

Chaetodon Reticulatus

A largish Butterfly fish sometimes as much as eight inches it is found around most of the Pacific Islands. It makes a break from the usual Butterfly colours by being dull grey in colour with attractive white spots on each scale which give it an almost black mother of pearl appearance. Its belly is yellow and the black eye concealment band is tinged with yellow.

Chaetodon Striatus

A West Indian fish which can reach six inches in the wilds but usually only manages three in the aquarium. A rather quarrelsome fish which spends most of its time rushing round looking for food or battles. It is white in colour with four black bands running vertically through the body. It has a coral picking snout. Its colour makes a break from the usual yellows of many Butterflies and, as its aggressive nature seldom leads to real damage, it is well worth consideration. It can easily be confused with the Damsel fish *Dascylus Aruanas* but the latter can be distinguished by its three black bars through the body instead of four.

Chaetodon Trifasciatus

Another yellow West Indian fish which looks very much like the *Unimaculatus* except for the horizontal black lines. It is also an expert in the backward swimming class and is difficult to remove from a well stocked aquarium.

Chaetodon Unimaculatus

Slightly smaller than the last fish and with a more foreward false 'eye' this is a much more desirable specimen being a brighter yellow and having no stripes except for the inevitable vertical 'eye' patch.

Chaetodon Vagabundus

This fish comes from the East Indies as do so many of our best coral fishes and is yet again yellow in colour and looking very much like the last fish. The scales however are much larger and more pronounced.

Chaetodon Xanthurus
This fish is found throughout the Pacific Ocean and is more or less an enlarged version of the *Vagabundus* but with even larger scales. The eye disguise is however not as strong and the dorsal finnage is more pronounced and flowing.

Chaetodonoplus Mesoleucus
This is rather a large fish for a Butterfly and one which appears not to have any fins. It looks in fact like one of the Holacanthus species. It has the usual eye concealment but not a false 'eye'. A placid fish but very nice.

Chelmon Rostratus
One of the most beautiful of all the Butterfly fishes commonly known as a Long-nosed Butterfly fish. It is found from Malaya to the Phillipines and often exceeds six inches in size and is seldom imported in the small sizes. It has an unusually long nose obviously adapted for foraging in the deeper crevices of coral in its natural surroundings. The *Chelmon Rostratus* is a well sought after fish and yet not the easiest to keep being very timid and quite uninterested in dried foods at first. The slightest disturbance in its tank sends it bolting for a place to hide and, if one is not available it will almost certainly die. Daphnia or brine shrimp is accepted and eaten in rather the same manner as sea horses and, after a while dried foods may be accepted but only if there are other inhabitants in the same tank for he simply will not learn what to eat without being taught by other fishes. A *Chelmon Rostratus* should never be purchased if he has even a slightly sunken belly. So many specimens arrive with wafer thin bodies and these never seem to overcome the fight for life which should follow. The *Chelmon Rostratus* is also prone to poisoning and Benedenia or even boils, the latter being most probably due to an incorrect diet. Another case where the fish seems to fare better when kept with living coral. It is a circular fish yellow in colour and with well marked orange trimmed stripes running the length of its body. It has a very well designed false 'eye' on the tail, making it a lovely fish.

Forcipiger Longirostris
This fish can be found all the way from East Africa to Hawaii and is

very similar in body shape to the *Chelmon Rostratus* except for the slightly more 'torpedo' shaped body made evident by a lack of forehead. It also gives the effect of having even a longer nose than the *Chelmon Rostratus*. It is a pale yellow and has a black and white face with a false 'eye' on the lower half of its tail. His real eye is very well hidden in the patch of black on its face. It is much rarer in Europe than the *Chelmon Rostratus* and is even more delicate.

Heniochus Acuminatus

This fish is not often referred to as a Butterfly for most aquarists believe it to be an Angel fish. It is however one of the largest of Butterflies, sometimes growing as large as ten inches. It is a striking fish and sometimes called the Poor Man's Moorish Idol, being very similar to the Zanclus due to its enormous dorsal fin. It is far easier to keep than the Zanclus being blessed with an insatiable appetite (Oh for such a condition with the Zanclus).The Heniochus inhabits the Indo Pacific and is a shoal fish. It is usually better to obtain this fish in an adult form for it soon settles down to community life although it may well chase off all comers at feeding time and, in a group, one will inevitably become a bully. A Heniochus is by far a better choice for the private aquarist than a Zanclus – and a much cheaper one.

Parachaetodon Ocellatus

A very similar fish to the *Chætodon Octofasciatus* both in complication of name and physical characteristics except that it only has four vertical bars and is a little less bold in colouration. It further displays a false 'eye'. The *Ocellatus* prefers the company of other specimens and, even in the largest of community tanks will seek out others of the same species. They are rather timid fish but will learn to take dried foods with a little patience.

Diodontidæ
(Porcupines)

THIS IS A rather small family composed of fishes that have an almost perfect natural safeguard against being eaten, they can puff up to large sizes like the puffer fishes but have the added deterrent of extremely sharp spikes, like porcupines, all over their bodies. These fishes are best kept alone for they like plenty of space and, like the puffers, require large quantities of oxygen. They grow very fast too. Most foods are accepted. These fishes are not for the amateur aquarist unless extremely large tanks are available.

Chilomycterus Schoepfi

Certainly a mouthful in more than one sense for this is the famous Spiny Boxfish and there are not many predators capable of digesting this weird creature. Specimens of twelve inches are not unknown so make sure you have enough room if you intend keeping one. He will accept nearly all foods. They are found in the Atlantic and Carribean. The horns above each eye grow shorter as the fish nears maturity but the spines never relax completely. It is coloured grey green. Usually seen in most public aquaria it is an eye catching fish and gathers large crowds.

Diodon Hystrix

This is another extremely large fish which occasionally exceeds three feet in length. It is only listed in this book because it is an interesting specimen but certainly not for the amateur. Unlike the last specimen the Hystrix swims with its spines in a relaxed position. It is found in most tropical seas. It is light green to brown in colour with various sized black splashes on the body. It can inflict nasty wounds with its spikes but lives quite well in a community tank with equal sized fishes. It is a more difficult fish to keep for the private aquarist than the last specimen but collects just the same crowds in public aquaria.

Gobiidæ
(Gobies)

GOBIES ARE VERY lively little fishes found in practically any salt or even brackish body of water. They are small fishes and relatively harmless although do not always suffer each other when more than a pair is kept in the same aquarium. The goby is a good cleaner fish and will even hop onto a passing fish and 'clean' it with the aid of its sucker. This sucker is in fact formed by the ventral fins and is not a mouth as in the fresh water sucker fishes. The Gobies are not unlike the fresh water catfishes in other aspects, preferring to spend their time on the ocean bed. They are however much more active in the aquarium. Furthermore the Gobies lack a lateral line.

They are not difficult to spawn but fry are seldom raised. They should not be left in an aquarium where breeding is being attempted by other fishes for their favourite food is fish eggs! Gobies are very tough and take a lot of killing and it is surprising that they are not more popular in Europe for they are not often seen here. Perhaps it is due to the lack of colour in most of the family. There are many fish in this family but, as there are only one or two kept in aquariums there is only one listed here.

Elacatinus Oceanops

This is undoubtedly the most popular Goby of all and its colours denote why. It is popularly known as the Neon Goby and is found from the Florida shores to the West Indies at a maximum size of three inches in length. A 'pair' will not always accept each other and a third is suicidal, they do however leave other fishes alone except of course for their cleaning activities. The Neon Goby is not unlike a fresh water black neon except for its more elongated body which is black with a bright blue streak running its whole length. They will accept most live foods but prefer the smaller worms.

Holocentridæ
(Squirrel fishes)

A NOCTURNAL FAMILY, evident by the extremely large eyes and not too dissimilar to some of the Cardinal fishes. They are even red in colour. They are rather small fishes and, being seldom imported, are less popular than one would expect for such pretty fishes. They are mostly found in the Pacific. Although small they are in fact predators and should be watched carefully with yet smaller fishes. They are most likely to devour these during the night.

Holocentrus Ascensionis

A Florida water fish and consequently more popular in the U.S.A. It grows to some four inches in the aquarium but often reaches twice this size in the sea. It is a striking fish, striped with pale pink stripes against a red body.

Holocentrus Erythräeus

This fish is found practically anywhere from the East Indies to Hawaii and is one of the largest of the Squirrel fishes growing in the wilds to at least ten inches. It has a dark crimson top half with a yellow belly. The stripes add an extra touch of velvet purple making it an extremely attractive fish.

Holocentrus Diadema

This fish comes from the Red Sea and can also grow very large in the wilds perhaps up to four inches, but aquarium specimens usually reach a mere two inches. It is a much deeper red than most of the Squirrels and is most attractive.

Labridæ
(Wrasse)

THE WRASSES ARE perhaps the hardiest of all the coral fishes and always seem to be the last to succumb when an epidemic occurs. They can stand wide ranges of temperatures and will feed on piactically anything. The wrasse family is another branch of the fish world which swims only by means of the pectoral fins. It is also a very confusing family for the changes suffered by some of the wrasses in colour and pattern are nothing less than astounding at times. Therefore it is not surprising to find that many members of the family are twice named through being totally different when mature to their younger appearance. The wrasses are interesting in many ways, the most interesting from the visual point of view being their apparent laziness and weird reclining positions in the aquarium. Do not be alarmed to see your prized wrasse folded upside-down and hanging onto a piece of coral by its eyelashes for this is quite normal. Nearly all wrasses adopt a 'going to bed' form which ranges from burying itself in the gravel at night to actually manufacturing a bubble which completely covers the body with a breathing and waste matter hole added. Most Wrasses have extremely sharp teeth which they fully realise the power of and considerable damage can be done to even the largest of fish if the Wrasse takes a dislike to it. This is a large family but once again we are only mentioning those fishes which are likely to be kept in aquariums.

Cheilio Inermis

This is the well known Cigar Wrasse and is so named for its shape. It is an Indo Pacific fish which is being imported widely at the moment. It would not be so popular however if buyers were aware of the adult size which is near the two foot mark! It varies in colour from light grey to green and brown. It cannot be kept with other fishes when even a quarter full size. All foods are accepted – fingers and all so keep well clear when feeding.

Coris Angulata

One of the prettiest and most delicately coloured of all the wrasses but, unfortunately, one of the largest, growing to the incredible size of four feet in the sea. I say incredible for these fish are never seen in the aquarium larger than twelve inches. A beautiful white with pale orange combined with jet black spots make it the most georgeous of all the wrasses to my mind. It is found mainly in the Red Sea and is relatively harmless although quite liable to outgrow even the largest of aquariums.

Coris Gaimard

This is one of most 'painted' looking of all the coral fishes second only to the Clown fish. It is dark brown with sharply contrasting white splotches like brush smears around the body. These white marks are 'pencilled' in with black edging making the colours look even more unreal. It can grow up to ten inches in length but is usually seen nearer the six inch mark when fully grown in the aquarium. The *Coris Gaimard* is often plausibly confused with the *Coris Formosa* but is identified by several differences, the main one being the dark patch on the dorsal fin of the Formosa. In time to come we will no doubt be able to breed coral fishes freely and I have little doubt that this will be one of the first pairs we will choose to abort science with. It is an easy fish to keep and a good one with which to start a coral fish collection. It is generally peaceful and will accept most foods. It is also a bit of a clown going through all sorts of peculiar motions and hanging at all angles from pieces of coral. He also likes to chew on the coral and should be provided with a soft sub sand in which to bury himself when the mood takes him. A very jolly fish and an easy one to keep.

Thalassoma Lunare

An extremely active fish resting not for a moment during the day but usually wearing itself out to such an extent with its continuous swift method of non-stop swimming that it flops out almost unconcious at night in a shell or similar sleeping quarters, and I *mean* flops out for he simply tips up onto his side and relaxes with one gasp apparently lifeless until dawn. No amount of banging on the tank will arouse him and even when he is poked into activity he can hardly see where he is going for some minutes. He must of course be provided with a large shell in which to sleep and a clam will always be preferred. The Lunare is incredibly greedy. Food should not be given to him by hand for his teeth are exceedingly sharp and fingers make a tasty extra to this glutton. He will eat absolutely anything from dog's meat to prize fish and should be carefully watched at all times with new specimens placed into his tank no matter how large they are for the Lunare is just as vicious as he is greedy and will attack a large fish in the eyes. First he tears the eyes out and then he will rip great pieces of flesh away whilst the fish is still swimming and will gradually devour it alive. With smaller fishes he simply takes hold of the head and shakes the poor creature senseless and then he dashes the half dead corpse against pieces of rock over and over until it is quite dead. His appetite is completely insatiable. It is also a very difficult fish to photograph being forever on the move in his constant search for food. It can however be trained to community with bigger fishes.

Monodactylidæ
(Malayan Angels)

THESE ARE FISHES well known to fresh water aquarists who attempt time and again to keep them in fresh water. This is perfectly alright for a while but I have never seen a truly nice 'Mono' kept in fresh water for very long. The Mono is often thought to be an angel fish and, although closely related, it is quite different when examined closely. Those kept in brackish water seem not to suffer too much but put one into its correct salt water and see just how lovely it can look. Monos are usually caught in fresh on brackish waters at the mouth of a river to which they have come to feed. They are greedy fish and are similar to the Scat in many ways – not physically of course – there always being a bully who takes over in the aquarium. They have almost tin foil skins and flash beautifully through the water as they turn. They have either one or two vertical black stripes passing down the front of their bodies' and very small heads in contrast to the Trigger fishes. They have the same appetites.

Monodactylus Argenteus

This is the most popular Mono and hails from the Indian ocean. They can grow to a length of six inches in the aquarium. Feeding is no problem for they are greedy fish and usually get to the food first. almost leaving the water in their excitement. Should there be more than one in the same tank then the stronger will not allow the others to approach the food and consequently one is always considerably larger than all the others.

Monodactylus Sebae

This West African Mono often goes to an estuary to feed, and is consequently caught in that area, and so it is accepted as a fresh water fish which in fact it is not. It will suffer brackish water for some considerable time but fresh water specimens do not usually last long. It differs from the Indian Ocean species in that it is slightly duller in colour and has one dark vertical band around its body instead of two. It is just as greedy but prefers tubifex.

Ostraciidæ
(Trunkfishes)

HERE IS ANOTHER example of nature's strange way. The trunk fish, although generally inoffensive is quite capable of dealing with enemies by emitting a lethal poison. The danger with aquarium kept Trunk Fishes is that none of the affected fish, including the Trunk fish, are able to leave the poisoned area and are all liable to die. The Trunk fish is therefore a suicide fish and great care should be taken not to upset him too much. They are usually left alone by other inmates of an aquarium for they are completely inedible being cloaked in a tough transparent shell. This shell is soft around the moving parts of the body such as the fin areas and mouth but is quite hard elsewhere. The colours of the fish on the skin shown through the shell. Due to the extreme rigidity of the shell, Trunkfishes swim in an almost bizarre manner by whirring their fins. This makes the fish look rather like a submarine with propellors whirring. It is easy to recognise a sick fish for the tail tends to curl when it is unhappy very much in the same way as puffer fishes. Feeding the Trunk fishes is not difficult for they are quite willing to try most foods, their favourite being Tubifex which they prefer in mid water so do not allow single worms to reach the bottom or they may be left there. New specimens must be left alone at first to get over the shock of being captured. If they are given a rock behind which to hide they will soon come out when hungry enough. Do not forget also to allow twice as much room for a trunk fish as you would for any other fish for they consume plenty of oxygen. They are jolly fish and will add a spice to your tank.

Lactophrys Cornutus

This weird little object has been aptly called the Cowfish and has no doubt received this nickname from the short pair of horns which sprout from the forehead. They also have a very comical expression permanently engraved on their faces looking for all the world as if someone has just stuck a pin into them. It is an·extremely bright yellow which unfortunately darkens with maturity. Added to the yellow and in great contrast is a sprinkling of blue spots covering the body. Another delightful feature lost with age is the very long tail. A good community fish it is sometimes molested by the attentions of a neon Goby. For some reason it objects to this and flees. We have yet to discover why.

Ostracion Cubicus

This fish is variable in colour from bright to dark yellow with brown or black spots all over the body. It is hornless and rather square in shape. It does not exceed four inches in size and usually dies if the sides are concave when it reaches the aquarist. Only specimens with dead flat sides should be purchased. It feeds well and will take all foods that do not float unless hungry.

Ostracion Lentiginosum

An Indo Pacific Trunk fish found in many colours and of various sizes, even up to twelve inches although the largest I have ever seen one in the aquarium was eight inches. It has a rather flat body and looks more fish-like than some other Trunk fishes do. It also has many more spots on its body. Its usual colours range from brown to green with whitish spots ringed with brown edging.

Ostracion Tuberculatus

Another Indo Pacific Trunk fish and again one which can reach twelve inches in size. It ranges in colour from purple to grey and has black and white spots. This is the longest bodied Trunk fish of all and is quite lively. It likes plenty of oxygen and will not last long in a crowded aquarium. As with many other Trunk fish it will commit suicide if attacked or frightened by the thoughtless aquarist so do not allow excessive shocks to occur.

Platacidæ
(Batfishes)

LARGE INDO PACIFIC fish greatly resembling the Angel fishes which are great favourites with the private collectors due to their extremely friendly natures. They are nearly all tameable and become great pets. This is mainly due to greed for any fish can be tamed if it is greedy enough. They are very hardy fish and usually fare better when not purchased too young. Feeding troubles are rare, they will soon get to know the aquarist and will eventually take food from his fingers. Like so many of the larger coral fishes the Bat fish soon loses its colours, which were not very bright to start with, and the markings are changed so much that it is often hard to recognise the same fish three months after first viewing it. They are interesting fish, hardy fish and quite good pets, never needing to be taken for a walk. They are all brown in colour.

Platax Orbicularis

This is the most popular of all the Bat fish and it certainly does not seem to mind being kept as a pet for it will soon know its owner and may even be tempted to roll over on its side to be tickled. It reaches the large dimensions of two foot in the sea but aquarium specimens are usually dwarfed into a fifteen inch maximum. The *Orbicularis* often arrives with very nasty looking white spots on the body but these should not cause alarm for they are merely pigment faults in the skin and can sometimes be seen on members of the Scorpion fish family. This fish suffers dramatic changes as it grows and the all brown body gives way to silver with stripes down the face. The immense fins extend and take up a third of the length of the fish each. Yougsters have hardly are fins at all.

Platax Pinnatus

This is a rather rare Bat fish. It is sometimes all black but with brown edging running all the way around the body and fins. It grows to around eighteen inches.

Platax Teira

This is the smallest of the Platax family for, when measuring a fish, only the body is referred to. It has however the largest fins of all the Bat fishes. Its dorsal fin alone is normally larger than the whole of its body. It is a more colourful fish than the others having a silver body with brown to black stripes on the face. It does not tame as readily as the *Orbicularis*.

Platax Vespertillio

This on the other hand is the shortest finned Platax of all. It is also the least colourful. It grows to the usual eighteen inches and loses its colours as with the others. It is the least popular of all the Bat fishes.

Pomacanthridæ
(Angel fish)

HERE IS THE most graceful and beautiful family of all the fishes of the world. Beautiful serene creatures with long flowing fins and unbelievable colours. Angel fish are found in both the Atlantic and Pacific and are in great demand. Some Angel fish can become quite vicious when upset and are equipped with very sharp spikes on the gill plates. These spikes are frequently used when tempers become frayed. Angel fishes are not good mixers especially when fish of equal sizes are mixed. A large Angel fish will no doubt feel quite safe from attack when his adversaries are smaller than he and can consequently relax. The smaller Angel on the other hand will avoid the larger one knowing full well he will not fare well in a scrap. Two equal sized fishes however will be constantly afraid of each other and always ready for an attack feeling perhaps that attack is the best form of defence. Angel fishes are mid water feeders in the aquarium and prefer food to drop from the surface and reach eye level before taking a snap at it. They may not feed too well when first introduced to the aquarium and should always be feeding in the quarantine tank before admission to a community tank. It is essential that the Angel fish is provided with a dark spot large enough to accommodate his long fins at night and, once a spot has been established it should not be changed for the Angel fish is a territorial creature and will allow no other fish to come near his 'place'. They are closely related to the Butterfly fishes and have much the same habits with perhaps a little more aggression. The Butterflies do not of course have the spiked gill plate nor the flowing finnage of the Angels. The Angel fishes are also much larger. Angels have a tendency to tip off centre when alarmed and a sudden flood of lights may cause this to happen. This is sheer shock and all lights should be gradually turned on, not suddenly flashed into brilliance. Shock is a very dangerous situation and Angel fishes are easily shocked so do take care.

Angelichthys Ciliaris

This is the fabulous Queen Angel fish and has been so dubbed due to the almost perfect 'crown' shaped marking in blue right on the apex of her head. It is a rather large fish when adult and can reach two foot in the sea. It passes through several colour phases during which it progressively brightens and then darkens off again to less vivid colours. It is blue faced and has long yellow flowing fins. It is of West Indian origin and is a highly prized exhibit.

Angelichthys Isabelita

This is the so called Blue Angel although I can think of more fishes entitled to this name. It is quite similar to the Queen Angel fish in body shape when adult but is slightly smaller at eighteen inches. It has the obvious difference of being blue but colour changes are so dramatic in this fish that young and old are far apart to look at. Its fins are reddish in colour and the crown is less pronounced than in the Queen Angel fish. Perhaps a little less desirable than the Queen Angel fish but nevertheless a very lovely and graceful fish.

Angelichthys Townsendi

This fish hails from Florida and, being rather rare, does not often reach Europe. It is a delicate fish almost sandy in colour with some original blue markings about the face and a completely blue circumference to touch off the whole effect. It lacks a 'crown' and grows to a lesser size than some of the Angels – about twelve inches. It is also rather expensive.

Holocanthurus Tricolour

This is the wonderful Rock Beauty so popular in public aquaria. It has of course three colours but only two of them are immediately noticeable, the third colour being a touch of red on the underside of the body. The fish is basically yellow with a large black patch towards the rear which takes up half the fish's body. It is generally a community fish but unfortunately grows to the size of twenty four inches. This may well make it an attraction in public aquaria but is is not a fish for the private man with limited space.

Pomacanthus Annularis

This fish is more often than not imported from Ceylon. It is an exceptionally beautiful fish having vivid, almost neon, blue streaks running at an angle through a dark brown body. No photograph can do true justice to these colours. Although it can reach eighteen inches in the wilds and is seldom imported less than six inches long it does not often grow larger than ten inches in the aquarium. It is an easy fish to feed and normally a quiet community dweller. Care however should be taken not to purchase an over thin specimen as it will probably die. The Annularis suffers considerable pattern and colour changes throughout its life but nevertheless keeps the powerful blue streaks, although perhaps a little less brilliant. A further attraction which is gained with age is a 'halo' just above the eye. An eye catching fish well worth the bother of locating.

Pomacanthus Arcuatus

The black Angel fish from the West Indies is a largish fish growing up to two foot in the sea. It is almost completely silver with delicate black spots on each scale. It also has fin extensions or filaments which branch back horizontally. When young this fish is striped in black and white. A rather delicate fish.

Pomacanthus Imperator

A small East Indian fish seldom exceeding twelve inches in size even in the wilds. It is a georgeous creature coloured an extremely bright mauve with undulating yellow lines crossing the body. It adds to its attraction with a white mouth and an orange tail. A real piece of mobile decoration fit for any aquarium. It can however become quite vicious and new specimens added to a tank which already houses an Imperator should be watched carefully for signs of aggression.

Pomacanthus Paru

This is the fabulous French Angel fish which is usually found between the West Indies and the coast of Florida. It grows to twelve inches and passes through many colour changes before doing so. When young it has three distinct yellow bars running vertically down its body. These vanish with time and end finally with a transformation to a black body with yellow flecked scales. The fins also extend to form filaments. It fights less with its own species than most Angels.

Pomacanthus Semicirculatus

This is the Koran fish and a suprisingly tough one it is too once it has been established in the aquarium. Of all the coral fish colour and pattern changes this fish takes the prize. It is a much more colourful fish when young but its growth is quite an interesting study, especially if you can be bothered to sketch the markings each month or so. A comparison at the end of a year will show interesting results and give you a great deal of enjoyment from the fish. When immature it is a bright blue with white circular lines all over the body. These white lines gradually move round until they are almost straight in places. They then begin to break up and are then reported to form the Arabic lettering of an old Mohammadan prayer. Needless to say this fish became quite popular among 'believers' in olden times. It usually reaches a size of fifteen inches and comes from the Indo Pacific. A very interesting fish indeed.

Pomacentridæ
(Damsels)

THIS IS ANOTHER of the larger coral fish families, comprised mainly of the smaller fishes. As most of these fishes can only grow to half their normal size in the aquarium this makes them small enough for the average amateur aquarist to keep in small tanks in his front room and for this reason they have become very popular. They tend to be scrappy fish but seldom do a great deal of physical damage and are more likely to cause a timid fish to be starved to death than actually wounded. They are small fish and very fast, a lively specimen being almost impossible to catch in the aquarium which is well stocked with coral. They also have a natural fear of being eaten all the time and in order to keep them reasonably happy and without a constant fear over their heads the aquarist must provide many hiding places for them. They will use these holes and crevices at night and will completely lose their colours when darkness falls. Their size alone indicates the need for an ability to rush at high speed to the nearest hole when danger threatens and if such retreats are not available then the unfortunate Damsel will never feel truly free to search for food and will be constantly fighting with other Damsels in a show of strength. Once again we are confronted with confusion regarding naming and classification for there are undoubtedly numerous specimens which have in fact been already named. The colour and marking differences during their lives causing extra names to be given to the same fish. The Damsels are perhaps the nearest the aquarist has come to spawning coral fishes. They will undoubtedly spawn under the correct conditions but the ability to raise to any size at all the fry will no doubt remain difficult until we are able to locate and supply the natural foods such as Plankton, which dies in the aquarium, with an artificial alternative. They are the nearest salt water fishes to the fresh water species being very similar to Cichlids in body anatomy, habits and size. They are very similar in shape and spawn in exactly the same manner with the parents guarding the young. Their temperament is also Cichlid-like and fights among neighbours, bullies and extra large feeders are common. They are active fish and a delight to watch. They are also very easy to feed. They therefore make excellent fishes with which to start your collection along with perhaps a wrasse or two.

Abudefduf Abdominalis

This fish is very similar to and often mistaken for the Sergeant Major (*Abudefduf Saxatillis*). The *Abdominalis* is a light to dark green with six dark bars running at intervals along its body. It can grow to six inches in length and comes from Hawaii. It is a scrappy fish and will tend to bully smaller specimens.

Abudefduf Aureus

An East Indian fish not often imported into the United Kingdom and a difficult fish to identify. Ranging in colour from brown on the back to yellow on the belly it is sometimes splashed with tiny white flecks. A rather expensive fish.

Abudefduf Leucosona

A Small Damsel fish usually adult at two inches found in most parts of the Indo Pacific. It is quite easy to identify having a perfect circle etched into the dorsal fin. It also has one vertical bar coloured a delicate blue. It is basically yellow to green with a lighter belly. It is not found in Europe very often.

Abudefduf Melas

This is a rather large Damsel fish being some six inches in length even in captivity. It goes through a strong colour change as it matures. The young are jet black with a sharply contrasting white patch on the side. It looks very similar to the Dominoe Damsel fish (*Dascyllus Trimaculatus*) but lacks the white patch on the forehead. It is also less boisterous than the Dominoe. The white body patch on the Melas sometimes leaves the fish quite early, especially if the salinity of the aquarium water is weak, but seldom does an adult fish show any colour except black or dark grey.

Abudefduf Saxatillis

This is a firm favourite for beginners and is often used as a 'teacher' for slow feeders. It is a tough hardy fish and rather more popular in America than in Europe. At meal times it is always the first to the food and will refuse nothing. For new fishes which do not know that the attentive aquarist is trying to feed them the 'Sergeant Major', as this fish is commonly called, soon shows them what to do with the food. There cannot be many hungry fish which are able to ignore the vast amounts of food the Sergeant Major gobbles up without their mouths' watering and an exploratory bite is eventually made, ending the fast. It is not a very colourful fish and one which will lose all its colours at the slightest disturbance, unless it is one concerning food. It can reach six inches in length and often does in the aquarium provided it is well supplied with food. It is a reasonably peaceful fish but can sometimes upset other specimens by pinching all the food, all the time. Colours range from dark blue to grey with black bars running down the body at regular intervals. These bars stop short at the face. A good fish to join your first selection list and a hardy one.

Amphiprion Group
(Clowns)

These are the well known Clown fishes and are grouped separately for they are almost a set on their own although they do of course belong to the *Pomacentridæ* family. They are most distinctive and have become the marine aquarists delight, usually taking first place in his collection. Seldom does a fresh water enthusiast, confronted with the unbelievable colours of a group of Clown fishes, not consider 'going Marine'. The Clown fishes are not however the easiest of coral fishes to keep although they are certainly not difficult provided the basic rules are adhered to. They are very prone to the disease Oodinium and are usually imported in a stricken state. The Clown fish is often seen in company with an anemone and is always happier when one is around. A frightened Clown fish will always rush into the nearest anemone and seems well aware of the fact that not many fish will follow it into the poisonous tentacles. The Clown fish is however *not* the only creature apparently immune to the poison. It is nevertheless the only fish which completely lives in an anemone, an honour shared by the anemone shrimp and the anemone crab. The difference being that the last two creatures are scavengers who feed on the anemone left overs, or vice versa, whereas the Clown fish actually supplies food to the Anemone in exchange for refuge and 'de-lousing'. Oodinium, the Clown fish, the anemone and their mutual interests are discussed in Chapter 11 (Diseases).

Most Clown fish come from the East Indies or Africa. This makes their cost rather cheap in Great Britain yet more expensive in the U.S.A. It is widely believed in the United States that the Clown fish will live safely in *any* Anemone. This is quite untrue but unfortunately a belief shared by the Clown fish itself for the English coastal anemones such as the Beadlet will sting to death *any* fish and the sting of a large Beadlet can actually penetrate the human hand. I once went to a great deal of trouble 'earthing' a 'live' tank thinking there was a slight electrical fault for I was always receiving jolts when cleaning it out. It was not for some time after the fruitless earthing that I realised it was in fact the anemone. It is not painful but merely a slight sensation usually only felt on the back of the hand.

The Clown fish in probably destined to be the first coral fish bred in captivity, and many people have reported spawnings and even hatchings. My personal experiences have always led to disaster and the nearest I have been to success was by accident in an experimental plant tank. It was so dirty and literally covered in algæ that I was unaware that fishes were there at all until I saw a very bedraggled and tattered Clown fish 'fanning' a small patch of eggs in the same way a Cichlid would, but unfortunately the eggs disappeared during the night. The Ph of the water was around ten! The greatest difficulty we must over come before expecting results from marine fishes is forced and dwarfed maturity. Although a number of Clown fishes exceed eight inches in the wilds it is extremely rare to find one anything like this size in an aquarium. This dwarfing, caused by lack of space, cannot possibly be natural and the fry resulting from parents which have been forced down to half their natural size must in turn be dwarfed and anaemic. It is almost like a human giving birth to a half sized baby, it hasn't much of a chance even with the meticulous care offered by modern science which is certainly not available to coral fishes. No, we must first learn to grow our specimens to a more natural size before attempting to breed them and then we must deal with the tricky problem of food for the fry to feed on. Added spice for the would be breeder however is the fact that the anal and ventral fins of Clown fishes which are ready for breeding darken considerably. I can only offer you my best wishes that you may be the lucky aquarist to first succeed where no other has – it's not far away.

Amphiprion Akallopsis

This is one of the least vivid of the Clown fishes and is a delicate yellow. It is also somewhat smaller than the others reaching only four inches in the wilds. It has a sharply contrasting white stripe running from nose to tail just like the marking of a skunk and is probably the most inoffensive of all the coral fishes. The *Akallopsis* seldom strays far from the anemone in which he lives constantly making only short sharp dashes out for food followed by even shorter, sharper dashes back into safety. It sleeps in the anemone and will not always live for very long without one. Foods are taken very carefully and patience may be needed.

Amphiprion Bicinctus

This is a rather rare fish and may well be one that has been incorrectly named, having been already named before whilst in a different colour condition. Colouration and markings change so much with environment and age that several of the following fish may even be each other! However, until we learn otherwise, we must bow to the Ichthyologists and treat them as directed.

Amphiprion Ephiprion

This is commonly known as the Tomato Clown fish and is quite large, up to six inches. It is the reddest of all the Clown fishes and is, needless to say, the colour of a ripe tomato. It normally has a white stripe behind each eye but can be found without this marking at all. It is unfortunately the Clown most liable to catching the disease Oodinium. This may be due to the fact that Clown fishes have the smallest scales in the Damsel family. Other small scaled fishes such as Triggers seem also to be more affected by skin diseases. The Ephiprion can be extremely aggresive if he feels that his territory has been invaded and his teeth are quite sufficient to cause damage to smaller fishes. An additional weapon common to all Clown fishes is a spike set to the rear of the mouth although this is seldom made use of, the aggression being limited to chasing away the intruder and returning to the territory.

Amphiprion Frenatus
This is almost indentical to the above fish except for the darker edges to lower fins and the bottom half of the caudal.

Amphiprion Laticlavius
This is an extremely Clown like fish being almost black-bodied, red faced at times and with a sharp white band encircling the face. It also has a white patch on its back. An easy fish to sex and spawn in the right aquarium and always a tough one. It is however rare and is not always readily obtainable.

Amphiprion Melanopus
This fish is almost a replica of the Tomato Clown and can seldom be distinguished from it. The main difference is one of size for it only reaches three inches in the sea and does not often see these proportions in the aquarium.

Amphiprion Percula
The most common of all the Clown, if not all the Coral fishes, this is a very cheap fish to buy in Europe and, if quarantined correctly, in a strong solution of Methylene Blue, reasonably easy to keep. It is safer always to assume that a newly imported Percula is infected and treat it accordingly.

It is the easiest of all fishes to identify being orange with almost unnatural painted white shapes curling around the body. Although reported by many specialists as only reaching four inches in size I have seen Perculas as large as eight inches in German aquariums and therefore assume that they can even exceed this size in the sea. The beautiful sharp band around the face makes this very attractive little fish look as though it has been bandaged and has been christened the Tooth-ache fish by my staff. My motor yacht has also been christened 'Percula' in honour of this beauty.

Although the Percula will always head for the nearest anemone when released into a tank it will be killed if allowed near an English coastal specimen.

Amphiprion Perideraion

This fish is similar in colour to the Akallposis but perhaps even less brilliant being a rather anaemic lemon with a sharp white vertical stripe just behind the eye. It is not a very strong fish and grows only to a maximum of two inches. It will not suffer the life of a very active community aquarium for very long.

Amphiprion Polymnus

This is one the rarest of Clown fishes and differs only slightly from the *Amphiprion Sebæ* for it is the same colour but with a different location of stripes.

Amphiprion Sebae

This fish is a deep brown with two thick white bars on the body. These white bars are very often luminous when the fish is first caught and this attractive colouring sometimes lasts for several weeks. A rather spiteful fish which will allow no other fish into its territory. It grows to at least four inches.

Amphiprion Xanthurus

This is very similar to the last fish and is often understandably confused with it. It seems however to hold a more erect dorsal fin (not always a bad sign in Damsel fishes) and looks generally heavier with more prominent scales. The fins, unlike the Sebæ, are black and the tail is yellow. It is the last of the Clown fishes.

Dascyllus Aruanas

This fish, like the Clown fish, will go into anemones with immunity. It does not however reside in one and only goes there when affected with some form of skin disease. It never goes to an anemone for protection against danger and will not go near one at all if healthy. Hiding places must be provided for the Aruanas or it will become quarrelsome. It is found throughout the Pacific and grows to about three inches. It is a contrasting black and white and, although sometimes aggressive when adult, seems happier in the company of other Aruanas. Like the Clown fishes it is a possible 'first to breed' and, also like the Clown fish, is susceptible to disease. The *Aruanas* seems to be prone to internal parasites and an attack is far more serious with this fish than with others. It is almost indentical to the *Dascyllus Melanurus* except for the continuous black on the dorsal fin. This is broken on the Melanurus and a black patch on the tail of this fish is missing on the other.

Dascyllus Reticulatus

This is a delicate pale blue to brown beauty which hails from the Red Sea and grows to a length of eight inches. It is a nervous fish which seldom travels well and should not be placed into an aquarium with over active fishes. Without a cave in which to recline at night it will die from sheer exhaustion. Once established it does quite well and prefers to feed with occasional live foods such as tubifex or white worms. Strangely enough it *can* turn and become quite spiteful.

Dascyllus Trimaculatus

This fish in commonly known as the Dominoe Damsel and is second in popularity only to the Clown fishes. It has a jet black body when young broken only by three white spots which give it its popular name. If anything it adds to its attractiveness with age by growing very prominent scales which take on a deep brown shade. It is a scrappy fish and, although reported to reach only three inches by many experts does in fact double this size in good conditions. In a community of Dominoes one will always outgrow the others and develop into a bully. Food of most kinds is accepted at all times for they are greedy fish.

Microspathodon Chrysurus

This is the well known West Indian Jewel fish aptly named for it has a deep blue body with pale blue spots all over it and a blue eye. Its white tail eventually turns yellow at the same time as the pale blue colouring vanishes. It is a large damsel and reaches six inches in the aquarium.

Pomacentrus Chrysus

Like most of the *Pomacentrus* species this is a small, active and sometimes spiteful fish. It has a yellowish body with a fake 'eye' on the tail. It has vivid electric blue flashes across the forehead which fade away with age. It grows only to two inches but makes up for its lack of size with courage.

Pomacentrus Coeruleus

This fish is an incredibly bright blue when young progressing to a paler sea blue when adult. It remains peaceful if grown to maturity along with plenty of other fishes but will become aggressive if kept alone for any length of time.

Pomacentrus Fuscus

This fish comes from the West Indies and has very little as far as I can see to warrant inclusion to an attractive aquarium. It is a dull brown which lightens with age. A difficult personality being unlikely to accept other fishes without a fight. It grows to three inches. Not a sensible fish for a community tank.

Pomacentrus Leucosticus

This is usually known as the Beau Gregory and starts life a brilliant yellow with an almost luminous blue top half. It is very popular at this age. Nature however not only removes most of the colour but adds a dash of viciousness later on.

Pomacentrus Melanochir

This fish is commonly known as the Blue Devil and is very similar to the last fish in many ways. It has a mauve to blue body with a yellow tail. It will live with other fishes provided it has never been isolated after maturity has been reached. A week without the company of other fishes at this time is sufficient to render it unfit for community life ever again. It loses most of its colours.

Scatophagidæ
(Scats)

A SMALL FAMILY - comprised of the so called 'muck eaters', probably because of their incredible greed which often takes them inland into fresh waters following a ship in the hope of refuse, rather like the sea gulls. Because of this habit of feeding in fresh or brackish water, and thus an obvious ability to stand vast water density changes, the Scat has become a popular fresh water aquarists' fish. The Scat however does not usually do too well in fresh water for any length of time and its colours are less bright. They are also aggressive fish but injuries are few when fighting does occur. There is always one Scat stronger than the others and he is usually a bully and will grow much larger than all the others. Scats like vegetation and should be provided with plenty of algæ, spinach, lettuce etc. Like all 'disc' shaped fishes they are prone to skin complaints and should be examined periodically for such infections. They are lively fish and are happier in the marine aquarium.

Scatophagus Argus

This fish has been named by someone as the 'Hundred eyed muck eater' and as the fish comes from the West Indies we can probably blame the natives. It can be a very spiteful fish and is not always the best of community dwellers. Colours vary from pale yellow to deep green depending upon diet and salinity of its water.

Scatophagus Tetracanthus

This is a very similar fish to the Argus but is tinged with red. It is sometimes known as the Tiger Scat and is triped with black lines. All foods taken.

Scorpænidæ
(Scorpion fish)

A FAMILY OF large predators which have been given more names than any other fish in the world namely: Scorpion fish, Lion fish, Zebra fish, Turkey fish, Feather fish, Tiger fish, Dragon fish, Cardinal scarpion, Cobra fish and possibly a few more. The trouble is that most of these names describe the Scorpion fishes perfectly. I have chosen to use 'Scorpion fish' mainly because it is the nearest to the correct Latin name. The Scorpion is to my mind the most breathtaking fish in the sea. It is beautiful to look at, graceful in movement, colourful, intelligent and a tough, *community* fish! It is also very poisonous and should never be handled without a net. Due to the long spines it should not be caught in an ordinary net at all or it may become entangled. A transparent all nylon bag with one or two holes makes a safer way to trap the Scorpion fish for removal from a tank. The Scorpion fish is extremely lazy and has therefore become a master of disguise. His camouflage consists basically of an ability to blend in with plant life by hanging, nose downwards, and by great flaps of skin which hangs all over the face making it look more like an algæ covered rock than a fish. The ability to hang motionless in this position for hours on end causes his breakfast no alarm and the unfortunate fish is engulfed with remarkable rapidity and choked to death in the stomach of the Scorpion fish. Even with this predatorial instinct the Scorpion fish becomes quite tame and soon learns which has been added to the aquarium for his benefit and which has not and those which are intended as decoration should not be too small for he is not infallible and may eat a small fish regardless. He will not attack a fish his own size and will keep to himself most of the time. Most of the Scorpion fishes are similar to look at and have the same habits. They are short-sighted fish on the whole and are subject to blindness on occasion. This can very often be cured by simply leaving the aquarium in darkness for a week or two. A Scorpion fish can be made to eat dried foods and I personally trained one to eat dogs meat broken into bite sized chunks. The Scorpion fish should be fed until he will eat no more, usually when his stomach is expanded to its fullest extent. He can be fed daily and will prefer this, but once a week will suffice.

Dendrochirus Zebra
This fish is found in the Red Sea and in East Africa. It is the darkest of all the Scorpion fish and is dark brown with white stripes running around its body. It grows to a size of twelve inches and is a reasonably lively fish.

Pterois Antennata
A very similar fish to the Volitans but with a much heavier body and shorter fins. It is a little aggressive at times and will only accept other fishes in the same aquarium if it has been brought up alongside them. It grows to eighteen inches.

Pterois Lunulatus
This is another fish closely resembling the Volitans and often mistaken for one. The spines however are shorter and there are no dermal flaps on the face. It is also smaller than the Volitans and only reaches twelve inches.

Pterois Miles
This is a very timid Scorpion fish with webbed fins and far less spines than the others. It is also small being only ten inches when adult.

Pterois Radiata
The most 'spiney' looking of all the Scorpion fish the Radiata has a white face and thin white streaks running vertically all over the body, The spines are very sharp and a special net is essential for removal. It has no dermal flaps at all and a short compressed body. With age the body darkens from brown to black.

Pterois Volitans
This is the most well known of all the family and most Scorpion fishes are called Volitans whether they are or not. It has the largest pectoral fins of any fish and these are sometimes as large *each* as the whole body of the fish. Contrary to belief the spines making up these fins are not poisonous, only the first dorsal fins are. The Pterois Volitans becomes very tame and will come to the glass wagging its tail for food when hungry. It will live contentedly in a community aquarium with fish of its own size but can engulf a fish only slightly smaller than itself having an elastic mouth. Colours of the Volitan change with mood many times during the day. It is very hungry most of the time and a lack of food interest should be regarded with great suspicion. It grows to the considerable size of two feet.

Syngrathidæ
(Sea horses and pipe fish)

THIS IS A family which consists of many types of Sea horse or Pipe fish. The connection between the two being very obvious and most unusual. They both have external skeletons and the males in each case suffers the burden of pregnancy!

Sea Horses

The smaller Sea horses are a more practical fish for the amateur aquarist for they require a considerable amount of live foods and fast moving water. The Sea Horse must also be provided with somewhere to 'anchor' and this can be best done with a piece of branch coral laid on the floor of the aquarium. Sea horses cannot be kept in the community aquarium for they are simply too slow to catch foods. They must therefore be kept alone or with pipe fishes. Their food requirements are demanding and the legend that brine shrimp will suffice should be forgotten for the newly hatched brine shrimp is almost useless as a food until it is at least one quarter of an inch long. Young Guppies or Mollies are ideal and should be offered when available. Although the Sea horse is slow swimming, due to the fact that it only has one fin, it can creep up on a prey and snap upwards with its mouth at a surprising speed.

The Sea horse is one of the few coral fishes which are breedable in the aquarium and their method of reproduction is remarkable. The male has a built in pouch on his abdomen. The female inserts her ovipostor into this pouch and emits her eggs which are then fertilised by the male. Incubation lasts between two weeks and two months. Sea horses are usually imported with an infestation of parasites and should be cleaned before admittance to the community aquarium.
(See Chapter eleven – Diseases).

Hippocampus Hudsonius

This is a large Sea Horse growing to a length of one foot and is not therefore the best of specimens for the amateur aquarist wihout very large tanks. It also needs a large amount of food and, consequently, excretes vast quantities of refuse and will need frequent water changes. As the eyesight of the Sea horse is not very good his aquarium will need to be quite bright or he will not be able to see his food. The Hudsonius incubates his young for about six weeks. He lives for around three years.

Hippocampus Zosterae

This is a dwarf Sea horse and is found in the warmer, shallow areas off the American coast. It grows to two inches and is variably coloured. It is much hardier than the common Sea horse and requires less space. Dwarf Sea horses breed only in the summer and a brood can exceed fifty. The incubation can be as short as ten days and the pair will be ready to mate again within a week. It lives for two years in the wilds but most aquarists have trouble keeping one alive for more than three months. This is probably due to mis-feeding.

Syngnathus Fuscus (pipe fish)

There is very little difference between the Sea horse and the Pipe fish which is only a stretched out version of the Sea horse anyway. Breeding, swimming and feeding are all identical the main physical difference being the fact that many Pipe fishes, due to their existence in bays and river mouths, are able to stand considerable salinity changes. They have even been kept in fresh water. I once placed a Pipe fish into a tank with a Scorpion fish by mistake. The Pipe fish no doubt regarded the Scorpion as an ideal hiding place and promptly swam into the spines and pretended to be another. His colours even changed to match those of the Scorpion and when the Scorpion fish moved the Pipe fish followed. When they were both stationary it took a good eye to find the Pipe fish at all.

Pipe fishes can be kept with Sea Horses but seldom with any other fish at all for, like the Sea horse, they will only accept live foods and are not much faster at catching it. When they die in the aquarium it is usually through malnutrition. They will accept fresh water but prefer salt.

Theraponidæ

ANOTHER SALT WATER fish which is able to accept severe changes in salinity. It also comes into the river mouths to feed and is sometimes classed as a brackish water fish. Colours vary from a golden or silver to a pale green. It has three brown lines which circumferate the body. It is an extremely fast fish and a somewhat disagreeable one spending most of its captive days fighting with other fish of its own family.

Thereapon Jarbua

One seldom sees a public aquarium lacking this fish and, in shoals they look an impressive sight. The private aquarist however should beware of keeping more than one specimen in the same tank for they can be most spiteful when cramped for space. They also exceed twelve inches in length. On the other hand they are very hardy and can stand great temperature drops and salinity changes. The Jarbua should be provided with sand on the floor of the aquarium for it may want to dig a hole into it for sleeping in. They are at their most aggressive when in one of these 'pits'.

Zanclidæ (Moorish idols)

THIS MUST SURELY be the Queen of the coral reef. The majestic, aloof, unbelievably stately and regal Moorish Idol is the delight, and misery, of coral fish fanciers the world over. Artists have painted this fabulous creature for years and the world zoos have been attempting to keep them for almost as long. They remain practically impossible to keep in the aquarium. This may seem a rather strong statement but it is nevertheless a fact. I have yet to find even a public aquaria which has managed to solve the puzzle. My personal experiences have led me to believe that the Moorish Idol is simply too highly strung to suffer the shock which racks its nervous system when it is first caught.

Typical of any 'difficult to keep' coral fish it has an elongated snout. It is a curious fact that *any* long-nosed fish will be found to be very fastidious about his food and tend to be difficult to keep. The Zanclus is no exception. Looking very much like an angel fish it is strange to find that it is in fact more closely related to the surgeon fishes. This indicates the need for algæ or other greenery but, as the Moorish Idol is more than difficult to feed at all, it should be kept completely alone and in a spacious, well lit, tank with several large hiding places which he can go into with his enormous dorsal fin. When obtaining a specimen, and they are expensive as well as difficult to obtain, make sure that the walls of his stomach are not concave and that the bone structures cannot be seen. A wafer thin Zanclus is unlikely ever to overcome this state. The difficulties are enough on their own without starting off with a half starved fish. Brine shrimp is often accepted at first but an immense quantity of this must be supplied. The Zanclus seems happier when surrounded with live coral or plant, all of which are in the same class as the Zanclus – difficult.

Zanclus Canescens

This is possibly a mis-named fish being perhaps simply an adult form of the Cornutus. The main differences are a fuller body, especially around the lower parts, and a lack of horns. It also grows larger and can reach twelve inches in the wilds.

Zanclus Cornutus

This fish is found in all parts of the Pacific and is white with delicate yellow patches and a long white dorsal fin. It has a black tail and two thick black bands running down the body. It also has two horns in front of the eyes. It is a smallish fish reaching only to nine inches when adult.

CORAL FISHES:
The majority of fishes are listed in the index of Coral Fishes which starts on page 94. They are in alphabetical order for easy reference and are therefore not repeated in this index.